High-Performance D/A-Converters

The Springer Series in Advanced Microelectronics provides systematic information on all the topics relevant for the design, processing, and manufacturing of microelectronic devices. The books, each prepared by leading researchers or engineers in their fields, cover the basic and advanced aspects of topics such as wafer processing, materials, device design, device technologies, circuit design, VLSI implementation, and subsystem technology. The series forms a bridge between physics and engineering and the volumes will appeal to practicing engineers as well as research scientists.

For further volumes:
http://www.springer.com/series/4076

Martin Clara

High-Performance D/A-Converters

Application to Digital Transceivers

Springer

Martin Clara
LANTIQ-A GmbH
Villach
Austria

ISSN 1437-0387
ISBN 978-3-642-31228-1 ISBN 978-3-642-31229-8 (eBook)
DOI 10.1007/978-3-642-31229-8
Springer Heidelberg New York Dordrecht London

Library of Congress Control Number: 2012954677

Printed on acid-free paper

Springer is part of Springer Science+Business Media (www.springer.com)

Preface

D/A-converters synthesize analog signals at the very heart of digital transceivers, thus linking the world of digital signal processing to the analog signal domain. The integration of the digital signal processor (DSP) together with data converters and analog signal processing functions in advanced CMOS technologies finally enables the realization of complete transceivers on a single silicon chip. This so-called system-on-chip (SoC) integration drastically reduces the cost and the form factor of the overall system, a highly desirable situation in cost-driven hardware development.

The topic of this work is the modeling and the implementation of high-performance current-steering D/A-converters for digital transceivers in nanometer CMOS technology. To comply with the target of SoC integration, the designed converter modules are fully embeddable in a mixed-signal system. They use only core devices, operate from a single supply voltage, and are optimized for small silicon area and low-power dissipation.

In the first part of this book the fundamental performance limitations of current-steering DACs are investigated. Based on simplified models, closed-form expressions for a number of basic nonideal effects are derived and tested. A special focus is put on the signal-dependent noise performance in multitone systems, showing that the traditional full-scale single-tone criteria are not adequate and may lead to substantial overdesign of biasing and/or clocking circuits. The knowledge of the basic performance limits allows to optimize the converter and system architecture already in an early design phase, essentially trading off circuit complexity, silicon area, and power dissipation for static and dynamic performance.

The second part of this book describes the design and experimental verification of four different current-steering DAC testchips, implemented in standard 130 nm CMOS. The converters use a single 1.5 V supply and have a resolution in the range of 12–14 bits for an analog bandwidth between 2.2 MHz and 50 MHz. Sampling rates between 100 MHz and 350 MHz are used. Dynamic element matching (DEM) and advanced dynamic current calibration techniques are employed to minimize the required silicon area. High-resolution converters with an active output stage to maximize the analog output signal swing, traditionally only used for low-frequency applications, are demonstrated to reach signal bandwidths of 30–50 MHz, while maintaining a dynamic linearity larger than 70 dB over the whole bandwidth.

This book is organized into seven chapters:

- Chapter 1 provides an overview of D/A-converter fundamentals, especially focusing on the current-steering architecture and segmentation strategies.
- In Chap. 2 the generally accepted performance figures used to characterize D/A-converters are reviewed. The impact of correlated bias noise and sampling jitter on the noise performance of a current-steering DAC is analyzed, and, based on simplified models, exact descriptions of the noise spectra for single-tone and multitone signals are derived. The jitter noise expressions obtained for multitone signals in an NRZ-DAC are also verified experimentally.
- In Chap. 3 the static linearity limited by random mismatch of the current sources is explored. Based on a statistical description of the fabrication yield, expressions for the required minimum current-source area fulfilling a given yield specification are derived. The code-dependent output resistance of the current-source array is identified as a further limiting factor for the static linearity. Expressions for the INL of the single-ended and the fully differential converter as a function of the finite output resistance of the unit current cell are given. The second part of Chap. 3 gives a general overview of two known methods to improve the static linearity: DEM and current calibration.
- Chapter 4 analyzes the three basic effects that limit the dynamic linearity of a current-steering D/A-converter. Under the idealizing assumption of perfectly matched current cells, which are also not influenced by any other large-scale imperfections, it is possible to derive closed-form expressions for the nonlinear distortion of a synthesized sine wave. The first effect is given by switching asymmetries that result in unsymmetrical current pulses. These are shown to generate even-order harmonic distortion in fully differential converters. The second effect is caused by the finite feed-through of the output voltage to the parasitic capacitance connected at the common tail node of the current switch pair. The resulting code-dependent error charge packets injected into the output generate odd-order harmonic distortion. The third effect that limits the dynamic linearity of a current-steering D/A-converter is the frequency-dependent distortion due to the code-dependent output impedance of the current-source array. It is shown that in a fully differential converter with standard current-cell architecture this effect is at least within the boundaries of practical sampling frequencies, not likely to become dominant compared to the switching errors. Nevertheless, it is a fundamental limitation to the achievable dynamic linearity, since it cannot be compensated straightforwardly, e.g., by applying a special switching algorithm. The second part of Chap. 4 describes various known circuit techniques to improve the dynamic linearity of current-steering D/A-converters.
- Chapter 5 describes the implementation of two $\Sigma\Delta$ D/A-converter testchips targeted at wireline communication applications. The first design is a 14-bit DAC targeted at the ADSL2+ downstream bandwidth of 2.2 MHz. It uses a second-order noiseshaper together with a very simple barrel-shift algorithm and

is optimized for low-power dissipation. The second design is a multi-mode $\Sigma\Delta$-converter for ADSL and VDSL. An interleaved current-cell architecture implements an effective return-to-zero (RZ) for the single current cell and also allows to use a modified data weighted averaging (DWA) algorithm. With a sampling rate of 350 MHz and an oversampling ratio of only 6, a dynamic range of 12 bits in a bandwidth very close to 30 MHz is demonstrated. Due to the effective RZ, the dynamic linearity around 30 MHz remains above 75 dB. The converter module is readily scalable—by register programming—to lower signal bandwidths, with not only drastically reduced power consumption but also increased resolution.

- In Chap. 6 two Nyquist-rate D/A-converter testchips using dynamic current calibration in the background are described. Both designs implement a segmented 13-bit converter core. The first module is a classical, resistively terminated single-polarity DAC, while the second converter uses a dual-polarity core with active transimpedance output stage to maximize the available voltage swing. In conjunction with an interleaved current-cell architecture this DAC achieves a signal bandwidth of 50 MHz with a dynamic linearity exceeding 70 dB over the full signal bandwidth. The periodicity of the background calibration, normally a source of low-frequency tonal disturbances, is destroyed by the introduction of a randomized calibration slot length. Thereby the calibration refresh tones are spectrally shaped and merged with the noise floor. This randomization also helps to suppress the image tone due to dynamic calibration effects in an interleaved architecture. Two novel strategies to trim the elements in converter segments with different weights are employed. The first design tries to match the segment boundaries by appropriately summing together uncalibrated and previously calibrated DAC elements for comparison with a unique reference element. The second design performs direct calibration of single DAC elements in differently weighted segments. In order to generate the required scaled reference currents with an accurate ratio, a reference cell array is introduced, which calibrates its constituting elements in a separate calibration loop, also running fully in the background.
- Finally, Chap. 7 summarizes the main results of this work, while trying to draw general conclusions. Also, it provides an outlook on future developments that are expected to happen in the area of CMOS current-steering D/A-converters targeted at digital communication systems.

Villach, Austria Martin Clara

Contents

Acronyms and Abbreviations

ADC	Analog-to-digital converter
ADSL	Asymmetric digital subscriber line
AFE	Analog front-end
CF	Crest factor
CLA	Clocked level averaging
CML	Current mode logic
CO	Central office
CPE	Customer premises equipment
CRFB	Cascade of resonators with distributed feed-back
DAC	Digital-to-analog converter
dBFS	dB relative to full-scale
DEM	Dynamic element matching
DMT	Discrete multi-tone
DNL	Differential nonlinearity
DPD	Digital pre-distortion
DR	Dynamic range
DSL	Digital subscriber line
DSP	Digital signal processor
DWA	Data weighted averaging
FDM	Frequency division muliplexing
FOM	Figure-of-merit
FS	Full-scale
FSR	Full-scale range
GBW	Gain-bandwidth
IF	Intermediate frequency
ILA	Individual level averaging
INL	Integral nonlinearity
LHP	Left-half-plane
LFSR	Linear feedback shift register
LSB	Least significant bit
LMS	Least mean square

MBPR	Missing band power ratio
MIM	Metal-insulator-metal
MMS	Modified mismatch shaping
MSB	Most significant bit
MTPR	Missing tone power ratio
NRZ	Non return-to-zero
NTF	Noise transfer function
OFDM	Orthogonal frequency division multiplexing
OSR	Oversampling ratio
PAR	Peak-to-average ratio
PGA	Programmable gain amplifier
PLC	Power line communication
PRDEM	Partial randomization dynamic element matching
PSD	Power spectral density
QCC	Quiescent current control
RAM	Random access memory
RX	Receiver
RZ	Return-to-zero
SAR	Successive approximation register
SFDR	Spurious free dynamic range
SNDR	Signal-to-noise and distortion ratio
SQNR	Signal-to-quantization noise ratio
SNR	Signal-to-noise ratio
SoC	System-on-chip
TX	Transmitter
VDSL	Very high-speed digital subscriber line
WLAN	Wireless local-area network
ZOH	Zero-order-hold

List of Symbols

Mathematical Symbols

Symbol	Description
$*$	Convolution
$\circ\!\!-\!\!\bullet$	Transformation symbol: $f(t) \;\circ\!\!-\!\!\bullet\; F(s), F(j\omega)$
$\mathscr{L}$	Laplace transform operator: $F(s) = \mathscr{L}(f(t))$
$\mathscr{L}^{-1}$	Inverse Laplace transform operator: $f(t) = \mathscr{L}^{-1}(F(s))$
$\mathscr{Z}$	Z-transform operator: $F(z) = \mathscr{Z}(f(n))$
$\overline{x}$	Average value of x
$\overline{x^2}$	Variance of x
$\arctan(x)$	Arctangent
$\delta(t)$	Continuous-time Dirac function
$\delta(k)$, δ_k	Discrete-time Dirac function
$E(.)$	Expectation operator
e^x	Exponential function
$\mathrm{erf}(x)$	Error function
$f^{-1}(x)$	Inverse of function f: $f^{-1}(f(x)) = x$
$\gcd(a, b)$	Greatest common divisor of integer numbers a, b
j	Imaginary unit, $j = \sqrt{-1}$
lim	Limit operator
$\ln(x)$	Natural logarithm
$\log_{10}(x)$	Decadic logarithm
$\max(x)$	Maximum operator
$\min(x)$	Minimum operator
mod	Modulus operator
$P[.]$	Probability operator
Π	Series of products
π	Ratio of circle's circumference to diameter, $\pi = 3.141592\ldots$
$\mathrm{rect}(x)$	Rectangular function: $\mathrm{rect}(x) = 1$ for $\lvert x \rvert \leq \frac{1}{2}$, 0 otherwise
$\mathrm{sinc}(x)$	Sinc-function: $\mathrm{sinc}(x) = \frac{\sin(x)}{x}$
Σ	Discrete sum
$\sigma(x)$, σ_x	Standard deviation of x

Physical Symbols

Symbol	Description	Unit
C	Capacitance	[F]
f	Frequency	[Hz]
k	Boltzmann's constant, $k \approx 1.38 \cdot 10^{-23}$	[J/K]
ω	Radian frequency	[rad/s]
q	Elementary charge, $q = 1.602 \cdot 10^{-19}$	[As]
T	Absolute temperature	[K]
t	Time	[s]
I	Current	[A]
R	Resistance	[Ω]
V	Voltage	[V]
Y	Complex admittance	[A/V]
Z	Complex impedance	[Ω]

Special dB-Units

Symbol	Description
dBc	dB relative to carrier
$dBFS$	dB relative to Full-Scale
dBm	dB relative to 1 mW signal power

MOS-Transistor Parameters

Symbol	Description	Unit
A_β	Current-factor mismatch constant	$[\% \cdot \mu\text{m}]$
A_{VT}	Threshold voltage mismatch constant	$[\text{mV} \cdot \mu\text{m}]$
C_{GD}	Gate-drain capacitance	[F]
C_{GDov}	Gate-drain overlap capacitance	[F]
C'_{GDov}	Gate-drain overlap capacitance per unit gate width	[F/m]
C_{GS}	Gate-source capacitance	[F]
C_{GSov}	Gate-source overlap capacitance	[F]
C_{ox}	Oxide capacitance per unit area	$[\text{F/m}^2]$
Γ	Combined current matching constant	$[\mu\text{m}^2]$
gm	Transconductance	[A/V]
gm_{MB}	Transconductance of bias transistor M_B	[A/V]
gm_{SRC}	Transconductance of current-source transistor M_{SRC}	[A/V]
gm_{SW}	Transconductance of current switch	[A/V]
gds	Drain conductance	[A/V]
gds_{SW}	Drain conductance of current switch	[A/V]
I_{DS}	Drain current of MOS-transistor	[A]
KF_F	Flicker noise voltage coefficient	$[\text{A}^2\text{s}^2/\text{m}^2]$
K_F	Flicker noise current coefficient	$[\text{F} \cdot \text{A}]$
L	Channel length of MOS-transistor	[m]
μ	Carrier mobility in the channel	$[\text{m}^2/(\text{Vs})]$
n	subthreshold slope factor	
ω_T	Transit frequency	[rad/s]
V_{GS}	Gate-source voltage of MOS-transistor	[V]
V_T	Threshold voltage of MOS-transistor	[V]
W	Channel width of MOS-transistor	[m]

Data Converter Theory

Symbol	Description	Unit
$A(j\omega)$	Open-loop gain of operational amplifier	
A_0	Open-loop gain of operational amplifier at DC	
A_1	Amplitude of fundamental	
A_k	Amplitude of kth order harmonic product DMT: intermodulation distortion at bin k	
A_{CAL}	Open-loop gain of calibration circuit at DC	
A_{CLK}	Clock signal peak amplitude	
A_{DMT}	Multitone peak amplitude	
A_{OFF}	Area of switch-off error	[Vs]
A_{ON}	Area of switch-on error	[Vs]
A_{peak}, A_{sin}	Single-tone peak amplitude	
$a_{LSB}(t)$	LSB-current step response	[V/s]
α	Ratio of full-scale voltage to gate overdrive	
B	Converter resolution in bits	
$B_j(\omega)$	Jitter noise frequency shaping function for DMT	
$B_n(\omega)$	Bias noise frequency shaping function for DMT	
C_B	Filter capacitor in DAC-biasing	[F]
C_C	Compensation capacitor of calibration loop	[F]
CF	Crest factor	
D	Duty factor of RZ-DAC, $D = T_S/T$	
DNL(k)	Differential nonlinearity at code k	
DNL_{max}	Worst-case DNL: $\text{DNL}_{\text{max}} = \max(\lvert\text{DNL}(k)\rvert)$	
DR	Dynamic range	[dB]
d_k, $d(k)$	Discrete-time converter input	
$d(t)$	Hypothetical continuous-time converter input	
Δ	LSB-size, unit DAC-element size	
δ_1	Deviation of calibrated cell current relative to I_{LSB}	
e_k	Error signal in sample k	
$e(t)$	Ideally reconstructed continuous-time error signal	
$E\left(e^{j\omega T}\right)$	Spectrum of sampled continuous-time error signal	
η_{BIAS}	Noise excess factor of DAC-biasing	
f	Frequency variable	[Hz]
f_B	Signal bandwidth	[Hz]
f_{CLK}	Clock frequency	[Hz]
f_{PLL}	PLL-bandwidth	[Hz]
$f_{refresh}$	Calibration refresh frequency	[Hz]
f_{sin}	Single-tone frequency	[Hz]
$\varphi(t)$	Instantaneous phase	[rad]
GBW	Unity gain-bandwidth product	[Hz]
GBW_{CAL}	Unity gain-bandwidth of calibration loop	[Hz]
GE	Gain-error	
HD_k	Harmonic distortion product of order k	
I_{cell}	Current of DAC-element	[A]

I_{FS}	Full-scale output current	[A]
I_{LEAK}	Leakage current	[A]
I_{LSB}	LSB-current	[A]
I_{OUTP}	Positive output current	[A]
I_{OUTN}	Negative output current	[A]
$\overline{I_{n,\mathrm{array}}^2(f)}$	Squared noise current density of DAC-array	[A^2/Hz]
$\overline{I_{n,\mathrm{array}N}^2(f)}$	Squared noise current density of NMOS-array	[A^2/Hz]
$\overline{I_{n,\mathrm{array}P}^2(f)}$	Squared noise current density of PMOS-array	[A^2/Hz]
$\overline{I_{n,\mathrm{Bias}}^2(f)}$	Squared noise current density of DAC-bias	[A^2/Hz]
$\overline{I_{n,\mathrm{MB}}^2(f)}$	Squared noise current density of bias diode	[A^2/Hz]
$\overline{I_{n,\mathrm{MBN}}^2(f)}$	Squared noise current density of NMOS bias diode	[A^2/Hz]
$\overline{I_{n,\mathrm{MBP}}^2(f)}$	Squared noise current density of PMOS bias diode	[A^2/Hz]
$\overline{I_{n,\mathrm{OS}}^2(f)}$	Squared noise current density of output stage	[A^2/Hz]
$\overline{I_{n,\mathrm{REF}}^2(f)}$	Squared noise current density of reference current	[A^2/Hz]
$\overline{I_{n,\mathrm{SRC}}^2(f)}$	Squared noise current density of M_{SRC}	[A^2/Hz]
I_Q	Quiescent current in output stage	[A]
I_{REF}	Reference current for calibration or biasing	[A]
I_{ULSB}	Cell current in ULSB-current	[A]
I_{unit}	Unit cell current	[A]
$\mathrm{INL}(k)$	Integral nonlinearity at code k	
$\mathrm{INL}_{\mathrm{max}}$	Worst-case INL: $\mathrm{INL}_{\mathrm{max}} = \max(\lvert\mathrm{INL}(k)\rvert)$	
$\mathrm{INL}_{\mathrm{max},s}$	Worst-case INL for single-ended converter	
$\mathrm{INL}_{\mathrm{max},d}$	Worst-case INL for differential converter	
$\Delta I_{\mathrm{MSB},0}$	Uncalibrated current error of MSB-cell	[A]
$\Delta I_{\mathrm{MSB,cal}}$	Calibrated current error of MSB-cell	[A]
K	Number of bits in intermediate converter segment	
K	Reference current noise scaling factor	
K_τ	Dynamic constant for active output stage model	
κ	Area overhead factor in DAC-bias	
L	Noiseshaper order	
L	Number of bands in multi-band DMT signal	
L	Number of bits in (L)LSB converter segment	
M	Number of bits in MSB converter segment	
$M(j\omega)$	Mirror factor of DAC-biasing	
M_0	Mirror factor of DAC-biasing at DC	
$\mathrm{MBPR}(f_k)$	Missing band power ratio at frequency f_k	[dB]
$\mathrm{MTPR}(f_k)$	Missing tone power ratio at frequency f_k	[dB]
N	Number of equal DAC-elements (in unary array)	
N_C	Number of carriers in multitone signal	
N_x	Integrated noise power of variable x	[W]
N_{eq}	Excess quantization noise power	[W]
$N_{j,\mathrm{NRZ}}$	Noise power due to sampling jitter	[W]
$N_{j,\mathrm{RZ}}$	Sampling jitter noise power for RZ-DAC	[W]
$N_{j,\mathrm{RZ},0}$	Sampling jitter noise power for RZ-DAC for $\omega_{\sin} \to 0$	[W]
N_k	Total sampling jitter noise power for carrier k	[W]
N_{FIX}	Minimum calibration slot length in number of clocks	
$N_{\mathrm{VAR}}(k)$	Output of random number generator in slot k	

$N_{\mathrm{VAR,max}}$	Maximum output of random number generator	
$\mathrm{NTF}(f)$	Noise transfer function	
OSR	Oversampling ratio	
$\omega_0, \omega_{\mathrm{CLK}}$	Sampling clock radian frequency	[rad/s]
ω_C	Corner frequency of correlated bias noise	[rad/s]
ω_L	Lower intercept frequency for charge-sharing effect	[rad/s]
$\omega_{\min}$	Lower band edge of multitone signal	[rad/s]
$\omega_{\max}$	Upper band edge of multitone signal	[rad/s]
ω_{PLL}	PLL-bandwidth	[rad/s]
ω_{p1}	Closed-loop bandwidth of calibration loop	[rad/s]
$\omega_{p1,\min}$	Minimum bandwidth of calibration loop	[rad/s]
$\omega_{p,\mathrm{BIAS}}$	Pole frequency of first-order bias noise model	[rad/s]
$\omega_{\sin}$	Single-tone radian frequency	[rad/s]
$\omega_{\mathrm{RZ,3dB}}$	Lowpass frequency for jitter-limited SNR in RZ-DAC	[rad/s]
ω_U	Upper intercept frequency for charge-sharing effect	[rad/s]
$\Delta\omega$	Carrier spacing of multitone signal	[rad/s]
P	Array pointer increment of barrel shifter algorithm	
P_{carrier}	Signal power of carrier in homogeneous DMT-signal	[W]
$P_{S,\mathrm{RZ}}$	Signal power of sine wave in baseband for RZ-DAC	[W]
$P_{\sin}$	Signal power of sine wave in baseband	[W]
$P_{\mathrm{ON}}(j\omega)$	Transformed impulse-response of switch-on error	
$P_{\mathrm{OFF}}(j\omega)$	Transformed impulse-response of switch-off error	
$p(k)$	Array pointer in sample k	
$p(t)$	DAC impulse response	
$p_{\mathrm{ON}}(t)$	Impulse-response of switch-on error	
$p_{\mathrm{OFF}}(t)$	Impulse-response of switch-off error	
$p_{\mathrm{NRZ}}(t)$	Impulse-response of NRZ-DAC	
$p_{\mathrm{RZ}}(t)$	Impulse-response of RZ-DAC	
$p_q(x)$	Probability density function of quantization noise	
$Q_{\mathrm{C0},n}$	Charge stored on tail capacitance, negative output	[As]
$Q_{\mathrm{C0},p}$	Charge stored on tail capacitance, positive output	[As]
$Q_{\mathrm{err}}(k)$	Charge error injected into differential output	[As]
Q_{ov}	Charge injected via C_{GDov} of current switch	[As]
Q_{unit}	Charge of unit current pulse	
$q(t)$	Quantization error signal	
$q(k)$	Discrete-time quantization error	
Δq_0	Charge error at end of calibration period	[As]
R_B	Open-loop resistance of calibration circuit	[Ω]
R_L	Load resistor, feedback resistor	[Ω]
R_{TRIM}	Relative trimming range of current cell: $\pm R_{TRIM}$	
R_{unit}	Output resistance of unit current cell	[Ω]
$\rho(\omega)$	Output voltage feedthrough factor	
$S_x(\omega)$	Power spectral density of x	
$S_B(\omega)$	PSD of correlated bias noise	[W/Hz]
$S_{\mathrm{eq}}(\omega)$	Excess quantization noise PSD	[W/Hz]
$S_i(\omega)$	Circuit noise PSD	[W/Hz]
$S_{i,\mathrm{DMT}}(\omega)$	Circuit noise PSD for DMT	[W/Hz]
$S_{i,\mathrm{SIN}}(\omega)$	Circuit noise PSD for sine-wave	[W/Hz]

$S_{j,\mathrm{DMT}}(\omega)$	Jitter noise PSD for DMT	[W/Hz]
$S_{j,\mathrm{NRZ}}(\omega)$	Jitter noise PSD for NRZ-DAC	[W/Hz]
$S_{j,\mathrm{RZ}}(\omega)$	Jitter noise PSD for RZ-DAC	[W/Hz]
$S_T(\omega)$	PSD of jitter sequence ΔT_k	[s^2/Hz]
SNR	Signal-to-noise-ratio	[dB]
$\mathrm{SNR}_{\mathrm{jitter}}$	Signal-to-noise-ratio due to sampling jitter	[dB]
$\mathrm{SNR}_{j,\mathrm{RZ}}$	Signal-to-noise-ratio of RZ-DAC due to jitter	[dB]
SNR_k	Signal-to-noise-ratio for carrier k	[dB]
$\mathrm{SR}_{\mathrm{LSB}}(t)$	LSB-step Slew-Rate as function of time	[V/s]
s	Laplace-domain complex frequency variable	
σ_F	Bias noise constant for 1/f-noise model	[A]
σ_{IB}	Integrated rms bias noise current	[A]
σ_{tj}	Absolute rms jitter	[s]
σ_{LT}	Long-term rms jitter	[s]
T	Sampling period	[s]
$T_{\mathrm{CAL}}(k)$	Calibration slot length for DAC-element k	[s]
$\overline{T_{\mathrm{CAL}}}$	Average calibration slot length	[s]
$T_{\mathrm{CAL,max}}$	Maximum calibration slot length	[s]
$T_{\mathrm{CAL,min}}$	Minimum calibration slot length	[s]
T_{fall}	Unit current pulse fall-time	[s]
T_{OFF}	Duration of switch-off error impulse response	[s]
T_{ON}	Duration of switch-on error impulse response	[s]
T_{refresh}	Refresh period	[s]
$T_{\mathrm{refresh,max}}$	Maximum acceptable refresh period	[s]
T_{rise}	Unit current pulse rise-time	[s]
T_S	Active period of RZ-DAC	
$T(s)$, $T(j\omega)$	Transimpedance transfer function	[V/A]
t_0	Absolute time at end of calibration period	[s]
t_1	Absolute time during normal operation period	[s]
ΔT_k	Timing uncertainty of sampling clock in sample k	[s]
τ_{CAL}	Closed-loop time constant of calibration loop	[s]
$u(t)$	Time-domain converter output signal	
$u_{\mathrm{NRZ}}(t)$	Time-domain output signal of NRZ-DAC	
$u_{\mathrm{RZ}}(t)$	Time-domain output signal of RZ-DAC	
V_0	Clock sine-wave peak voltage	[V]
V_{FS}	Full-scale output voltage range	[V]
V_k	Converter output voltage at code k	[V]
$V_{k,\mathrm{ideal}}$	Ideal converter output voltage at code k	[V]
$V_{k,\mathrm{meas}}$	Measured converter output voltage at code k	[V]
V_{LSB}	LSB-voltage	[V]
$\overline{V_{n,\mathrm{DAC}}^2(f)}$	Squared noise voltage density at DAC-output	[V^2/Hz]
$\overline{V_{n,\mathrm{FS}}^2(f)}$	Squared noise voltage density at full-scale	[V^2/Hz]
$\overline{V_{n,\mathrm{MS}}^2(f)}$	Squared noise voltage density at midscale	[V^2/Hz]
V_{OFFSET}	Zero-code or midscale offset voltage	[V]
$V_{\mathrm{OUT}}(t)$	Converter output voltage	[V]
$V_{\mathrm{OUT,ideal}}(t)$	Ideal converter output voltage	[V]
$V_{\mathrm{OUT},N}(t)$	Voltage at negative converter output	[V]
$V_{\mathrm{OUT},P}(t)$	Voltage at positive converter output	[V]

$V_S(t)$	Voltage at switching node	[V]
$v(t)$	Clock signal waveform	[V]
ΔV_{FSN}	Negative full-scale voltage error	[V]
ΔV_{FSP}	Positive full-scale voltage error	[V]
ΔV_G	Gate voltage difference in switch pair	[V]
$\Delta V_{G,\min}$	Minimum gate voltage difference in switch pair	[V]
ΔV_k	Deviation of converter output voltage at code k	[V]
$\Delta V_{\max}$	Maximum deviation of converter output voltage	[V]
$w_{\mathrm{OFF}}(k)$	Discrete-time switch-off sequence	
$w_{\mathrm{OFF}}(t)$	Equivalent continuous-time switch-off signal	
$W_{\mathrm{OFF}}(j\omega)$	Spectrum oft continuous-time switch-off signal	
$w_{\mathrm{ON}}(k)$	Discrete-time switch-on sequence	
$w_{\mathrm{ON}}(t)$	Equivalent continuous-time switch-on signal	
$W_{\mathrm{ON}}(j\omega)$	Spectrum of continuous-time switch-on signal	
Y_X	Fabrication yield with respect to parameter X	
Y_{INL}	Fabrication yield with respect to worst-case INL	
Y_{DNL}	Fabrication yield with respect to worst-case DNL	
Y_L	Load admittance	[A/V]
$Y_{\mathrm{OUT},N}$	Admittance seen into negative converter output	[A/V]
$Y_{\mathrm{OUT},P}$	Admittance seen into positive converter output	[A/V]
Y_{unit}	Output admittance of unit current cell	[A/V]
z	z-domain complex frequency variable	
Z_{unit}	Output impedance of unit current cell	[Ω]
Z_L	Load impedance, feedback impedance	[Ω]
$Z_{L,\mathrm{eff}}$	Effective load impedance	[Ω]

Chapter 1
Introduction

1.1 Integrated D/A-Converters

A digital-to-analog converter (DAC) is designed to perform an accurate summation of a given number of electrical unit quantities. This number is specified by the *digital* input code applied to the converter. The result of the summation appears as an *analog* electrical signal at the converter output. Thus, the output signal of the DAC represents, with more or less accuracy, the digital input code in the analog domain and can be further processed by subsequent analog circuitry, e.g., filters, amplifiers, mixers, or transducer interfaces. Fundamentally, it is the summing operation of appropriate electrical unit quantities that marks the actual transition from the digital into the analog domain. This data conversion process is also subject to a number of error sources that introduce distortion, as well as additional noise, into the analog output signal. Since these nonideal effects tend to limit the performance of practical converters, they should be adequately considered already during the design phase. The modeling of conversion inaccuracies is believed to be of central importance for close-to-optimum, system-specific DAC design. Indeed, a large part of this work is dedicated to providing theoretical insight into fundamental nonidealities of integrated D/A-converters with a focus on the current-steering architecture and, wherever appropriate, linking it to the system-level perspective.

The DAC unit cell comprises one or more primary circuit elements that generate the summable electrical unit quantity in the first place, as well as a couple of analog switches that allow to connect the unit cell to common circuit nodes within the converter. Primary circuit elements in integrated converters are capacitors, resistors, or transistors. Important types of DAC unit cells used in integrated converters are shown in Fig. 1.1. The complete D/A-converter is eventually constructed using a suitable number of such unit cells, together with a couple of auxiliary circuits required for proper operation of the DAC. Auxiliary circuits include, e.g., biasing and clocking circuits, an output stage interfacing the converter to subsequent analog processing blocks, decoders and control circuitry, as well as digital interfaces.

M. Clara, *High-Performance D/A-Converters*, Springer Series in Advanced Microelectronics 36, DOI 10.1007/978-3-642-31229-8_1,

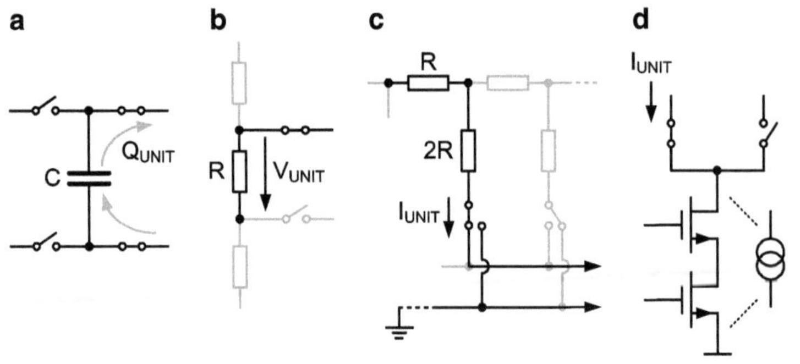

Fig. 1.1 DAC unit cell types: (a) charge, (b) voltage, (c) current, and (d) current

In general, a D/A-converter can have a voltage or a current output, depending on the type of input the circuitry connected to the converter provides. In high dynamic range AFEs the preferred method for analog signal interfacing is usually voltage mode, which in turn requires the D/A-converter to perform an accurate conversion from the previously summed up primary electrical quantity into the corresponding output voltage. In other circumstances, the electrical unit quantity is already available as a voltage, but a direct connection to the load would destroy the accuracy. In such cases, the data converter must include a sufficiently low-noise and linear voltage buffer.

Charge-based D/A-conversion is performed by storing electrical unit charges Q_{UNIT} on an array of integrated capacitors (Fig. 1.1a) and appropriately redistributing them onto a summation capacitor. Such converters are called Switched-Capacitor DACs and are especially popular as subsystems in more complex integrated MOS A/D-converters [1]. The main reason is that MOS technologies offer high-quality and easy-to-drive switches. Equally important, many integrated capacitor types are very linear[1] and can be laid out to exhibit good matching behavior [2]. Switched-capacitor DACs are also found as stand-alone building blocks, especially for high-resolution, low-frequency applications, e.g., in audio front ends [3, 4]. Figure 1.2 shows the principle of a multibit Switched-Capacitor DAC [5]. The charge stored on the unit capacitors C_i in phase Φ_1 is transferred to capacitor C_F during phase Φ_2, provided that the corresponding input bits are set, i.e., $b_i = 1$. The charges of those capacitors C_i, for which $b_i = 0$, are instead dumped to ground in phase Φ_2 and do not contribute to the output signal. The low-impedance node required for complete charge transfer from C_i to C_F is provided by the virtual ground node of the operational amplifier. The output voltage at the end of phase Φ_2 is additionally stored on the holding capacitor C_H, which is then put in the feedback of the operational amplifier during the subsequent phase Φ_1.

[1] An exception are diffusion- and MOS-based capacitors, which display a relatively strong voltage dependency. High-precision converters therefore tend to use poly–poly, poly–metal and metal–metal structures to form integrated capacitors.

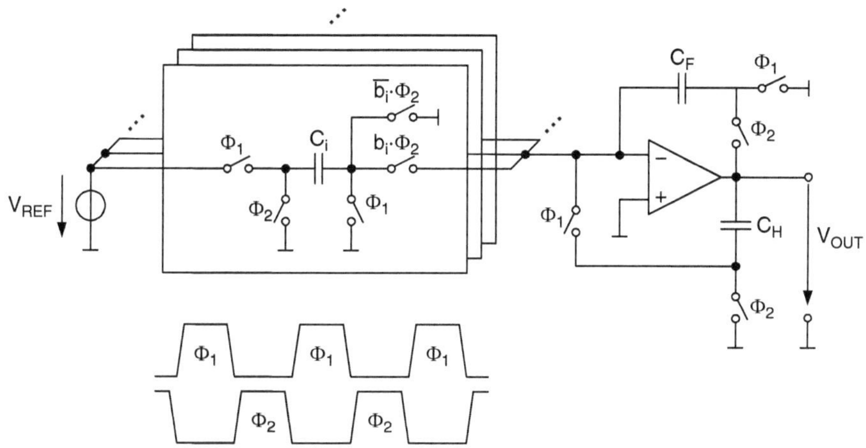

Fig. 1.2 Multibit parallel switched-capacitor DAC

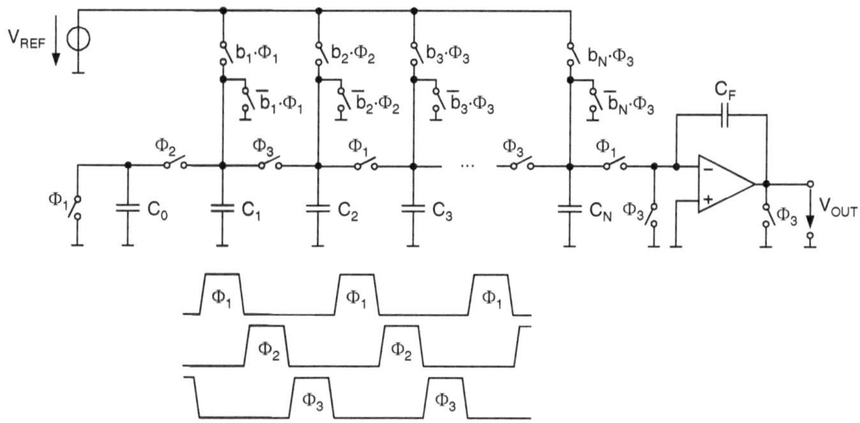

Fig. 1.3 Multibit pipelined switched-capacitor DAC

This clock phase is also used to precharge the unit capacitors C_i to V_{REF} and discharge the summing capacitor C_F, in effect preparing the converter for the next conversion cycle, while the actual output voltage is held on C_H.

Another approach is the pipelined switched-capacitor DAC [6] shown in Fig. 1.3. Using only $N+1$ identical capacitors C_0-C_N, an N-bit converter can be constructed, which is very area efficient. The operating principle is in some sense similar to a charge-coupled device (CCD), but with additional input-code-dependent precharging of the single capacitors. Using three clock phases and appropriately delayed input bits, the binary-weighted output charge corresponding to the digital input word is building up by successive charge redistribution along the capacitor pipeline. The reconstructed "analog" charge on the last capacitor C_N can finally be transferred to the output via a SC-amplifier. Drawbacks of this DAC architecture

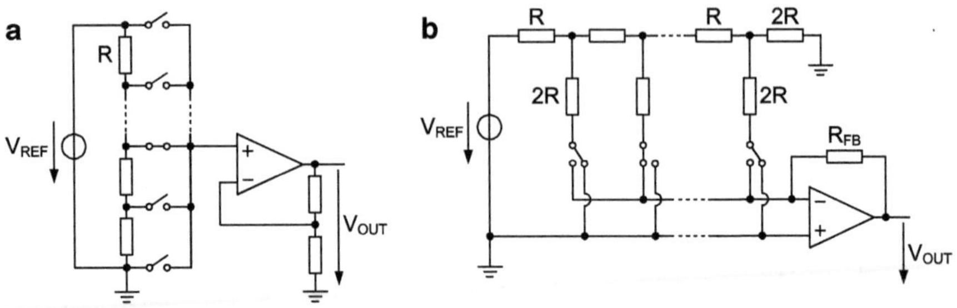

Fig. 1.4 Basic resistor-DAC types: (**a**) R-string and (**b**) R-2R multiplying DAC

are the latency of $N+2$ clock cycles and, possibly, a certain limit on the achievable resolution due to error progression along the capacitor pipeline. Nevertheless, due to the pipelining of the conversion process, it allows operation at respectably high sampling rates, as demonstrated for example in [7].

Integrated matched resistors can be arranged either as a programmable voltage divider (Fig. 1.1b), also called resistor string, or as a programmable current divider in the form of an R-2R ladder (Fig. 1.1c). In the voltage divider case, the output node must display a sufficiently high impedance compared to the total resistance of the string. Since the output impedance of a tapped resistor string depends on the actual position of the tap, the voltage distribution within the resistor stack is (code-dependently) disturbed, in case a significant current is flowing into the output. In the current divider case, on the other hand, the output node must be of very low impedance, otherwise the current distribution within the R-2R ladder will depend on the actual setting of the switches, i.e. on the digital input code. In practical implementations, both types of resistor-based D/A-converters require not only a precise voltage reference driving the resistor array but also a high-quality operational amplifier as output stage.[2] This amplifier either serves as a high-impedance voltage amplifier/buffer in the string-type converter, or it provides the low-impedance current-summing node for the current divider, along with the implicit current to voltage conversion using resistive feedback.

Figure 1.4 shows the two basic resistor-type DACs in greater circuit-level detail. Figure 1.4a shows the resistor string DAC or string-DAC [8]. The converter in Fig. 1.4b with accessible reference voltage V_{REF} is called R-2R multiplying DAC [9], owing to the fact that the output voltage is the product of the reference voltage and the digital input code [10], both applied externally. With the amplifier configured as a voltage buffer, the R-2R ladder can also be tapped in voltage mode, as shown in Fig. 1.5. In this case, the ladder behaves like a programmable voltage divider. Such an arrangement is called voltage-mode R-2R ladder DAC [10], sometimes also R-2R back-DAC [11].

[2]This is certainly true for general purpose instrumentation applications. However, for on-chip control applications with known suitable loading a dedicated op-amp is often not necessary.

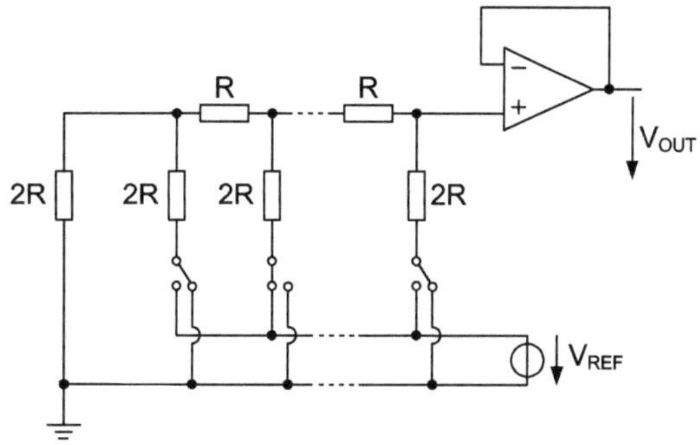

Fig. 1.5 Voltage-mode R-2R DAC (R-2R back-DAC)

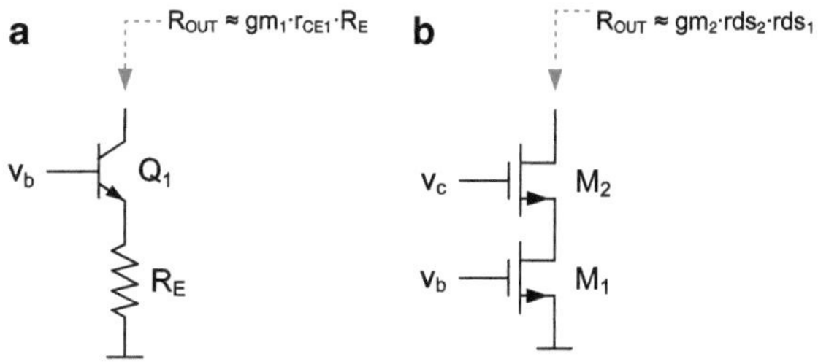

Fig. 1.6 Transistor-based current sources: (**a**) BJT and (**b**) MOS

As stand-alone converters, resistor-based DACs are heavily used not only in instrumentation and control applications but also in high-end digital audio products [12–14]. In conjunction with resistor (and reference) trimming, these converters can achieve exceptional accuracy, linearity, and long-term stability at very low power dissipation [15, 16]. The signal bandwidth covered by resistor-based D/A-converters is normally quite small, nevertheless sufficient for the usual target applications. A notable exception that covers video bandwidths, albeit displaying also considerably less resolution, can be found in [17].

Integrated current sources based on active elements require at least one transistor operated in the active region (Fig. 1.1d). In most cases, either resistive degeneration or cascoding is employed to further boost the output resistance to sufficiently high values. Traditionally, bipolar designs tend to use resistive emitter degeneration, while MOS current sources are normally in favor of cascoding. Both techniques are shown in Fig. 1.6. Extensive treatments of integrated current sources can be found in reference texts on integrated analog circuit design; see, e.g., [5, 18–20]. D/A-converters built with active current sources are called current-steering DACs because the unit cell currents must not be turned on and off. Instead, they have to be *steered*, rather cautiously, to the appropriate summing node(s) in order not to disturb the potentially delicate biasing required for these circuits. Current-steering DACs implemented in CMOS-technologies are the main subject of this book and described in more detail starting from Sect. 1.4.

In general, every unit cell contains at least one high-accuracy switch. This switch connects the unit cell, when triggered by the digital input code, to an appropriate common node within the converter circuitry, where, e.g., the summation of the electrical unit quantities occurs. Differential architectures (see below) have two

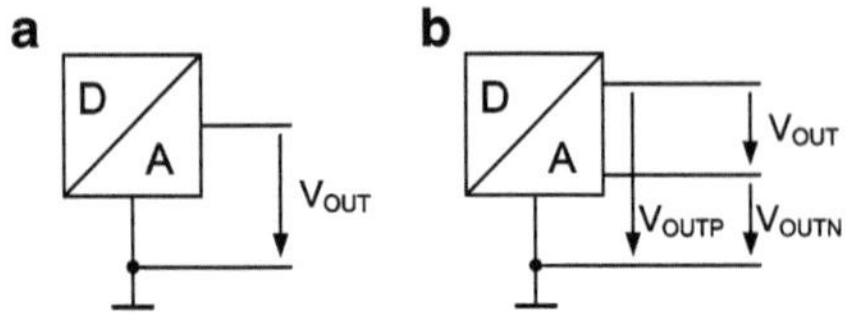

Fig. 1.7 Single-ended (**a**) and differential output (**b**) DAC

output nodes, obviously requiring at least two high-accuracy switches within each unit cell. Additional switches may be needed to establish a proper behavior of the unit circuit element during subsequent conversion cycles. For example, in charge-based converters, as described above, the unit capacitors must be correctly precharged prior to each summation operation.

In a D/A-converter, the unit cells need not necessarily be all of the same type. Nevertheless, true hybrid D/A-converters are not very common, an example using voltage division plus charge redistribution is reported in [21]. On the other hand, resistor-based current division of transistor current sources appears to be quite popular in bipolar designs [22–24], because it allows the construction of a true binary-weighted current-cell array, while employing only equal-sized current source transistors. This architecture has already been used[3] in the feedback DAC of EPSCO's 1954 11-bit 50 kS/s SAR-ADC (the "DATRAC"), an all-valve 500 W (!) A/D-converter system [10, 25].

On a more architectural level, we can distinguish between single-ended output and differential output D/A-converters; see Fig. 1.7. Single-ended output DACs are mostly found in low-frequency instrumentation and control applications, whenever the output node is single-ended by definition. Differential output DACs, on the other hand, are nowadays the primary choice for dynamic signal synthesis, whenever the spectral purity, i.e., the signal quality with respect to noise and distortion, is of primary importance. According to common understanding, this has to do with the doubled signal swing of differential architectures, leading to 3 dB better signal to noise performance. Additionally, defining the output signal as being the difference of two single-ended, opposite polarity signals[4] has the tendency to suppress even-order harmonic distortion, as well as providing a certain immunity against disturbance and noise present on common circuit nodes, e.g. the supply rails. Note, also, that any differential output DAC can be converted into a single-ended DAC by using only one of its two output nodes and discarding the other. Of course, by doing so, any previously present common-mode rejection is given up on as well.

On system level we can distinguish between Nyquist-rate and noiseshaped D/A-converters. Nyquist-rate converters exhibit a flat quantization noise spectrum (see Sect. 2.3.1), and, in principle, allow to exploit the full converter bandwidth for signal synthesis. Since virtually all communication systems require a certain

[3] According to [10] for the first time in data converter history.

[4] The two single-ended signals in a differential configuration are opposite in polarity relative to a fixed common-mode level above on-chip ground. Ideally, the differential output appears floating with respect to any fixed ground-referenced potential.

amount of oversampling, e.g., to facilitate the necessary analog filtering, in practical applications, only a certain fraction of the full Nyquist bandwidth, as defined by half the sampling rate, is actually used.

Noiseshaped D/A-converters, on the other hand, try to concentrate the major part of the quantization noise outside of the signal bandwidth of interest, thereby greatly improving the in-band resolution. As a consequence, they can only be used for signal synthesis within a relatively small fraction of the Nyquist bandwidth. Noiseshaped D/A-converters are also called $\Sigma\Delta$-DACs.

The MOS-transistor based, current-steering, and differential output converter topology (see Sect. 1.4) has become the most prominent DAC architecture for signal synthesis in communication systems, because it offers high-speed and high-accuracy capability at moderate power consumption. Whether oversampling, or even noiseshaping is employed, largely depends on the overall system specification, because these decisions usually have a rather big influence on the overall AFE architecture. In any case, current-steering D/A-converters can be realized quite straightforwardly, even in standard digital CMOS processes, and are easily combined with a fair amount of digital circuitry, e.g., to control DAC-element shuffling or calibration algorithms. This "digital assistance" is used extensively in the hardware examples in later chapters of this book, with the goal to improve the converter performance, while keeping the required silicon area small.

1.2 DACs for Highly Integrated Transceivers

Digital communication systems have become an integral part of the infrastructure in our modern information society. Although the bulk of the signal processing is already done in the digital domain, still the transmission medium, or channel, across which the connection of the so-called data link is finally established, is analog by nature [26]. Therefore, the digital data must be converted into an appropriate analog signal by the transmitter and back into digital data samples by the receiver.

Figure 1.8 shows a generic (unidirectional) digital communication link. In the transmitter (TX) a digital signal processor (DSP) performs the necessary coding and modulation of the input bit stream. In the AFE the encoded digital data is then converted into an analog signal by a DAC. After suitable filtering and power amplification, the analog transmit signal is finally applied to the channel.

The input signal of the receiver (RX) consists of the attenuated and spectrally distorted wanted signal together with noise and interferers. This composite signal is first amplified by a programmable gain amplifier (PGA), then filtered, and subsequently converted back into a digital signal by an analog-to-digital converter (ADC). Only then can the DSP in the receiver recover the original bit stream sent by the transmitter down through the channel. For bidirectional (duplex) operation a complete receiver and transmitter module must be implemented at both data terminals. This combination of *trans*mitter and re*ceiver* is often called *transceiver*. If transmitter and receiver are active at the same time, i.e.,

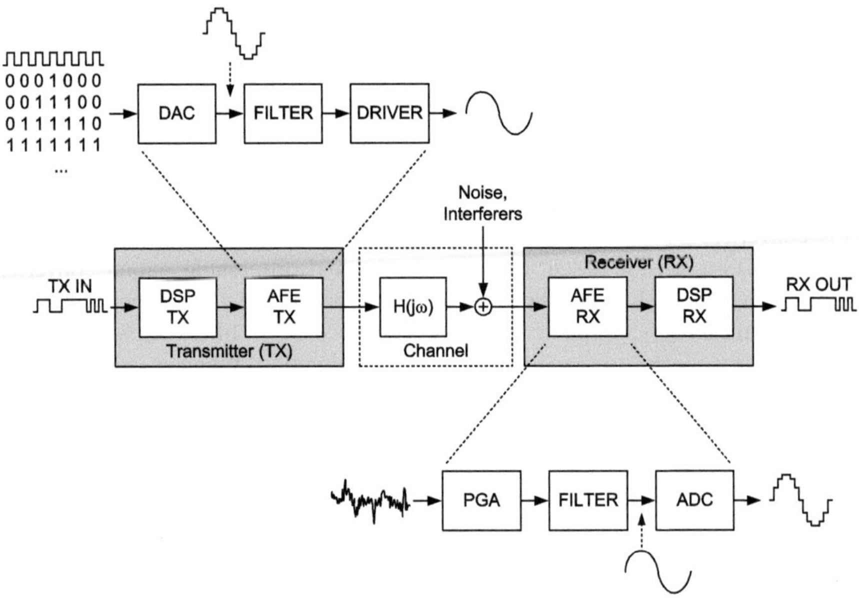

Fig. 1.8 Digital communication link

a bidirectional data link is established simultaneously, we speak of full-duplex operation, whereas in so-called half-duplex systems transmitter and receiver are active and transmitting/receiving only alternately. A typical example for full-duplex systems is the current xDSL-families using DMT modulation, while today's WLAN-standards and most PLC-systems operate in half-duplex.

In wireless systems we additionally find an analog mixer in the transmitter for spectral upconversion of the signal to the desired center frequency. In the corresponding receiver there is at least one analog mixer for downconversion of the antenna signal, either directly to the baseband frequency, or first to some intermediate frequency for further amplification and filtering, before the signal is downconverted to the baseband and digitized by the ADC [27]. In case the A/D-converter is able to sample the signal at an intermediate frequency (IF), the final downconversion can also be performed in the digital domain [28].

In any case, the two inverse data conversion operations in the analog front-end (AFE), D/A in the transmitter and A/D in the receiver, constitute very central signal processing functions that can be found in every digital communication system. The fidelity with which the data conversion can be accomplished on either side of the analog channel has an immediate impact on the quality of the data link. Therefore, considerable effort is spent in optimizing the performance of the data converters in the context of a limited power budget and silicon area constraints. These boundary conditions tend to become especially stringent in multi-channel integration, where multiple transmit and receive channels must be integrated together on the same silicon die.

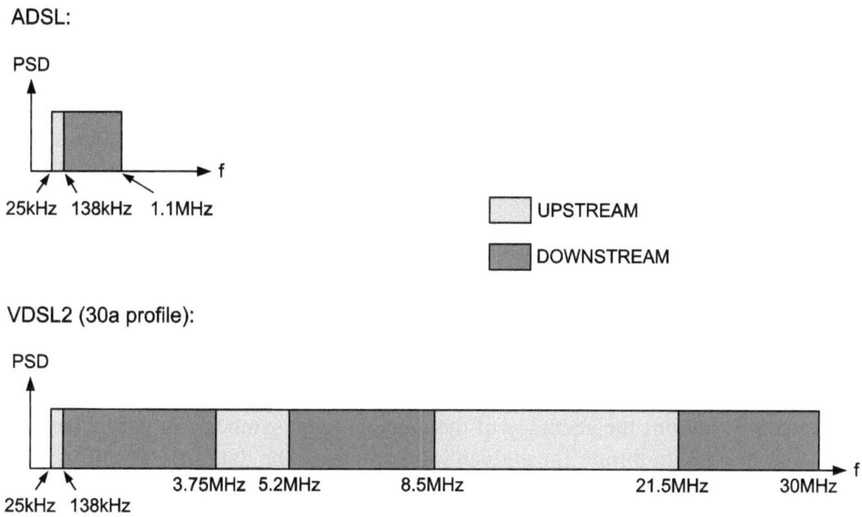

Fig. 1.9 DSL frequency-band plan examples

Special difficulties also arise when a single AFE-design must cover multiple transmission standards with considerably different bandwidth and linearity specifications. This feature is sometimes called multi-mode capability. As an example, Fig. 1.9 shows exemplary frequency band plans for two popular digital subscriber line (DSL) standards. The asymmetric digital subscriber line (ADSL) standard employs a single upstream and downstream[5] band from 25 kHz up to 138 kHz and from 138 kHz up to 1.1 MHz, respectively. On the other hand, the very high-speed digital subscriber line (VDSL) revision 2 system (30a profile in this example) uses a spectrum that is split into several interleaved upstream and downstream bands in the frequency range between 25 kHz and 30 MHz [29]. In practice, a myriad of different, even country-specific DSL-standards and substandards exist, all with different frequency band plans and sometimes even different transmit power masks.

The resolution and linearity requirements for the ADSL and VDSL system are also strongly differing. While ADSL requires around 14-bit performance, 11–12 bits are generally sufficient for VDSL, depending on the amount of out-of-band filtering and echo attenuation achieved for a given loop. Including enough flexibility into the analog building blocks to cover strongly differing operating modes, while still maintaining close-to-optimum silicon area and power consumption, considerably complicates the design of a multimode AFE.

[5]In wireline systems "upstream" is the data transfer from customer premises equipment (CPE) to central office (CO). "Downstream" is the reverse direction.

Table 1.1 D/A-converter requirements for digital communication systems

System	Signal bandwidth	Resolution	Sampling rate
WLAN 802.11a/g	10 MHz	8 bit	80 MS/s
WLAN 802.11n	20 MHz	9 bit	160 MS/s
Bluetooth 2	2 MHz	10 bit	13 MS/s
Cable modem	8 MHz	10 bit	160 MS/s
1000Base-T	62.5 MHz	5 levels	125 MS/s
ADSL	1.1 MHz	14 bit	100 MS/s
ADSL2	2.2 MHz	14 bit	100 MS/s
VDSL2	30 MHz	11–12 bit	200 MS/s
PLC	30–100 MHz	11 bit	400 MS/s

In many situations the accuracy of the analog signal synthesis in the transmitter limits the maximum bitrate for a given loop length of the data link. In full-duplex systems with frequency division muliplexing (FDM) noise and distortion generated in the transmitter can eventually leak into the near-end receive path and limit the achievable signal-to-noise ratio (SNR) in the ADC, which is trying to digitize the signal from the far-end transmitter. This problem is especially encountered in systems with interlaced or even partly overlapping transmit and receive spectra, e.g., in DSL. Consequently, the D/A-converter in the transmitter of a digital transceiver is a very important building block with considerable impact on the overall system behavior.

In Table 1.1 coarse performance specifications for the TX-DAC of different digital communication systems are reported. Note, that the actual choice of the physical resolution and the sampling rate of the data converters may vary in different implementations. For example, it may be advantageous to run the D/A-converter at a higher clock rate, since this generally relaxes the order of the analog reconstruction filter that follows the DAC in the transmit chain. Although a higher sampling rate means a somewhat higher power consumption in the DAC, this is eventually overcompensated by a much simpler analog filter. On the other hand, the last stage of the digital interpolation filter preceding the DAC then also has to run at a higher clock rate. Likewise, it is in general more difficult to achieve a certain performance level, if the DAC is run at a higher sampling rate, because nonideal effects generated by the switching of DAC-elements will occupy a larger relative portion of the sampling interval. A careful optimization of the AFE-architecture is therefore mandatory to optimize the area and power consumption of the overall transceiver for a given performance target.

D/A-converters for modern digital transceivers are almost exclusively designed in advanced CMOS technologies. The main reason for this choice is that CMOS allows the data converters to be integrated together with analog and RF circuits, as well as a powerful DSP. This highest possible level of integration is often termed system-on-chip (SoC). To simplify the system integration, DAC-modules designed for the transmit path of digital transceivers must be fully embeddable macros and preferably run from a single supply with lowest possible power dissipation.

1.3 The Ideal D/A-Converter

As already stated previously, a DAC generates an analog output signal by summing together as many electrical unit quantities as represented by the actual digital input code. The electrical unit quantity can be a voltage, current, or an electrical charge, and it is usually derived from a reference circuit, e.g., a bandgap voltage. We can write a B-bit binary digital input code vector d as a weighted sum of the input bits b_i that constitute the digital word:

$$d = \sum_{i=0}^{B-1} 2^i \cdot b_i \qquad\qquad b_i \in [0, 1]. \tag{1.1}$$

The bit with the smallest weight (b_0) is called the least significant bit (LSB), while the bit with the largest weight (b_{B-1}) is called the most significant bit (MSB). An ideal B-bit D/A-converter, fed with the digital input code of Eq. (1.1), produces an output signal u given by

$$u = \Delta \cdot \sum_{i=0}^{B-1} 2^i \cdot b_i . \tag{1.2}$$

Δ is the smallest output step that the D/A-converter can perform. It is commonly called the unit quantity or LSB-quantity. The maximum and minimum value that the output signal can assume is called positive and negative full-scale (FS) value, respectively. The difference between these two values is called the full-scale range (FSR) of the converter:

$$\text{FSR} = \Delta \cdot \left(2^B - 1\right). \tag{1.3}$$

According to Eq. (1.2) the output of the D/A-converter can only take on integer multiples of the unit-quantity Δ. This discretization of the amplitude range is a fundamental property of digital signal processing and commonly called quantization. Likewise, the difference between a continuous-amplitude signal and its quantized counterpart is called the quantization error. For an ideal converter the quantization error is always in the range of $-\frac{1}{2}\Delta$ and $+\frac{1}{2}\Delta$ and is essentially a nonlinear error [30]. However, if the amplitude range of a quantized signal comprises many LSB-levels and the signal is not correlated with the quantization error, then we can treat it, to a good approximation, as noise [31]. This quantization "noise" obviously becomes smaller when the LSB-size is decreased relative to the full-scale range, meaning that the resolution of the converter is being increased. Note that the quantization error is not introduced by the ideal D/A-converter itself, still it is a fundamental limitation to the accuracy with which an analog signal can be reproduced. The relationship between the amplitude quantization and the achievable converter resolution is further explored in Sect. 2.3.1.

Digital signals are not only discrete in amplitude, but also discrete in time. As such, they are only defined at equidistant sampling instants, spaced T seconds apart.

T is called the sampling clock period, sampling period, or clock period. In practice, a D/A-converter must hold the corresponding output value for a significant fraction of the sampling period, in order to deliver sufficient signal power to the load.

1.3.1 The Non Return-to-Zero DAC

We define the unit impulse response of an ideal non return-to-zero (NRZ) D/A-converter, $p_{\mathrm{NRZ}}(t)$, to the (digital) Dirac sequence δ_k as a perfectly rectangular pulse with the height of 1 LSB and a duration $T = 1/f_{\mathrm{CLK}}$. In the following, we normalize the LSB-size Δ to 1, such that

$$p_{\mathrm{NRZ}}(t) = \mathrm{rect}\left(\frac{t - \frac{T}{2}}{T}\right). \tag{1.4}$$

The standard rect()-function is defined to be equal to 1 if its argument is between $-\frac{1}{2}$ and $+\frac{1}{2}$, and 0 otherwise. The DAC unit pulse in Eq. (1.4) is additionally shifted by half a clock period to the right on the time-axis in order to make the sampling instants fall on integer multiples of T. The ideal time-domain analog output $u_{\mathrm{NRZ}}(t)$ is given by multiplying the input data sequence d_k with the ideal unit impulse response $p_{\mathrm{NRZ}}(t)$ of the D/A-converter:

$$u_{\mathrm{NRZ}}(t) = \sum_k d_k \cdot p_{\mathrm{NRZ}}(t - k \cdot T) = \sum_k d_k \cdot \mathrm{rect}\left(\frac{t - k \cdot T - \frac{T}{2}}{T}\right). \tag{1.5}$$

The digital input code sequence d_k can assume only integer values and is updated with the sampling clock of frequency $f_{\mathrm{CLK}} = 1/T$. The output of the ideal NRZ D/A-converter is a staircase signal whose value stays constant within the whole sampling period T; see Fig. 1.10.

Equation (1.5) can also be written as the convolution of the continuous-time signal $d(t)$, which has been "sampled" (multiplied) with the Dirac comb, and the unit impulse response $p_{\mathrm{NRZ}}(t)$ [26]:

$$u_{\mathrm{NRZ}}(t) = \left[d(t) \cdot \sum_k \delta(t - k \cdot T)\right] * p_{\mathrm{NRZ}}(t). \tag{1.6}$$

At first, $d(t)$ is a purely hypothetical continuous-time signal that corresponds exactly to the digital input code sequence d_k at the sampling instants $t = kT$, but could take on arbitrary values elsewhere. A more likely scenario, however, consists in $d(t)$ being a real-valued continuous-time signal obeying the Nyquist criterion, which has been digitized by an ADC. In this case Eqs. (1.5) and (1.6) formalize the reconstruction process using an ideal NRZ-DAC.

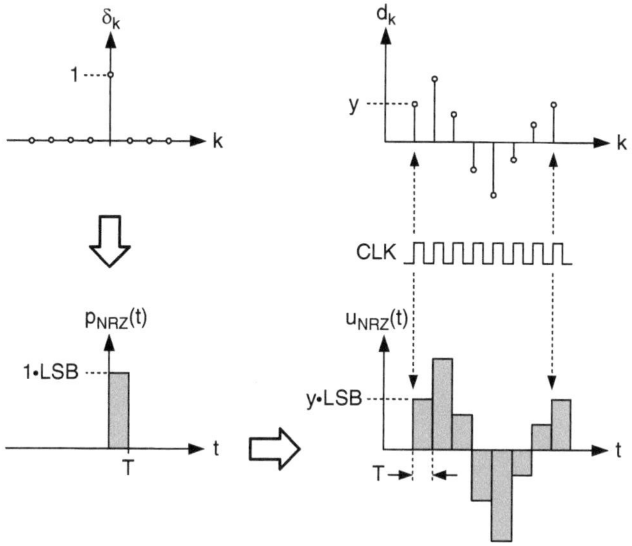

Fig. 1.10 Ideal DAC output signal

In the frequency domain, Eq. (1.6) is equivalent to the product of the repetitive spectrum of the input signal $d(t)$, as obtained by the ideal sampling process, and the Fourier transform of the ideal unit pulse $P_{\text{NRZ}}(j\omega)$.

$$U_{\text{NRZ}}(j\omega) = \frac{1}{T}\sum_{m} D\left(j\left(\omega - m\frac{2\pi}{T}\right)\right) \cdot P_{\text{NRZ}}(j\omega). \tag{1.7}$$

The Fourier transform of the ideal rectangular unit impulse response of Eq. (1.4) is given by

$$p_{\text{NRZ}}(t) = \text{rect}\left(\frac{t - \frac{T}{2}}{T}\right) \quad \circ\!\!-\!\!\bullet \quad P_{\text{NRZ}}(j\omega) = T \cdot \frac{\sin\frac{\omega T}{2}}{\frac{\omega T}{2}} \cdot \mathrm{e}^{-\frac{j\omega T}{2}}. \tag{1.8}$$

The ideal NRZ-DAC thus represents a zero-order hold (ZOH) [31]. Its transfer function is the well-known sin(x)/x or sinc(x) function, with zeros at multiples of f_{CLK}. Figure 1.11a shows the spectrum of the original signal $d(t)$, before and after sampling. According to the Nyquist–Shannon theorem [26], only signals with a bandwidth smaller than $\frac{1}{2}f_{\text{CLK}}$ can be reconstructed unambiguously after being sampled with f_{CLK}. Therefore, $\frac{1}{2}f_{\text{CLK}}$ is also called the Nyquist frequency. Figure 1.11b displays schematically the ZOH transfer function $P_{\text{NRZ}}(j\omega)$, and Fig. 1.11c shows the spectrum of the reconstructed signal at the output of an ideal NRZ DAC, $U_{\text{NRZ}}(j\omega)$. At the Nyquist frequency $\frac{1}{2}f_{\text{CLK}}$ the sinc function introduces an amplitude error of $2/\pi$, corresponding to -3.92 dB.

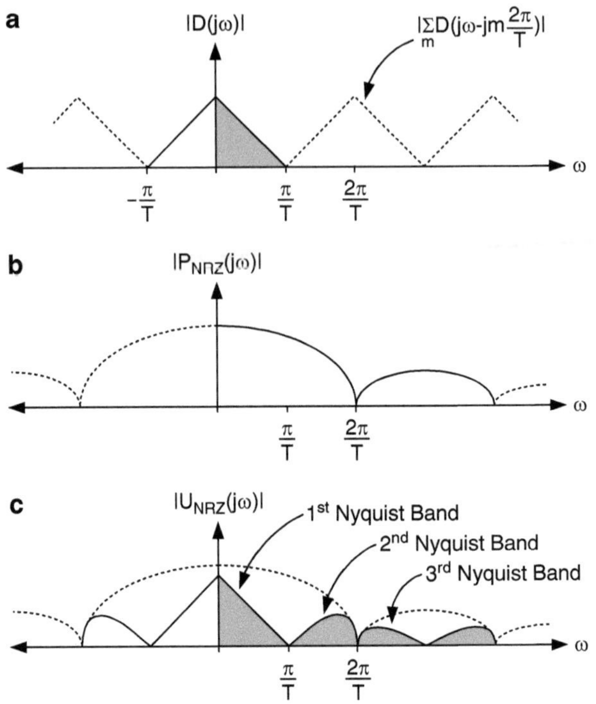

Fig. 1.11 Spectrum of ideal NRZ DAC: (a) (Repetitive) input spectrum, (b) NRZ DAC unit pulse spectrum, and (c) NRZ DAC output spectrum

1.3.2 The Return-to-Zero DAC

If the unit impulse response is made shorter than the sampling period T, we speak of a return-to-zero (RZ) DAC. As shown in Fig. 1.12a, the output of the DAC is only active for $T_S < T$ and reset to zero for the rest of the sampling period. We call the fraction $D = T_S/T$ with $D \in]0..1[$ the duty-factor of the RZ-DAC. The unit impulse response and the corresponding Fourier transform (shown in Fig. 1.12b) are given by

$$p_{\text{RZ}}(t) = \text{rect}\left(\frac{t - \frac{T_S}{2}}{T_S}\right) \quad \circ\!\!-\!\!\bullet \quad P_{\text{RZ}}(j\omega) = T_S \cdot \frac{\sin \frac{T_S}{2}}{\frac{\omega T_S}{2}} \cdot \text{e}^{-\frac{j\omega T_S}{2}}. \tag{1.9}$$

The first zero in the frequency response is now shifted out to $1/T_S = 1/D \cdot f_{\text{CLK}}$, such that the relative amplitude error introduced by an RZ-DAC is smaller, as long as the reset period, or gap, $(T - T_S)$ consumes a significant fraction of the clock period T (Fig. 1.12c).

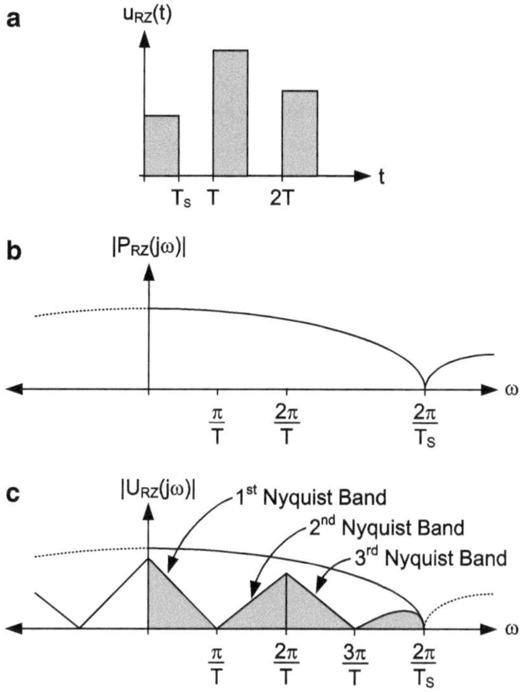

Fig. 1.12 Spectrum of ideal RZ-DAC: (a) ideal time-domain output waveform, (b) RZ DAC unit pulse spectrum, and (c) RZ DAC output spectrum

As discussed in Sect. 4.2.6, the primary motivation to use RZ is to improve the dynamic performance of D/A-converters. A major drawback is, however, that at the same time, the signal power residing in the baseband that can be delivered to the load is also reduced by the factor $(T_S/T)^2 = D^2$. In an oversampled converter employing an active transimpedance stage, the resulting high slew-rate at the output of an RZ-DAC may be problematic (see Sect. 1.4.3). In addition, the increased jitter sensitivity of the RZ-DAC, especially for low signal frequencies, requires in general a much higher-quality clock as compared to NRZ-implementations (see Sect. 2.3.3).

The frequency band from DC to $\frac{1}{2} f_{CLK}$ is sometimes called the first Nyquist band. Here, we find the reconstructed signal $d(t)$ that has been filtered with the unit impulse response of the DAC. Likewise, the frequency band from $1/2 \cdot f_{CLK}$ to f_{CLK} is called the second Nyquist band, from f_{CLK} to $3/2 \cdot f_{CLK}$ the third Nyquist band, and so forth. The majority of D/A-converter designs is intended for signal synthesis in the first Nyquist band, while the signal components residing in the higher Nyquist bands are eliminated by an analog filter.

An example of a D/A-converter intended for operation in higher Nyquist bands is described in [32]. The implemented half-clock RZ moves the first zero of the unit impulse response to $2 f_{CLK}$, and the extraction of the second or even the third Nyquist band by an analog bandpass filter is proposed. Although this method potentially allows the translation of the baseband signal to very high frequencies without the use of an analog mixer, it requires a precise and sufficiently linear analog bandpass filter.

Moreover, with signal content near DC the filter order must be correspondingly high to suppress the adjacent image. Note, also, that the signal spectrum in the even Nyquist bands is mirrored with respect to the baseband spectrum. This must be taken into account when generating the digital input sequence for the DAC, in case an even-order Nyquist band is intended for signal transmission.[6]

1.4 The Current-Steering DAC

Current-mode D/A-converters represent the digital input code by summing together the corresponding number of unit currents. Because the major part of analog signal processing still takes place in the voltage domain, the output current of the DAC is in most cases converted into an output voltage. The output stage, i.e., the circuit that provides the I-V conversion, must be viewed as an integral part of the D/A-converter because its properties also do influence the behavior of the current sources. In the simplest case, the output stage is a pair of resistors connected to a supply rail.

1.4.1 General Description

At a very basic level, a current-steering DAC consists of an array of current sources. A pair of current switches is connected to each current source to steer the current to either the positive or negative output node. The current switches are controlled by the digital input code—depending on the array coding—via a suitable decoder (see Sect. 1.5.1).

In single-ended implementations (see Fig. 1.13a), one switch connects the current source to the output, while the second switch, when switched on, diverts the current to a low-impedance node (e.g. a supply rail), without putting it to further use.

Most modern implementations, especially for transceiver applications, are differential designs that use both complementary output nodes (see Fig. 1.13b). One reason is, that the analog circuits, to which the D/A-converter in a digital transceiver must interface, are usually differential as well. A single-ended DAC-output would then require a single-ended to differential conversion, which is never trivial in high-performance applications. More importantly, differential implementations double the available signal swing, while at the same time the resulting structural symmetry improves other properties, like immunity to certain common-mode disturbances.

Current sources are built with active circuit elements, i.e., MOS or bipolar transistors operated in the active region. MOS-implementations are generally preferred in transceiver applications because they can be integrated in the most basic digital CMOS process together with a large DSP. However, modern digital

[6]Alternatively, the flipped spectrum must be handled correctly in the receiver DSP.

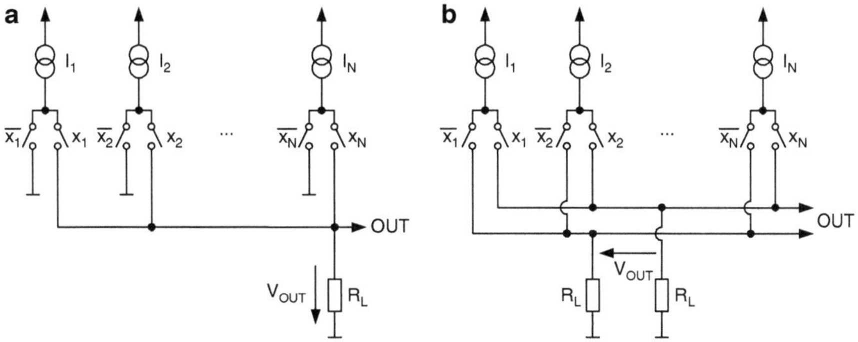

Fig. 1.13 Single-ended (**a**) and differential current-steering (**b**) DAC

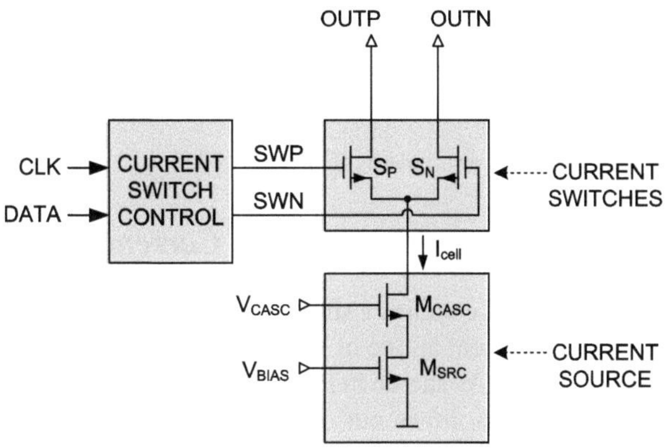

Fig. 1.14 Principle of MOS-based current cell

transceivers are usually fabricated in "analog" CMOS technologies, featuring also high-quality passive circuit elements like linear resistors and capacitors. Such processes offer a much broader range of possibilities to integrate together digital signal processing, data converters and linear analog signal processing circuits.

Figure 1.14 shows the principle of a (single-polarity) current-cell in MOS technology. It consists basically of three parts. The current source (M_{SRC}, M_{CASC}) generates the cell current I_{cell}, which is then steered to either OUTP or OUTN with the current switches S_P and S_N, depending on the input data (DATA). Because a D/A-converter contains many current cells, the input data must be locally resynchronized with a high-quality clock (CLK), to ensure that all current cells switch at exactly the same time. Also, the switching signals SWP and SWN have to be properly aligned to minimize the dynamic voltage change at the drain node of the cascode transistor M_{CASC} during switching. These tasks are accomplished by the current switch control.

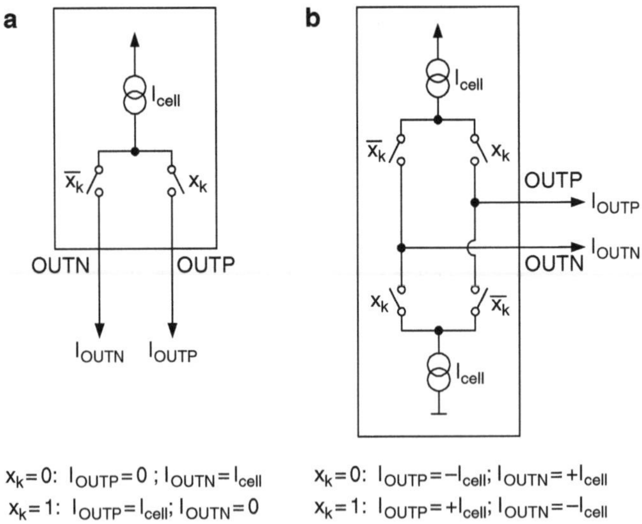

Fig. 1.15 Single-polarity (**a**) and dual-polarity (**b**) current cell

1.4.2 *Single-Polarity and Dual-Polarity Current Cells*

A single-polarity current cell contains a single current source I_{cell} and can therefore only source (or sink) its current at one of the two output branches, while the other output is left open-circuited. As shown in Fig. 1.15a, when the right switch is closed (x_k = high) the current I_{cell} flows out of OUTP, while OUTN is disconnected. With x_k = low, the left switch connects the current source with OUTN. The impedance seen at the output branches thus depends on the value of the switching bit x_k. Because single-polarity current-steering DACs are usually working into a low impedance load, this fundamental asymmetry is not problematic.

A dual-polarity current cell (Fig. 1.15b) can source *and* sink current. The two complementary current sources in the upper and lower half nominally carry the same current I_{cell}. When x_k = high, the upper current source is connected to OUTP and the lower current source to OUTN, the differential output current $I_{\text{OUTP}} - I_{\text{OUTN}}$ being $2 \cdot I_{\text{cell}}$. With x_k = low the connections are reversed and the currents in both output branches change polarity. The differential output current is now $-2 \cdot I_{\text{cell}}$. Provided that both current sources have identical properties, the impedance seen at both output branches is equal and independent of the switching bit x_k. This symmetry is important to guarantee code independent settling when an active output stage is used.

1.4.3 *Passive and Active Output Stage*

Single-polarity and dual-polarity current-cell arrays also require a different type of output stage that converts the signal current into a voltage. Assuming a differential

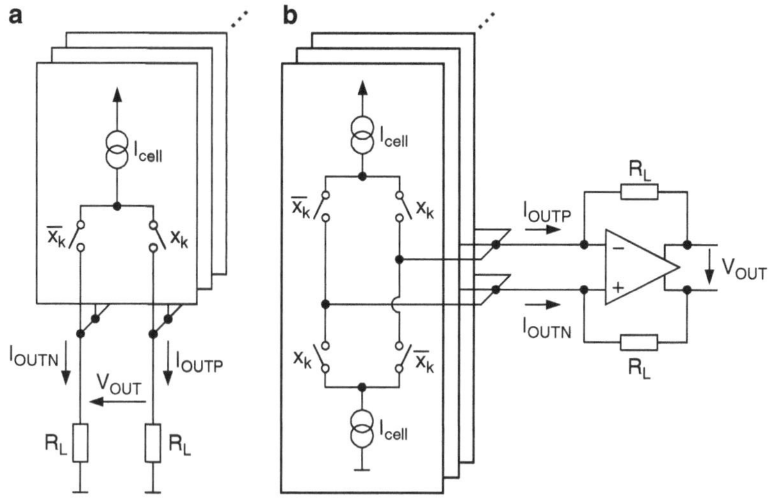

Fig. 1.16 Current-mode DAC output stages

output, the simplest I-V conversion is provided by a pair of resistors connected to one of the supply rails; see Fig. 1.16a. This output stage is preferred in conjunction with a single-polarity current-cell array. If I_{FS} is the maximum output current, the so-called full-scale current, then the available voltage range at each of the single-ended outputs is $[0; R_L I_{\mathrm{FS}}]$, while the differential output voltage range is $[-R_L I_{\mathrm{FS}}; +R_L I_{\mathrm{FS}}]$. The common-mode voltage of the differential output is at $\frac{1}{2} \cdot R_L I_{\mathrm{FS}}$, which is quite close to one of the supply rails.

The maximum output voltage of a single-polarity current-steering DAC with resistive loading is limited by the compliance voltage range of the current cell that still allows it to achieve a high accuracy. As a rule of thumb, in a single-supply environment with core supply voltage VDD, the maximum single-ended output voltage for high-linearity applications is typically limited to one quarter of VDD. The maximum differential FSR is thus 50 % of the supply voltage. Op-amp-based analog circuits on the other hand typically use a differential voltage range of 100–120 % of VDD, with a common-mode voltage about halfway between the supply rails. Therefore, if the D/A-converter of Fig. 1.16a interfaces to an active analog circuit, e.g., a reconstruction filter, then this block must provide a voltage gain of typically 6 dB. Unless a high input impedance buffer circuit is used, a common-mode current I_{CM} will flow into the load resistors R_L. This common-mode current offsets the single-ended output voltage by $R_L I_{\mathrm{CM}}$ and further decreases the linear differential voltage range by $4 \cdot R_L I_{\mathrm{CM}}$. Additionally, the input impedance of the following stage also appears in parallel to the load resistors R_L. As an advantage, the single-polarity current-cell array together with a passive output stage is an open-loop structure, and therefore allows the highest conversion speed and signal bandwidth.

Dual-polarity current-steering D/A-converters usually employ an active output stage, as shown in Fig. 1.16b. The I-V conversion is performed in the feedback branch of the operational amplifier by the resistors R_L. If I_{FS} is the maximum output current of each half of the current-cell array, then the full-scale differential output voltage range is $[-2R_L I_{FS}; +2R_L I_{FS}]$, with the common-mode voltage set by the operational amplifier. The output nodes of the current-cell array are also maintained at the same common-mode level and the voltage swing at these nodes is very small due to the large gain of the amplifier. The active output stage thus decouples the current-cells from the output nodes and the full-scale voltage can be optimized. Typically, the output voltage range of the DAC can be made equal to the voltage range of the other op-amp-based building blocks connected to the converter. Thus, no further amplification is required, and no loading effects occur, since the output voltage is buffered by the operational amplifier. However, the dual-polarity current-steering DAC with active output stage is a closed-loop system and therefore not suited for very-high-bandwidth applications.

In summary, the choice of the current cell structure and associated output stage for the transmit-DAC of a digital transceiver heavily depends on overall system considerations.

1.5 Array Coding

As already mentioned, a D/A-converter relies on the ability of accurately summing together an appropriate number of unit quantities and providing the result at its output port. In every sampling period the number of unit quantities to be summed is given by the digital input word. We call the way in which the possible digital input codes are represented internally by arrays of circuit elements and their respective weights relative to the full-scale value array coding. The following three array types are commonly encountered in D/A-conversion:

- Unary array
- Binary array
- Segmented array

1.5.1 Unary Array

A B-bit unary array consists of $2^B - 1$ identical unit elements, each having LSB-size, as shown in Fig. 1.17. To represent a binary digital code, the unary array requires a digital decoder that maps the binary input data to an equivalent number of DAC-elements. Because the unit elements are not distinguishable, this mapping function is not unique. This means that a specific input code can be represented by more than one set of DAC-elements. Only the full-scale values, i.e., all zeros or ones, have a unique representation, all other codes do have at least $2^B - 1$ different combinations of unit elements that generate nominally the same output value. Dynamic element matching (DEM) techniques (see Sect. 3.2) try to average out the unit element mismatch over time and make extensive use of this property of unary arrays.

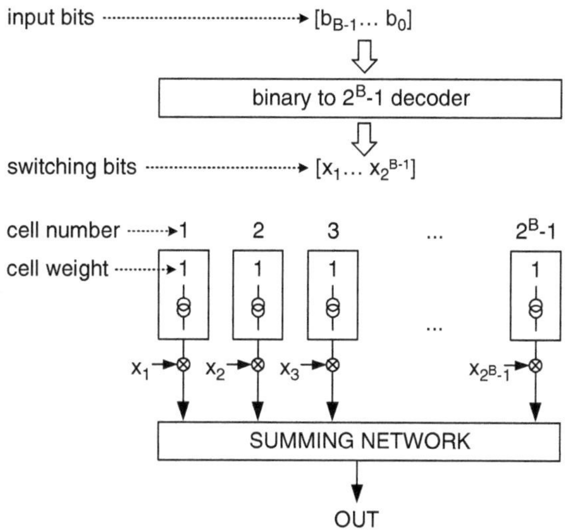

Fig. 1.17 Unary array

The drawback of unary arrays is that the complexity of the digital decoder is exponentially related to the resolution. In practical implementations the resolution of a unary array is therefore limited to typically 4–8 bits.

On the other hand, unary arrays offer the best possible performance. Because of their strictly incremental nature, monotonicity is always guaranteed. As shown in Sect. 2.1.2, the differential nonlinearity (DNL) of unary arrays is unequaled, because for every code transition it depends only on the accuracy of one single DAC-element of LSB-size. The same is true for a limited class of dynamic effects. When switching from one code to the next, certain transitional errors are linearly related to the code step. If the DAC-elements and the associated switching errors are additionally well matched, then, supposedly, only linear errors can occur in the output signal [33]. However, it will be shown in Sect. 4.1, that even perfectly matched unary current-steering arrays are subject to certain switching imperfections that cause nonlinear distortion.

Unary arrays are also the architecture of choice for multibit current-steering $\Sigma\Delta$-DACs. Two design examples using a 6-bit unary array consisting of 64 current sources are described in Chap. 5.

1.5.2 Binary Array

In a binary array the elements are binary weighted and thus directly correspond to the respective weight of the bits of a binary digital input code. The LSB-element has unit weight, while the MSB-element has a weight of 2^{B-1}. Because of the direct correspondence between input bits and single DAC-elements, digital decoding is not necessary—the input bits can directly control the respective DAC-cells. For a binary array the mapping function is always unique: each input code is represented by exactly one combination of DAC-elements (Fig. 1.18).

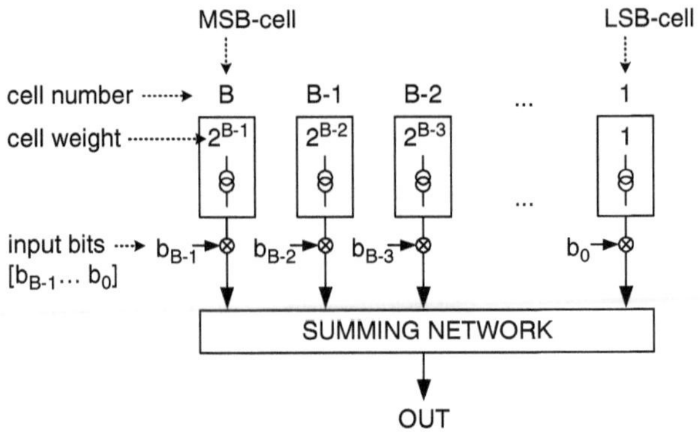

Fig. 1.18 Binary array

Binary arrays have the simplest possible implementation and minimum overhead for digital logic. In terms of performance, however, they are inferior to unary arrays. Monotonicity is not guaranteed, and the DNL of binary arrays is generally worse, because major carry transitions rely on addition *and* subtraction. For the same reason, dynamic errors related to the switching of the DAC-elements, commonly called glitches, are also much more pronounced. For these reasons, binary arrays are believed to be limited to the lower resolution range. As a counter-example, a 250 MS/s 10-bit binary-coded current-steering DAC having a SFDR>60 dB within the first Nyquist band is described in [34].

1.5.3 Segmented Array

Segmented arrays consist of different sub-arrays, or segments, each with a potentially different array coding. Figure 1.19 shows an example of a segmented converter with two sub-arrays. The MSB-segment is a unary array with $2^M - 1$ elements and represents the upper M bits. The LSB-segment realizes the lower L bits in a binary array. The overall resolution of the converter is $B = M + L$. The MSB-elements all have an equal weight of 2^L and are controlled by a binary to $2^M - 1$ decoder. The LSB-elements on the other hand can be directly controlled by the lower L bits of the digital input code. In order to equalize the delay of the MSB-decoder, usually a dummy decoder is inserted to align the timing for the LSB-segment [35].

Segmented arrays offer an attractive compromise between the superior static and dynamic performance, but increased complexity of unary arrays and the simplicity, but inferior performance of binary arrays. Therefore, in a segmented architecture the MSB-segment is virtually always a unary array, while the LSB-segment is typically a binary array. The higher the amount of segmentation, i.e., the resolution of the unary MSB-array relative to the overall converter resolution, the better the dynamic

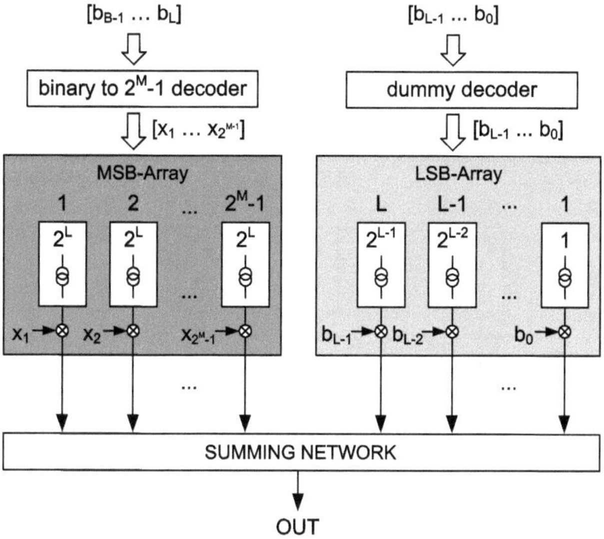

Fig. 1.19 Segmented D/A-converter with two segments

performance will be. On the other hand, the complexity and silicon area needed for the binary to thermometer decoder is exponentially related to the resolution of the MSB-segment.

In [33] the optimum amount of segmentation for current-steering DACs is empirically stated to be at the point where the analog area (current sources) and the digital area (decoder) are equal. Along this line of reasoning, with the continuing migration to finer geometry technologies the optimum amount of segmentation should be shifting towards fully unary arrays, because analog and digital circuits do not scale equally. At least in uncalibrated D/A-converters the analog area will be dominated by the current source transistors, and these are subject to accuracy and noise requirements. Although the matching generally improves with smaller minimum channel length, various process related effects tend to counteract this trend in deep-submicron technologies [36–38]. Moreover, the continuous reduction of the supply voltage decreases the gate overdrive voltage that can be applied to the current source transistors. This generally worsens the drain current matching for a given transistor area [39]. Because the output voltage swing is also reduced along with the supply voltage, the full-scale current must be increased accordingly, in order to maintain a constant SNR. In a given technology with already maximized gate overdrive this can only be achieved by increasing the transistors' aspect ratio. All these effects taken together do not allow the analog portion of the converter to follow the aggressive "digital" scaling rules.

However, in present technologies segmentation is still generally used for converters having a resolution above 8 bits. The unary MSB-segment in these designs typically realizes 4–8 bits. For still higher resolutions, typically starting

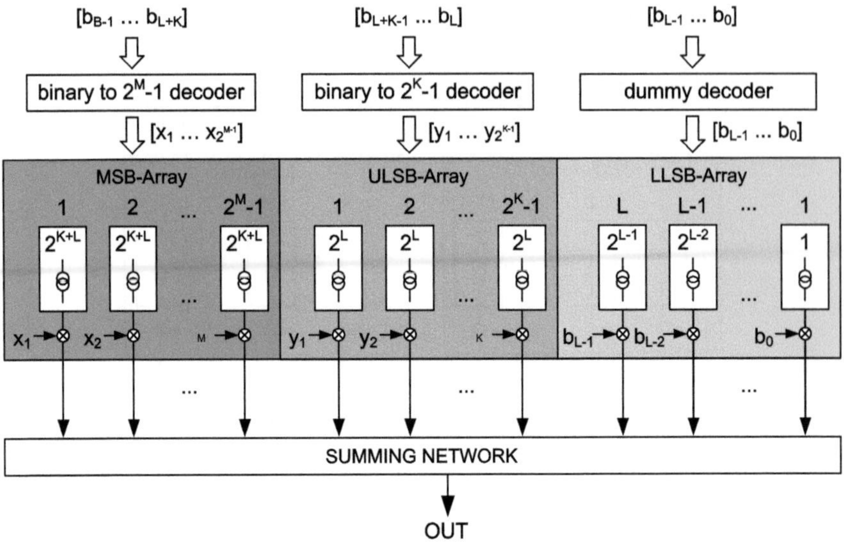

Fig. 1.20 Segmented D/A-converter with three segments

around 12 bits, an additional low-resolution (2–3 bits) intermediate segment is inserted between the MSB-array and the LSB-array [40–42]. Although the performance is slightly compromised by splitting the current-cell array into more segments, the total area required for the digital circuitry is greatly reduced. In Fig. 1.20 a segmented converter with three sub-arrays is shown. The MSB-segment again represents M bits in a unary array with $2^M - 1$ equal elements and thus requires a binary to $2^M - 1$ decoder. The middle segment, sometimes called the upper-LSB array (ULSB), codes the next K bits, also in a unary array with $2^K - 1$ elements. Since K is usually small, the ULSB-decoder will not add significantly to the overall digital complexity. Finally, the lower-LSB array (LLSB) represents the remaining L bits in a binary array. The overall resolution of the converter is thus $B = M + K + L$. The MSB-elements and ULSB-elements each have a weight of 2^{K+L} and 2^L, respectively.

Two examples of segmented current-steering D/A-converters are described in Chap. 6. Both designs realize a 13-bit converter using three segments:

- MSB-segment: 6-bit unary array
- ULSB-segment: 2-bit unary array
- LLSB-segment: 5-bit binary array

This choice of the segmentation mainly depends on practical considerations. A 6-bit MSB-segment is a compromise between a moderate amount of decoding logic, but considerably improved dynamic performance. The 2-bit thermometer-coded intermediate segment provides a somewhat smoother transition to the binary-coded 5-bit LLSB-array, while the required decoder complexity for the ULSB-segment remains practically negligible.

Chapter 2
Performance Figures of D/A-Converters

2.1 Static Accuracy

Static accuracy parameters describe the DC-characteristic of the D/A-converter. In all cases the difference between the actual characteristic and the presumed ideal converter behavior is quantified. The static performance figures are mostly referred to the LSB-size of the converter, in some cases also to the full-scale range. In the following, we assume a D/A-converter with voltage output.

2.1.1 Gain and Offset Error

If the converter resolution is sufficiently large, the granularity of the output voltage is not visible on a macroscopic level when looking at the static output characteristic. In this case the converter output voltage, as a function of the input code, appears as a smooth curve. The linear deviation of the measured output characteristic from the ideal output is expressed as gain and offset error.

In the differential output characteristic of Fig. 2.1 the offset error V_{OFFSET} is measured as the output voltage deviation at the midscale code (MS), where the ideal characteristic passes through 0.

The gain error of the D/A-converter is defined as the difference in the full-scale range $V_{\mathrm{FSP}} - V_{\mathrm{FSN}}$ between the measured and the ideal characteristic, after the offset error is removed [5]. Referring to Fig. 2.1, the gain error GE, normalized to the LSB-size, can be expressed as

$$\mathrm{GE} = \frac{\Delta V_{\mathrm{FSP}} - \Delta V_{\mathrm{FSN}} - V_{\mathrm{OFFSET}}}{V_{\mathrm{LSB}}} \quad [\mathrm{LSB}]. \tag{2.1}$$

Gain and offset error are linear errors that affect the whole code range of the converter in the same way. Both errors can relatively easily be trimmed to zero

M. Clara, *High-Performance D/A-Converters*, Springer Series in Advanced Microelectronics 36, DOI 10.1007/978-3-642-31229-8_2,

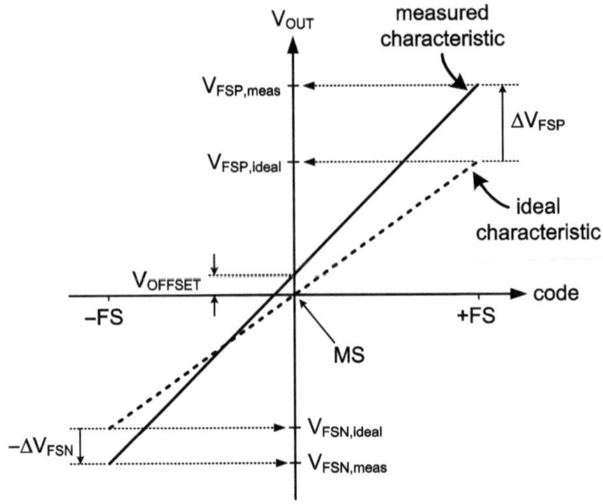

Fig. 2.1 Gain and offset error

by changing the value of the reference voltage or current (gain) and by adding a constant voltage or current at the converter output (offset). Note that the linearity of the converter is not affected by gain and offset deviations.

2.1.2 *Differential Nonlinearity*

The differential nonlinearity (DNL) of a D/A-converter for a given code transition $(k-1) \rightarrow k$ is defined as the difference between the actual step height and the ideal output step, i.e., the LSB-step V_{LSB}. Normally, the DNL is referred to the LSB-size, such that (see Fig. 2.2).

$$\mathrm{DNL}(k) = \frac{V_k - V_{k-1} - V_{\mathrm{LSB}}}{V_{\mathrm{LSB}}} \quad [\mathrm{LSB}] \qquad k \in [1, 2, \ldots, 2^B - 1]$$
$$\mathrm{DNL}(0) = 0. \tag{2.2}$$

The DNL is defined for all codes k by additionally setting $\mathrm{DNL}(0) = 0$. Often, the DNL of a converter is characterized by reporting only the maximum and minimum value of $\mathrm{DNL}(k)$.

If the absolute value of the DNL is smaller than 1 LSB for all codes, the converter is guaranteed monotonic [5, 31]. In this case the output is strictly increasing for increasing input codes, meaning that the slope of the output characteristic does not change sign for any given code transition. In a binary-weighted structure non-monotonicity can occur, when the DNL exceeds ± 1 LSB.

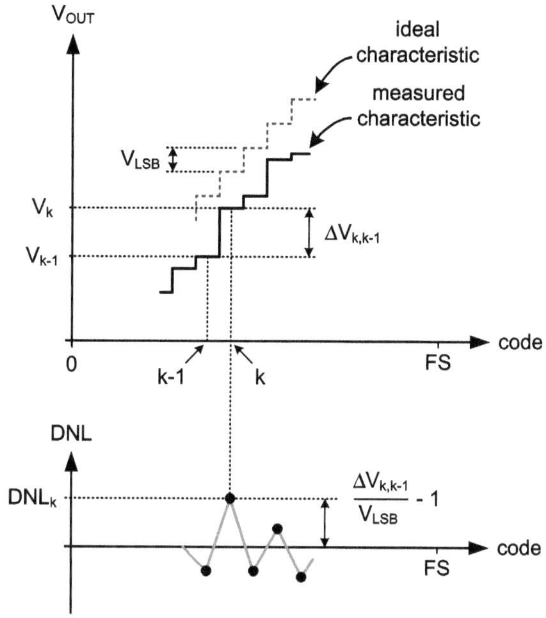

Fig. 2.2 Differential nonlinearity (DNL)

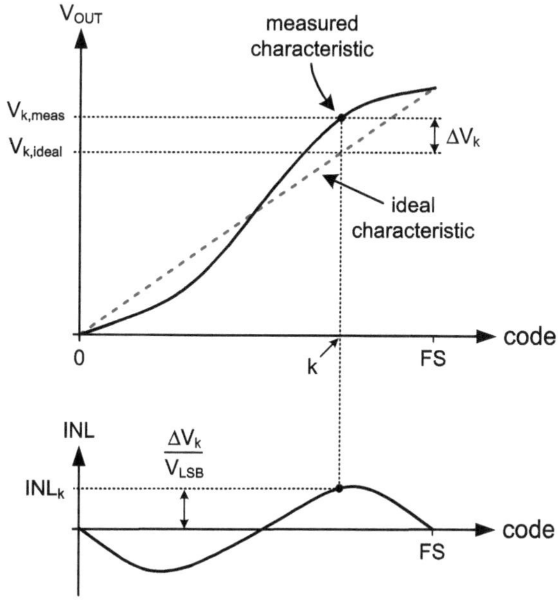

Fig. 2.3 Integral nonlinearity (INL)

2.1.3 Integral Nonlinearity

The integral nonlinearity (INL) is defined for every code k as the deviation of the actual output characteristic from a straight line, see Fig. 2.3. The INL is normally referred to the LSB-level:

$$\mathrm{INL}(k) = \frac{V_{k,\mathrm{meas}} - V_{k,\mathrm{ideal}}}{V_{\mathrm{LSB}}} \qquad [\mathrm{LSB}] \qquad k \in [0, 1, \ldots, 2^B - 1]. \tag{2.3}$$

Also the INL of a converter is often specified by the maximum and minimum value of $\mathrm{INL}(k)$. While the DNL is a measure for the local step-size deviation at single code transitions, the INL represents the cumulative effect of errors in the output step size, and thus is a measure for the macroscopic deviation from the ideal converter characteristic.

In Fig. 2.3, the straight line that represents the ideal output characteristic is drawn between the "endpoints" of the actual output characteristic given by the full-scale output values. This so-called endpoint-INL is easy to measure and is therefore preferred in the characterization of D/A-converters.[1]

The endpoint-INL and the DNL of an arbitrary converter are related by:

$$\mathrm{INL}(n) = \sum_{k=0}^{n} \mathrm{DNL}(k) \qquad n \in [0, 1, \ldots, 2^B - 1]. \tag{2.4}$$

A sufficient, but not necessary condition for the monotonicity of a converter is that the endpoint-INL is smaller than or equal to $\pm\frac{1}{2}$ LSB [31].

2.2 Dynamic Performance

Because the transmit-DAC of a digital transceiver is used for signal synthesis, its dynamic performance is even more important than the static accuracy. Although the latter gives a lower bound on the achievable linearity near DC, at higher signal frequencies other effects, which tend to further deteriorate the linearity of the converter, can become dominant. Also, spurious tones that are not directly related with the synthesized signal may be injected into the DAC output.

2.2.1 *Harmonic Distortion*

The output signal of a nonlinear circuit under sinusoidal excitation contains not only the fundamental sine wave, also called the carrier, but also frequency components at integer multiples of the fundamental frequency, also called carrier frequency. These spectral components generated by the nonlinearity are called harmonic distortion

[1] Another INL-definition uses the straight line corresponding to a least mean square (LMS) fit as ideal characteristic [43]. In practice, however, the difference is sufficiently small, such that the actual choice of the "best-fit" straight line is not of highest importance for the static linearity characterization of a D/A-converter.

Fig. 2.4 Harmonic distortion

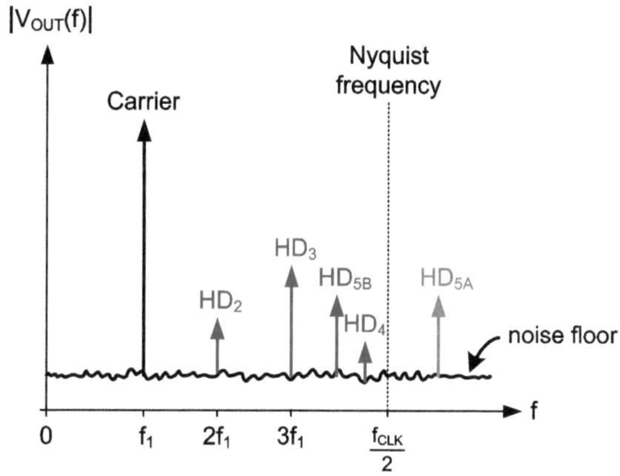

components, or harmonics. Given the carrier frequency f_1, the kth-order harmonic distortion at frequency $k \cdot f_1$ is given by

$$\mathrm{HD}_k = 20\log_{10}\left(\frac{A_k}{A_1}\right), \tag{2.5}$$

where A_1 is the magnitude of the fundamental sine wave and A_k the magnitude of the kth harmonic. Note that nonlinear distortion components generated in a sampled-data system also experience aliasing, i.e., the kth order harmonic distortion is found at $|kf_1 - nf_{\mathrm{CLK}}|$, with n the smallest integer fulfilling $2kf_1 > nf_{\mathrm{CLK}}$.

In a current-steering D/A-converter, distortion components that experience aliasing are necessarily generated before the current switches, e.g., HD_5 in Fig. 2.4. Nonlinearities in the output stage, located after the current switches, can only generate frequency multiples of the existing spectral components, without folding them back at the Nyquist frequency. In Nyquist-rate DACs, this effect can sometimes help to localize the source of the nonlinearity.

2.2.2 *Intermodulation Distortion*

When two sine waves, usually with similar frequencies, are injected into a weakly nonlinear system, intermodulation distortion products are generated, as shown in Fig. 2.5. Like harmonic distortion components, intermodulation products are subject to aliasing, provided they are generated before the current switches.

With the carrier frequencies f_1 and f_2, a quadratic nonlinearity generates the second-order intermodulation products IMD_2^- at $f_2 - f_1$ and IMD_2^+ at $f_2 + f_1$, the latter eventually aliased back at the Nyquist frequency. A cubic nonlinearity generates third-order intermodulation products $\mathrm{IMD}_{3,R}^-$ at $2f_2 - f_1$ and $\mathrm{IMD}_{3,L}^-$ at $2f_1 - f_2$. Furthermore, also third-order intermodulation products $\mathrm{IMD}_{3,R}^+$ at $2f_2 + f_1$

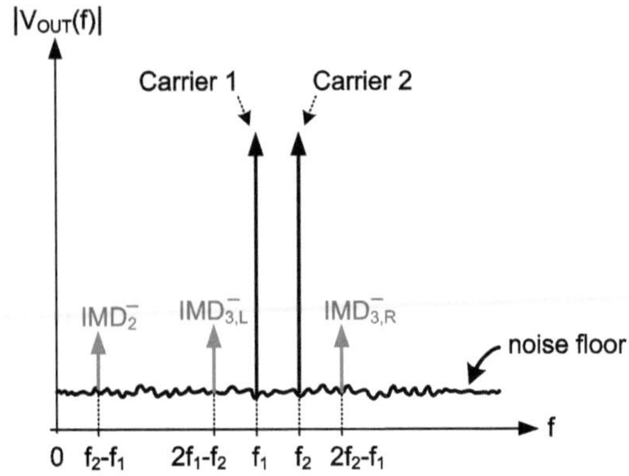

Fig. 2.5 Intermodulation distortion

and $\text{IMD}_{3,L}^{+}$ at $2f_1 + f_2$, of course eventually subject to aliasing, are generated by a cubic nonlinearity.

Assuming $f_1 < f_2$, the subscripts L and R denote the left- and right-hand position of the corresponding nonlinear distortion component on the frequency axis, at least as long as aliasing is avoided. If generated by a memoryless polynomial nonlinearity, the L and R components have equal magnitude. The $\text{IMD}_{3,LR}^{-}$ components appear close to the carriers and are therefore more important for narrowband applications, since they are perceived as in-band distortion. The same is true for the higher odd-order intermodulation components.

In general, a nonlinearity of order p produces intermodulation products up to order p, thus located at $|mf_1 \pm nf_2|$ with $m + n \leq p$. In differential circuits the odd-order nonlinearities are generally dominant.

2.2.3 *Spurious Free Dynamic Range*

Data converters are mixed-signal circuits and therefore can easily pick up digital disturber signals, that subsequently show up at the output. When these digital disturbers are not directly related with the input signal, they can appear in the output spectrum at different frequency locations than the normal nonlinearity products. Another potential source of disturbance is a continuously running background calibration that necessarily influences the behavior of the DAC-elements (Sect. 3.3). Figure 2.6 shows an example, where, besides second and third harmonic distortion, other spectral lines that are not (aliased) multiples of the carrier frequency appear.

Independent of the origin of unwanted spectral components, the spurious free dynamic range (SFDR) is defined as the distance from the carrier to the largest spectral component, after the carrier has been removed. However, good mixed-signal design practice should take care to minimize the injection of digital disturbances into the DAC output signal, such that the SFDR is mostly limited by nonlinear distortion.

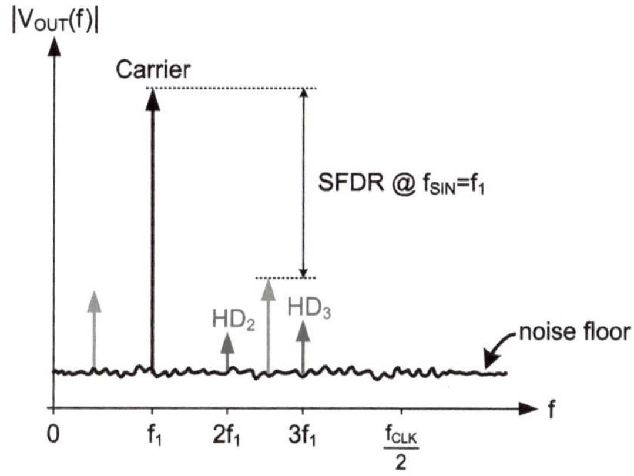

Fig. 2.6 Spurious free dynamic range

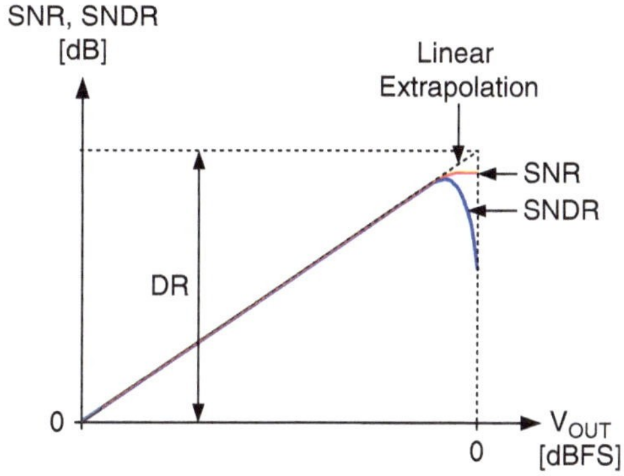

Fig. 2.7 Dynamic range

2.2.4 Dynamic Range

The signal-to-noise ratio (SNR) of a converter is defined as the ratio of the signal power divided by the total noise integrated over the signal bandwidth. Unless defined otherwise, harmonic distortion components are usually treated separately and therefore not included in the noise summation for the SNR. When the integrated noise power deliberately includes harmonic distortion, then we often speak of signal-to-noise and distortion ratio (SNDR). For practical purposes these definitions typically assume a single-tone signal when applied to data converters.

Mainly in the context of $\Sigma\Delta$-converters also the term signal-to-quantization noise ratio (SQNR) is used. It includes the in-band noise and harmonic distortion generated by the quantization operation in an otherwise ideal (digital) noiseshaper [44]. SQNR is thus identical to the SNDR in this special case.

Figure 2.7 shows the typical SNR and SNDR characteristic of a $\Sigma\Delta$-converter. For low signal levels[2] distortion products are usually not dominant and the two

[2]The x-axis is in dB relative to full-scale (dBFS). This means that the input signal is normalized to the converter full-scale 0 dBFS.

Table 2.1 Multi-carrier modulation systems

System	Modulation	Bandwidth	Carriers	Spacing
ADSL2+	DMT	2.2 MHz	512	4.3125 kHz
VDSL2-17a	DMT	17.66 MHz	4,096	4.3125 kHz
VDSL2-30a	DMT	30 MHz	3,479	8.625 kHz
802.11a/g (WLAN)	OFDM	20 MHz I+Q	48+4	312.5 kHz
DVB-T	OFDM	7.6 MHz	6817	1.116 kHz

curves are on top of each other. At higher signal levels, however, the $\Sigma\Delta$-modulator gets increasingly overloaded and the SNDR-characteristic peaks and falls off rapidly when approaching the full-scale level. In extreme overload also the SNR-characteristic typically saturates.

The dynamic range (DR) of the converter is defined as the maximum value of the linearly extrapolated SNR/SNDR-curves up to the full-scale value [45]. This is equivalent to stating that the smallest signal that can be reliably detected has the same power as the integrated in-band noise. The DR is primarily used to characterize the resolution of noiseshaped converters.

2.2.5 Multitone Linearity

With more than two carriers, the possible number of cross-modulation products increases drastically, even when only low-order nonlinearities are considered. Many wideband digital communication systems use a multitude of individually modulated carriers to optimize the data throughput over strongly varying channels. Digital subscriber line (DSL)-systems, e.g., use the so-called discrete multi-tone (DMT) modulation. Systems communicating over time-varying and rapidly fading channels favor a special multi-carrier modulation technique called orthogonal frequency division multiplexing (OFDM). In the following we will simply speak of DMT-signals, or multitone signals, irrespective of the actual modulation format.

Since the number of carriers in all these systems is quite high see Table 2.1, the direct calculation of a significant number of nonlinear products is computationally not feasible. Therefore, a different definition for the effect of a nonlinear transfer characteristic on multitone signals must be adopted.

We define a homogeneous single-band DMT signal (Fig. 2.8) to consist of N_C+1 carriers, each with the same amplitude A_{DMT}. The carriers are equally distributed within the frequency band $\omega_{\min} \leq \omega \leq \omega_{\max}$. The carrier spacing $\Delta\omega$ can thus be expressed as

$$\Delta\omega = \frac{\omega_{\max} - \omega_{\min}}{N_C}. \tag{2.6}$$

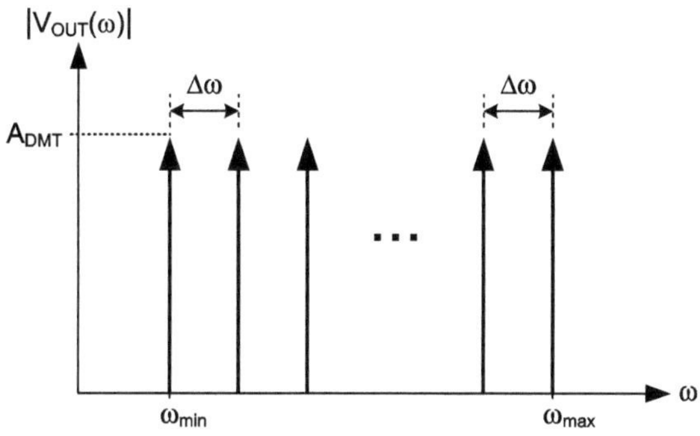

Fig. 2.8 Homogeneous multitone signal

The Crest Factor CF, also called peak-to-average ratio (PAR), is defined for an arbitrary signal as the ratio of its peak value A_{peak} to the rms value.[3] In a homogeneous DMT-signal the signal power is equally distributed among all carriers, therefore:

$$A_{\text{DMT}} = \frac{A_{\text{peak}}}{\text{CF}} \cdot \sqrt{\frac{2}{N_C + 1}} \approx \frac{A_{\text{peak}}}{\text{CF}} \cdot \sqrt{\frac{2\Delta\omega}{\omega_{\text{max}} - \omega_{\text{min}}}}. \tag{2.7}$$

A multiband DMT signal consists of a number of distinctive frequency bands ($n = 1, 2, \ldots, L$) populated with carriers. If the carrier amplitude A_{DMT} and the carrier spacing $\Delta\omega$ are equal in each band, we speak of a homogeneous multiband DMT signal.

The L frequency bands have a minimum and maximum frequency $\omega_{\text{min},n}$ and $\omega_{\text{max},n}$ with $n \in [1; L]$. The carrier amplitude A_{DMT} can be calculated as:

$$A_{\text{DMT}} \approx \frac{A_{\text{peak}}}{\text{CF}} \sqrt{\frac{2\Delta\omega}{\sum_{n=1}^{L} (\omega_{\text{max},n} - \omega_{\text{min},n})}}. \tag{2.8}$$

Shown in Fig. 2.9 is an example of a homogeneous multiband DMT signal. All carriers have the same amplitude A_{DMT} and the same frequency spacing $\Delta\omega$ within each band.

Missing Tone Power Ratio

To measure the effect of a nonlinear characteristic on a multitone signal, single carriers at specific frequency locations f_k are left out during signal synthesis. The resulting cross-modulation products that fall into these gaps are observed.

[3]The Crest Factor of a single sinusoid is $\sqrt{2}$; a two-tone signal has CF $= 2$. For more than two tones the Crest Factor depends on the relative phases of the single carriers.

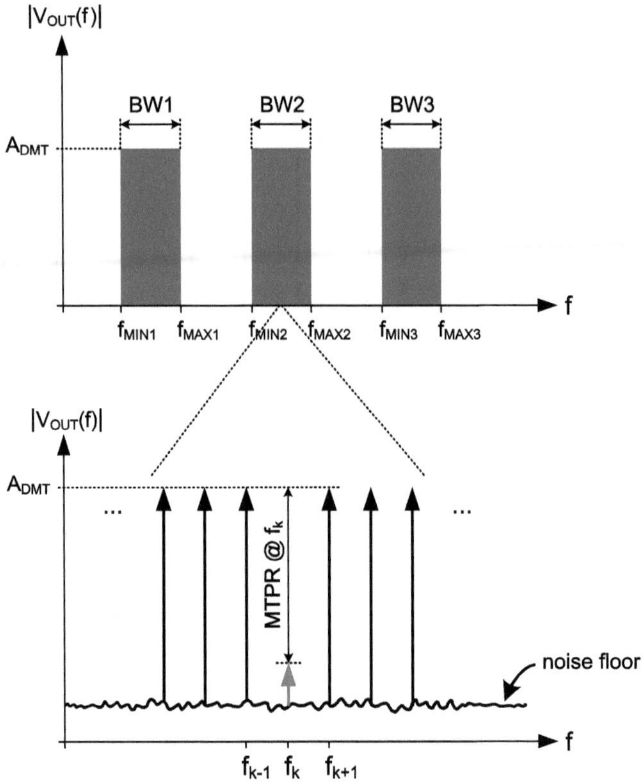

Fig. 2.9 Missing tone power ratio

The characteristic of the multitone signal is not altered significantly, at least as long as the number of missing tones is small compared to the total number of tones. The missing tone power ratio (MTPR) at the frequency of the missing tone (bin) f_k is then defined as the ratio of the power of the nonlinearly generated spectral component at frequency f_k and the power of the neighboring tone:

$$\text{MTPR}\,(f_k) = 20\log_{10}\left(\frac{A_k}{A_{k\pm 1}}\right) = 20\log_{10}\left(\frac{A_k}{A_{\text{DMT}}}\right). \tag{2.9}$$

Note that the MTPR is measured with unmodulated carriers.

Missing Band Power Ratio

Cross-modulation products of a multitone signal typically also fall outside of the frequency band(s) populated by the carriers, see Fig. 2.10. Due to the limited isolation between transmitter and receiver, these nonlinear modulation products

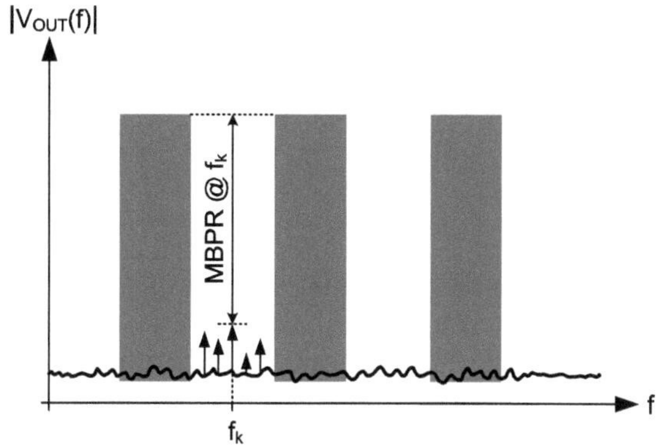

Fig. 2.10 Missing band power ratio

generated by the transmitter eventually appear in the receive bands and can limit the dynamic range of the receiver in a full-duplex system.

The missing band power ratio (MBPR) at a certain frequency f_k, typically a tone location in the receive-band, is defined as the ratio of the power of the nonlinearly generated spectral component at frequency f_k and the power of a single tone in the neighboring transmit band.

$$\text{MBPR}\,(f_k) = 20\log_{10}\left(\frac{A_k}{A_{\text{DMT}}}\right). \tag{2.10}$$

2.3 Noise Performance

Besides the linearity, also the noise performance of a converter is of importance, since it limits the dynamic range of the signals that can be synthesized or digitized. Noise in a data converter comes in three basic flavors:

- Quantization "noise"
- Circuit noise
- Jitter noise

2.3.1 Quantization "Noise"

As already stated in Sect. 1.3, quantization sets the fundamental limit for the accuracy with which a signal can be represented in the digital domain. When digitizing an analog signal in an A/D-converter the quantization error is irreversibly introduced due to the granularity of the decision levels. A D/A-converter reconstructing an analog signal preserves the quantization error already present in its input data. It may, however, add additional errors due to several nonideal effects. This will be discussed in the following sections and later chapters.

If the converter resolution is sufficiently large, the variance of this quantization "noise" $q(t)$, equivalent to its power, can be calculated as [31]:

$$\sigma_q^2 = E(q(t)^2) = \int_{-\infty}^{\infty} x^2 p_q(x)\mathrm{d}x = \frac{\Delta^2}{12}, \tag{2.11}$$

where $E(\cdot)$ is the expectation operator, $p_q(x)$ the probability density function of the quantization error, and Δ the LSB-size.

Equation (2.11) represents the total noise power of the quantization process in the frequency range from 0 to $\frac{1}{2}f_{\mathrm{CLK}}$, i.e., the first Nyquist band. The spectral shape of the quantization noise depends on the type of the data converter.

Nyquist-Rate Converter

A so-called Nyquist-rate DAC exhibits a flat quantization noise spectrum. The (one-sided) power spectral density (PSD) of the quantization noise is given by

$$S_q(f) = \frac{\Delta^2}{12} \cdot \frac{2}{f_{\mathrm{CLK}}}. \tag{2.12}$$

The amount of quantization noise, the quantization noise power, within the bandwidth of interest, denoted by f_B, can be calculated as

$$N_q = \int_{f_B} S_q(f)\mathrm{d}f = \frac{\Delta^2}{12} \cdot \frac{2 \cdot f_B}{f_{\mathrm{CLK}}} = \frac{\Delta^2}{12} \cdot \frac{1}{\mathrm{OSR}}. \tag{2.13}$$

The oversampling ratio (OSR) is defined as the Nyquist frequency $\frac{1}{2}f_{\mathrm{CLK}}$ divided by the signal bandwidth f_B:

$$\mathrm{OSR} = \frac{f_{\mathrm{CLK}}}{2 \cdot f_B}. \tag{2.14}$$

If we divide the power of a full-scale sine wave by the quantization noise power residing in the signal bandwidth of interest, we can calculate the maximum Signal-to-Noise Ratio (SNR) for the Nyquist-rate converter [31]:

$$\mathrm{SNR_{max}} = 10\log_{10}\left(3 \cdot 2^{2B-1}\mathrm{OSR}\right) = 6.02B + 1.76 + 10\log_{10}\mathrm{OSR} \;\; [\mathrm{dB}]. \tag{2.15}$$

The last term in (2.15) is also called the oversampling gain. With a constant integration bandwidth, every doubling of the clock frequency improves the effective resolution by 3 dB, or half a bit. Figure 2.11 shows the effect of different sampling rates on the quantization noise floor in a Nyquist converter.

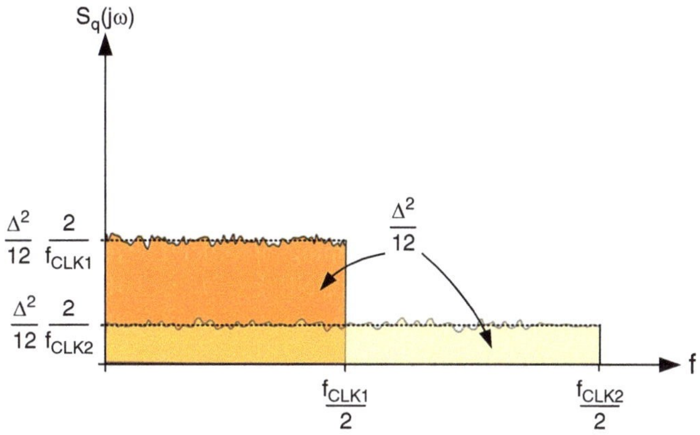

Fig. 2.11 Quantization noise of Nyquist-DAC

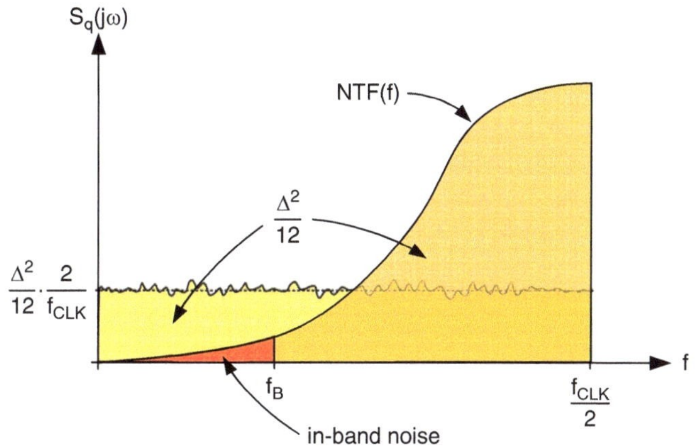

Fig. 2.12 Noiseshaped quantization noise

Noiseshaped Converter

A much more efficient use of oversampling can be achieved by spectral shaping of the quantization noise, commonly called noise shaping [44]. By suitable filtering, the major part of the quantization noise power is pushed outside the bandwidth of interest, leading to a much higher in-band SNR as compared to using oversampling alone. The principle of (low-pass) quantization noise shaping is shown in Fig. 2.12.

The spectral shaping of the quantization noise is described by the so-called noise transfer function (NTF). Since the total quantization noise power is always given by Eq. (2.11), the resulting noise in the frequency band of interest $[0; f_B]$ can be calculated by integrating the modulus squared of the NTF:

$$N_q = \frac{\Delta^2}{12} \cdot \frac{2}{f_{\text{CLK}}} \cdot \int_0^{f_B} |\text{NTF}(f)|^2 \text{d} f. \tag{2.16}$$

An ideal noiseshaper of order L has a noise transfer function in the z-domain given by $\mathrm{NTF}(z) = \left(1 - z^{-1}\right)^L$ with L zeros at DC ($z = 1$). The theoretical maximum SNR of a Lth-order noiseshaper is given by [46]

$$\mathrm{SNR}_{\mathrm{max},L} = 6.02B + 1.76 + (20L + 10)\cdot\log_{10}\mathrm{OSR} - 10\log_{10}\frac{\pi^{2L}}{2L+1}\ [\mathrm{dB}]. \quad (2.17)$$

According to Eq. (2.17), with a noiseshaping of order L every doubling of the OSR increases the maximum achievable SNR by $6L + 3$ dB, or $L + 0.5$ bits. Unfortunately, a NTF of the form $\mathrm{NTF}(z) = \left(1 - z^{-1}\right)^L$ with $L \geq 2$ cannot be realized, because the resulting noiseshaper is not unconditionally stable. Practical implementations must therefore use a less aggressive noise-shaping function by introducing poles into the NTF [44]. To optimize the noiseshaper performance, some of the zeros can additionally be moved away from the origin. In any case, the theoretical SNR predicted by (2.17) cannot be reached in actual designs for $L \geq 2$.

A hardware example of a D/A-converter displaying a remarkable performance boost due to noiseshaping is described in [47]. It describes an audio-DAC with a physical resolution of only 6 bits. Nevertheless, using a second-order noiseshaper and an OSR of 128, it achieves a theoretical in-band SNR of 125 dB, equivalent to 20.4 bits, while Eq. (2.17) predicts a theoretical, but not realizable maximum of 130.34 dB, or 21.3 bits. The actually measured in-band SNR, limited by the output stage noise, is still 111 dB in a 20 kHz bandwidth, equivalent to over 18 bits.

2.3.2 *Circuit Noise*

Data converters are affected by circuit noise, just like any other building block in the analog signal processing domain. The amount of circuit noise visible at the output of the converter puts an upper limit on the achievable dynamic range. For system-level considerations it must therefore be put into context with the other two noise-generating mechanisms, namely quantization noise and jitter noise.

In a current-steering D/A-converter we can identify three sources for the circuit noise:

- Current sources
- Output stage
- Biasing circuit

Figure 2.13 shows a single-polarity current-steering DAC with resistive load, together with the relevant noise sources. The (one-sided) PSD[4] of the current noise

[4] In this text the power spectral density (PSD) is almost always assumed one-sided. However, to avoid ambiguity, the qualifier "(one-sided)" is added at times. Examples of a two-sided PSD are found in Eqs. (2.41) and (B.2).

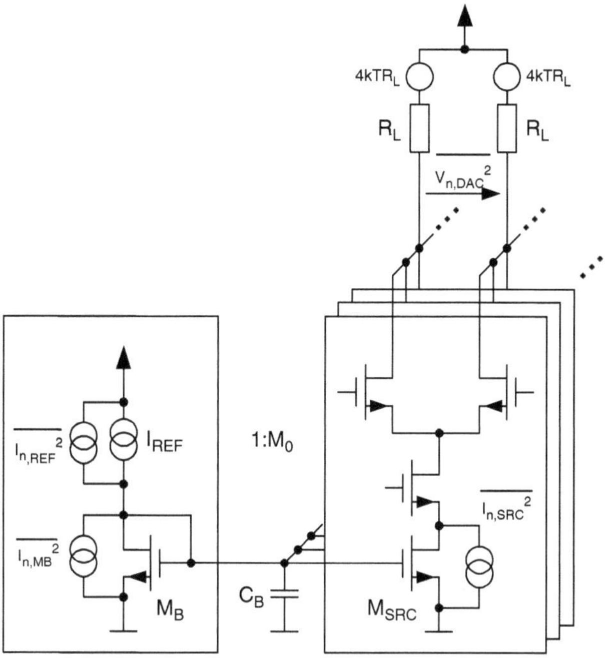

Fig. 2.13 Noise sources in a single-polarity current-steering DAC

in the long-channel MOS transistors M_{SRC} in strong inversion can be approximated by [5, 20, 48, 49]:

$$\overline{I_{n,\mathrm{SRC}}^2(f)} \approx 4kT \cdot \frac{2}{3} gm_{\mathrm{SRC}} + \frac{\mathrm{KF}_F}{WLC_{\mathrm{ox}}^2 f} gm_{\mathrm{SRC}}^2 = 4kT \cdot \frac{2}{3} gm_{\mathrm{SRC}} + \frac{K_F I_{\mathrm{DS}}}{C_{\mathrm{ox}} L^2 f}. \tag{2.18}$$

L is the channel length of the current-source transistor M_{SRC}, C_{ox} the oxide capacitance per unit area, KF_F and K_F are (not bias independent) technology constants related by $K_F = 2\mu \cdot \mathrm{KF}_F$, with the carrier mobility μ. The factor 2/3 in the thermal noise component is an approximation for long-channel transistors operating in strong inversion [20, 49]. More exact models include at least one additional, bias dependent factor; see, e.g., [50].

Note that in a single-ended implementation the total output noise of the current-source array is linearly dependent on the input code, such that the maximum output noise is generated when all current sources are switched to the output node, i.e., at positive full-scale. On the other hand, in a differential implementation, the noise power of all current sources is always present in the differential output signal, independent of the input code. Therefore, for the noise calculation in a differential converter we can assume that the current-source array consists of N identical sources, each carrying a fraction $\frac{1}{N}$ of the full-scale current I_{FS}:

$$\overline{I_{n,\mathrm{Array}}^2(f)} = N \cdot \overline{I_{n,\mathrm{SRC}}^2(f)} \approx 4kT \cdot \frac{2}{3} \cdot \frac{2I_{\mathrm{FS}}}{V_{\mathrm{GS}} - V_T} + \frac{K_F I_{\mathrm{FS}}}{C_{\mathrm{ox}} L^2 f}. \tag{2.19}$$

In (2.19) we use the following expression for the transistor transconductance in strong inversion [20, 50]:

$$gm = \frac{2I_{\mathrm{DS}}}{V_{\mathrm{GS}} - V_T}. \tag{2.20}$$

Equation (2.19) shows that the thermal noise of the current-cell array can be minimized by choosing the gate overdrive of the current-source transistors M_{SRC} as large as possible, thus making M_{SRC} very long. This in turn also reduces the 1/f-noise of the array, at least as long as the flicker noise current parameter K_F does not change significantly and the transistor does not enter velocity saturation.

The (differential) noise current of the output stage is simply given by the contribution of the two load resistors:

$$\overline{I_{n,\mathrm{OS}}^2(f)} = 2 \cdot \frac{4kT}{R_L}. \tag{2.21}$$

The current-source array is biased by the MOS-diode M_B and a reference current I_{REF}. M_B and M_{SRC} are "matching" transistors, thus they have the same channel length and gate overdrive voltage. The current noise of M_B and I_{REF} is mirrored to each of the N current sources and is steered via the current switches toward the positive or negative output, depending on the digital input code. Because the mirrored noise current from the biasing is correlated, it has the same amplitude and phase in all current sources. Therefore, the noise contribution of the biasing at the output of the DAC depends on the absolute value of the digital input code.

At the positive and negative full-scale output the noise from the biasing is maximum, while at midscale an equal number of current sources carrying the correlated bias noise is switched to either output node. Therefore, the noise contribution from the biasing is canceled in the differential output current at the midscale code.[5]

If we denote the (frequency-dependent) ratio between the full-scale current I_{FS} and the reference current I_{REF} with $M(j\omega)$, then the maximum bias current noise injected into the differential output is

$$\overline{I_{n,\mathrm{Bias}}(f)^2} = |M(j\omega)|^2 \cdot \left(\overline{I_{n,\mathrm{MB}}^2(f)} + \overline{I_{n,\mathrm{REF}}^2(f)}\right). \tag{2.22}$$

[5]Standard DAC-architectures implement an odd number of LSB-steps, see Eq. (1.2). Thus, at midscale the differential output current is not exactly zero but equal to the LSB unit current. If the resolution is reasonably large, though, the bias induced noise in the differential output will be practically negligible at midscale, and we may therefore safely ignore this subtlety.

The total noise voltage PSD at the output of the single-polarity current-steering DAC of Fig. 2.13 is given by

$$\overline{V_{n,\mathrm{DAC}}^2} = R_L^2\left(\overline{I_{n,\mathrm{Array}}^2} + \overline{I_{n,\mathrm{OS}}^2} + |d|\overline{I_{n,\mathrm{Bias}}^2}\right) = R_L^2\overline{I_{n,\mathrm{DAC}}^2}. \tag{2.23}$$

The digital input code d in Eq. (2.23) is normalized to the full-scale value, $|d| \leq 1$. The frequency dependent mirror factor $M(j\omega)$ in Eq. (2.22) displays a first-order roll-off:

$$M(j\omega) = \frac{M_0}{1 + j\omega\dfrac{C_B}{gm_{\mathrm{MB}}}}. \tag{2.24}$$

M_0 is the ratio of I_{FS} and I_{REF} at DC and is directly given by the mirror ratio, equal to the sum of the widths of the current-source transistors M_{SRC} divided by the width of the bias transistor M_B. The pole frequency of $M(j\omega)$ is the ratio of the transconductance of the MOS-diode g_{MB} and the capacitance C_B connected to its gate node.

We can approximate the transconductance of the MOS-diode M_B with dimensions W and L in the long-channel square-law model [5] as

$$gm_{\mathrm{MB}} \approx \mu C_{\mathrm{ox}}\frac{W}{L}(V_{\mathrm{GS}} - V_T). \tag{2.25}$$

With an additional MOS-capacitor M_C of area $\kappa(1+M_0)WL$, i.e., an area overhead of κ, the total capacitance at the gate node of M_B is approximately given by

$$C_B \approx (1 + M_0)\cdot\frac{2}{3}WLC_{\mathrm{ox}}\left(1 + \frac{3}{2}\kappa\right). \tag{2.26}$$

Assuming $M_0 >> 1$, the pole frequency $\omega_{p,\mathrm{BIAS}}$ of the mirror that biases the current sources of the D/A-converter becomes

$$\omega_{p,\mathrm{BIAS}} = \frac{gm_{\mathrm{MB}}}{C_B} \approx \frac{1}{M_0}\cdot\frac{3\mu}{2L^2}\cdot\frac{V_{\mathrm{GS}} - V_T}{1 + \frac{3}{2}\kappa} = \frac{\omega_T}{M_0\left(1 + \frac{3}{2}\kappa\right)}. \tag{2.27}$$

The typical frequency dependence of $V_{n,\mathrm{DAC}}$ for a midscale and full-scale input code is shown in Fig. 2.14. At low frequencies the noise from the DAC-bias becomes dominant for $|d| > 0$.

To simplify the analysis with the goal to achieve a more general insight, we neglect the 1/f-noise contribution. Furthermore, let the reference current noise be a scaled version of the noise of the MOS-diode M_B, $\overline{I_{n,\mathrm{REF}}^2} = K\cdot\overline{I_{n,\mathrm{MB}}^2}$, such that:

$$\overline{I_{n,\mathrm{Bias}}^2} = |M(j\omega)|^2\cdot\overline{I_{n,\mathrm{MB}}^2}\cdot(1 + K). \tag{2.28}$$

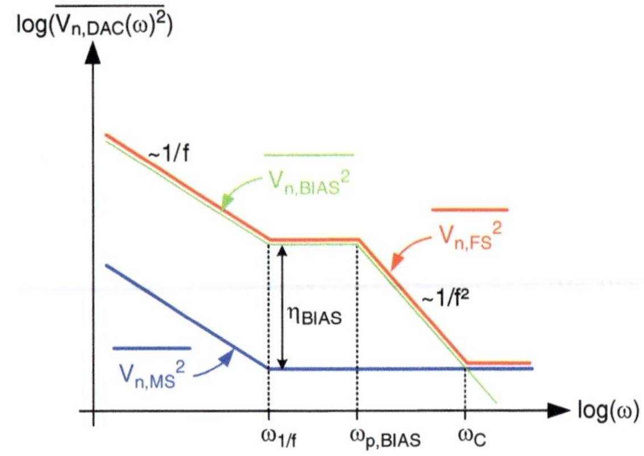

Fig. 2.14 Noise PSD of a current-steering DAC

In this case we can estimate the excess noise corner frequency ω_C:

$$\omega_C = \omega_{p,\text{BIAS}} \cdot \sqrt{\eta_{\text{BIAS}}}$$

$$\eta_{\text{BIAS}} = M_0^2 \cdot \frac{R_L^2 \overline{I_{n,\text{MB}}^2}\,(1+K)}{R_L^2 \overline{I_{n,\text{Array}}^2} + 8kTR_L}. \tag{2.29}$$

η_{BIAS} is the excess noise factor of the DAC-biasing for frequencies below $\omega_{p,\text{BIAS}}$. It is the ratio of the maximum noise current injected from the biasing and the noise of the current sources plus output stage. With the long-channel square-law approximation for the transistor transconductance (2.20), the thermal noise current of the current-source array and the MOS-diode M_B becomes

$$\overline{I_{n,\text{Array}}^2} \approx 4kT \cdot \frac{2}{3} \cdot \frac{2I_{\text{FS}}}{V_{\text{GS}} - V_T}$$

$$\overline{I_{n,\text{MB}}^2} \approx 4kT \cdot \frac{2}{3} \cdot \frac{1}{M_0} \cdot \frac{2I_{\text{FS}}}{V_{\text{GS}} - V_T} = \frac{\overline{I_{n,\text{Array}}^2}}{M_0}. \tag{2.30}$$

By introducing the ratio α of the maximum peak output voltage to the current-source transistors' gate overdrive[6]:

$$\alpha = \frac{I_{\text{FS}} R_L}{V_{\text{GS}} - V_T} = \frac{1}{2} \cdot \frac{V_{\text{FS}}}{V_{\text{GS}} - V_T} \tag{2.31}$$

[6] α reflects, at least partly, the relative partitioning of the supply voltage between output range and current-source headroom.

we can write Eq. (2.29) as

$$\omega_C = \omega_{p,\mathrm{BIAS}} \cdot \sqrt{\eta_{\mathrm{BIAS}}}$$
$$\eta_{\mathrm{BIAS}} = \frac{2}{3} \cdot M_0 \cdot (1 + K) \frac{\alpha}{1 + \frac{2}{3}\alpha}. \tag{2.32}$$

Below ω_C the excess noise of the DAC-biasing becomes dominant in the output noise of the D/A-converter for $|d| > 0$. With (2.27) we get

$$\omega_C = \frac{1}{\sqrt{M_0}} \cdot \frac{\mu}{L^2} \cdot \frac{V_{\mathrm{GS}} - V_T}{1 + \dfrac{3}{2}\kappa} \sqrt{\frac{3}{2} \cdot \frac{\alpha\,(1 + K)}{1 + \dfrac{2}{3}\alpha}} \propto \omega_T. \tag{2.33}$$

The excess noise factor η_{BIAS} can only be minimized by choosing the mirror factor M_0 as small as possible and at the same time minimizing the reference current noise factor K. In general, both measures lead to an increase of the power dissipated by the DAC-biasing. Less effect has the minimization of α by maximizing the gate overdrive voltage of the current source transistors relative to the output full-scale voltage.

The excess noise corner frequency ω_C is inversely proportional to the square root of the mirror factor, suggesting that a very large M_0 would be beneficial. There are however practical restrictions that do not allow to choose the mirror factor arbitrarily large. An additional filter capacitor at the gate node of the current source transistors ($\kappa > 0$) can be used to decrease ω_C, but at the cost of additional silicon area. For the same reference noise, i.e., equal K, PMOS implementations apparently perform better due to their lower mobility, $\mu_p < \mu_n$.

Figure 2.15 shows the noise sources in a dual-polarity current-steering DAC with active output stage. Assuming the same mirror factor $M(j\omega)$ for both polarities, the total output noise is still given by Eq. (2.23) with the following contributions:

$$\overline{I^2_{n,\mathrm{Array}}} = \overline{I^2_{n,\mathrm{Array}N}} + \overline{I^2_{n,\mathrm{Array}P}}$$
$$\overline{I^2_{n,\mathrm{OS}}} = 2 \cdot 4kT\frac{1}{R_L} + \frac{\overline{v^2_{n,\mathrm{OP}}}}{R_L^2} \tag{2.34}$$
$$\overline{I^2_{n,\mathrm{Bias}}} = |d| \cdot |M(j\omega)|^2 \cdot \left(\overline{I^2_{n,\mathrm{MB}N}} + \overline{I^2_{n,\mathrm{MB}P}} + \overline{I^2_{n,\mathrm{REF}N}} + \overline{I^2_{n,\mathrm{REF}P}}\right).$$

The results obtained from the simplified analysis of the single-polarity current-steering DAC are still valid in principle. However, the biasing of a dual-polarity architecture requires at least one additional current mirror and is thus inherently more noisy.

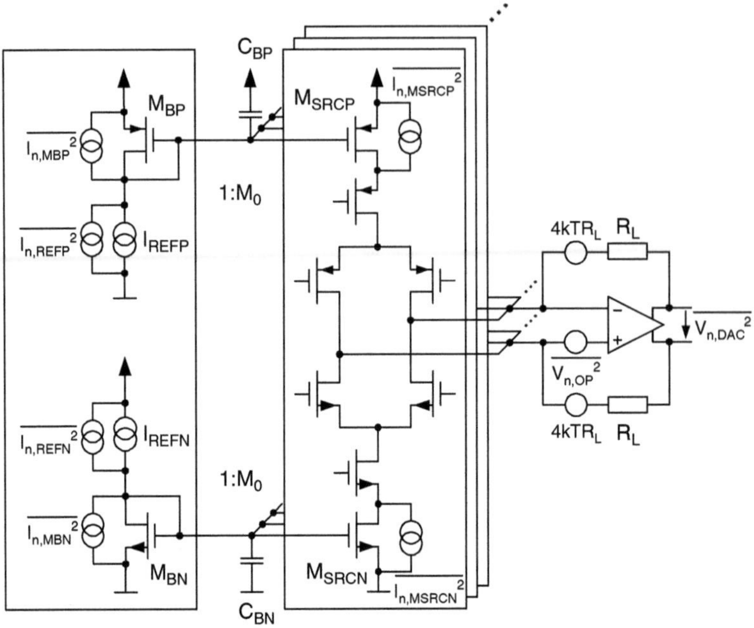

Fig. 2.15 Noise sources in a dual-polarity current-steering DAC

Fig. 2.16 DAC noise model

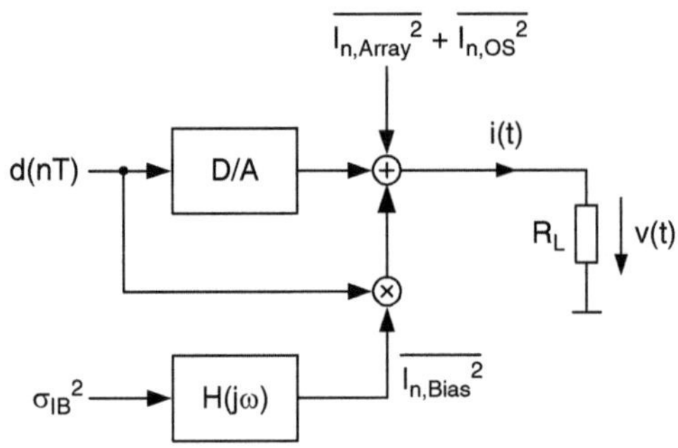

A general circuit-noise model of a fully differential current-steering D/A-converter is shown in Fig. 2.16. In this model the correlated bias noise $\overline{I^2_{n,\text{BIAS}}}$ is generated with a (hypothetical) white current noise source having power σ^2_{IB} and filtered by $H(j\omega)$.

The uncorrelated noise of the current sources and the output stage, $\overline{I^2_{n,\text{Array}}} + \overline{I^2_{n,\text{OS}}}$, is directly added to the output current $i(t)$. The correlated bias noise $\overline{I^2_{n,\text{BIAS}}}$, however, is first multiplied with the input code sequence $d(nT)$ before being added to the output current. The conversion to the voltage domain is performed by the multiplication with R_L.

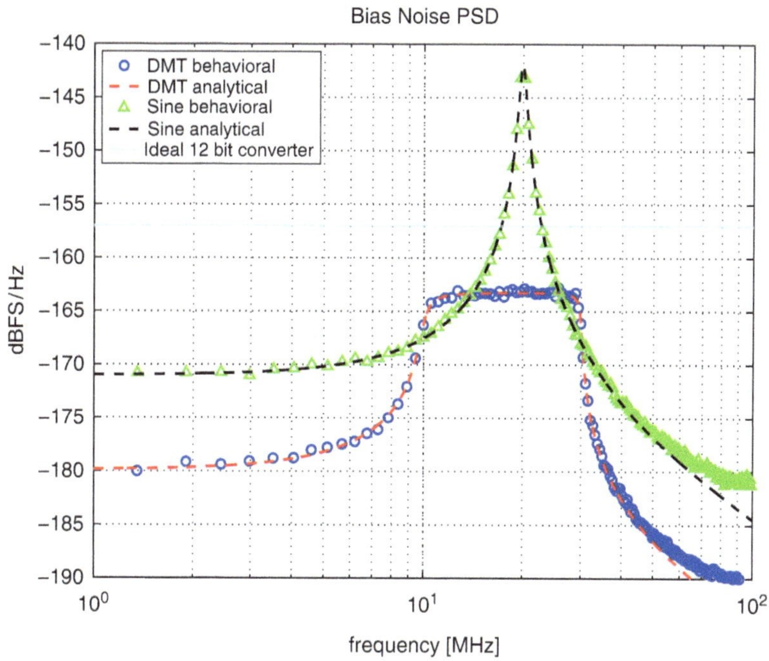

Fig. 2.17 Sine wave and DMT-signal with correlated bias noise

Neglecting the sampling operation, the circuit noise PSD contained in the output current can be calculated as

$$S_i(\omega) = \overline{I_{n,\text{Array}}^2} + \overline{I_{n,\text{OS}}^2} + D(j\omega) * \overline{I_{n,\text{Bias}}(\omega)^2}. \tag{2.35}$$

$D(j\omega)$ is the (baseband) spectrum of the input signal and "$*$" denotes the convolution operation. Appendix A.1 derives the noise for a DMT-signal using a single-pole model for the bias noise, thus neglecting 1/f-noise. In the case of a sufficiently dense multitone signal with $\Delta\omega << \omega_{p,\text{Bias}}$ the (one-sided) output noise PSD can be approximated by

$$S_{i,\text{DMT}}(\omega) \approx \overline{I_{n,\text{Array}}^2} + \overline{I_{n,\text{OS}}^2} + \frac{2}{\text{CF}^2} \cdot \frac{A_{\text{peak}}^2 \sigma_{\text{IB}}^2}{\omega_{\max} - \omega_{\min}} \left(B_n(\omega) + B_n(-\omega)\right)$$

$$B_n(\omega) = \arctan\left(\frac{\omega_{\max} - \omega}{\omega_{p,\text{Bias}}}\right) - \arctan\left(\frac{\omega_{\min} - \omega}{\omega_{p,\text{Bias}}}\right). \tag{2.36}$$

Figure 2.17 shows a comparison of the calculated bias noise PSD and the result of a time-domain behavioral simulation for a single-tone and a multitone signal with equal peak value, corresponding to the converter full-scale. Both signals are sampled

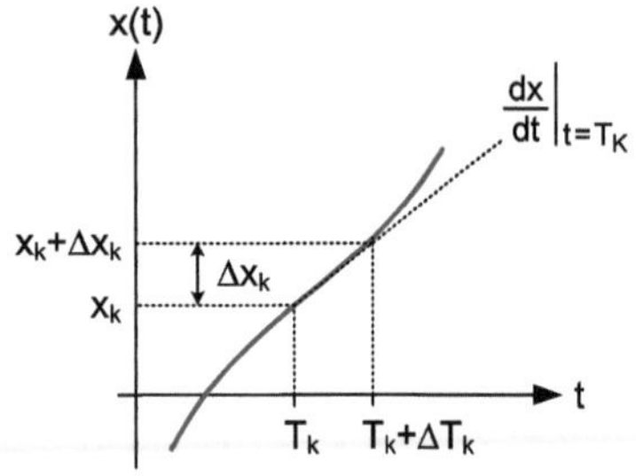

Fig. 2.18 ADC aperture error

with a clock of 200 MHz and affected by a correlated bias noise with a total noise power corresponding to a full-scale resolution of 12 bits and a noise bandwidth of 0.5 MHz. The multitone signal has the following parameters:

- Crest Factor CF $= 4.5$
- Minimum frequency $f_{\text{min}} = 10\,\text{MHz}$
- Maximum frequency $f_{\text{max}} = 30\,\text{MHz}$
- Number of carriers $N_C + 1 = 925$

The difference of the peak noise density between single-tone and multitone signal in this example is 20.2 dB. Broadband multitone systems are much less susceptible to correlated bias noise, because the effect is distributed over many carriers. Specifying the bias circuit of a converter intended for multitone systems based on the assumption of a full-scale single-tone signal can thus lead to significant overdesign of the DAC-biasing.

2.3.3 *Jitter Noise*

Data converters perform the transition from the analog domain to the sampled data domain in the receiver (ADC), or from a digital code sequence to an analog signal in the transmitter (DAC). The conversion is triggered by a dedicated clock signal, the so-called sampling clock. Any timing uncertainty that affects the sampling has an influence on the signal being digitized or synthesized. Figure 2.18 shows the effect of a small timing error that slightly displaces the moment in which a sample of the continuous-time signal $x(t)$ is taken. In ADC's this effect is also called aperture error, indicating the uncertainty about when exactly the converter front-end stops tracking the analog input and starts the quantization process.

If the timing error is small compared to the clock period, $|\Delta T_k| << T$, then the amplitude error can be linearly approximated by the slope of the signal at time T_k multiplied by the timing error ΔT_k:

$$\Delta x_k = \left.\frac{\mathrm{d}x(t)}{\mathrm{d}t}\right|_{t=T_k} \cdot \Delta T_k. \tag{2.37}$$

Because the amplitude error is proportional to the slope, signals with high-frequency content are much more affected by timing errors than slowly changing low-frequency signals.

The timing uncertainty in a clocked system is commonly called clock jitter. Different definitions to quantify jitter are in use, depending on which property is most critical in the system at hand [51]. For data converters, the introduction of (jitter-) noise into the sampled signal is of main interest. We assume that the timing jitter ΔT_k is a stochastic process with rms value σ_{tj}, also called the absolute timing jitter. Then the resulting SNR, evaluated over the first Nyquist band, when sampling a sine wave of frequency f_{sin} is given by [52]:

$$\text{SNR}_{\text{jitter}} = -20 \cdot \log_{10}\left(2\pi f_{\text{sin}} \sigma_{tj}\right). \tag{2.38}$$

The quality of the sampling clock for data converter systems, normally generated by a PLL-type clock generator, is usually specified in terms of its long-term rms jitter σ_{LT}. The long-term rms jitter is the standard deviation of the clock edges measured relative to a starting edge, at a distance where the accumulated jitter has already settled [53]. Equation (2.38) then becomes

$$\text{SNR}_{\text{jitter}} = -20 \cdot \log_{10}\left(2\pi f_{\text{sin}} \frac{\sigma_{\text{LT}}}{\sqrt{2}}\right). \tag{2.39}$$

The division by $\sqrt{2}$ in (2.39) is necessary, because the reference edge of the accumulated jitter measurement is not ideal, but also subject to the same jitter-generating mechanism. From (2.38) and (2.39), the relation between the absolute timing jitter σ_{tj} and the long-term rms jitter σ_{LT} is obviously given by [53]:

$$\sigma_{tj} = \frac{\sigma_{\text{LT}}}{\sqrt{2}}. \tag{2.40}$$

Figure 2.19 shows the obtainable full-Nyquist SNR as a function of the signal frequency according to (2.39) with σ_{LT} as a parameter.

Practical clock sources exhibit a low-pass phase noise spectrum. For a first-order PLL with noiseless input the (two-sided) PSD of the phase noise can be described by a so-called Lorentzian spectrum [54]:

$$S_\varphi(\omega) = \left(\frac{2\pi}{T}\right)^2 \frac{\sigma_{\text{LT}}^2}{\omega_{\text{PLL}}} \frac{1}{1 + \left(\frac{\omega}{\omega_{\text{PLL}}}\right)^2}. \tag{2.41}$$

In hardware implementations, the phase noise spectrum observed at the output of an integrated PLL has a much more complex behavior, due to a multitude of different noise sources and their respective noise transfer functions [54–57]. Mathematically, the sampling process can be described by a convolution of the phase noise PSD with

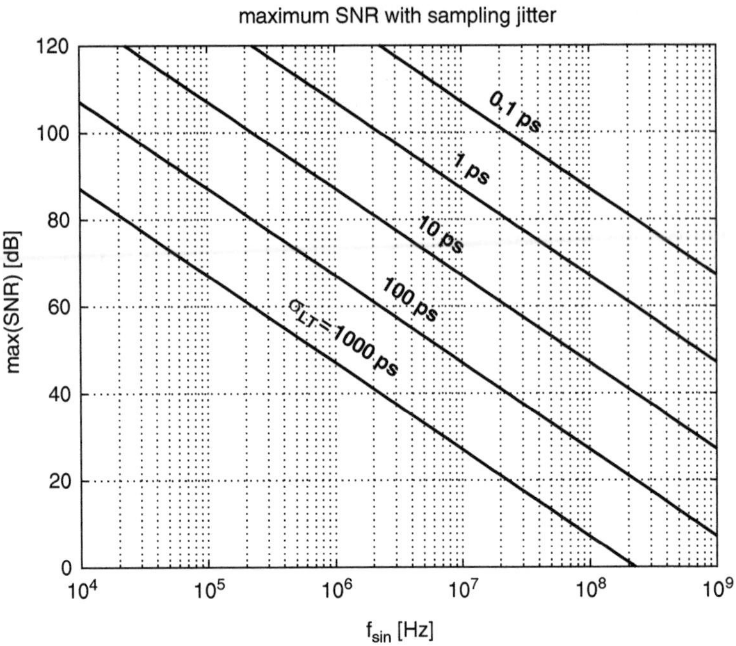

Fig. 2.19 Nyquist SNR with sampling clock jitter

the signal PSD, the latter additionally weighted with the square of the frequency. In time domain this is equivalent to a multiplication of the jitter sequence with the derivative of the signal [53]. Since in the following we are not interested in the exact shape of the jitter noise spectrum in the proximity of the signal, we can neglect the $1/f^n$-terms appearing close to the clock carrier and use Eq. (2.41) to approximate the phase noise of the sampling clock. The -3 dB bandwidth ω_{PLL} of the Lorentzian spectrum also roughly corresponds to the PLL-bandwidth [51].

The (one-sided) jitter noise PSD generated when sampling a sine wave $A_{\text{sin}} \sin(\omega_{\text{sin}} t)$ with a clock affected by Lorentzian phase noise (2.41) with bandwidth ω_{PLL} and long-term rms jitter σ_{LT} can be approximated by (see also Appendix B):

$$S_j(\omega, \omega_{\text{sin}}) \approx \frac{A_{\text{sin}}^2 \omega_{\text{sin}}^2 \frac{1}{2} \sigma_{\text{LT}}^2}{\omega_{\text{PLL}}} \left[\frac{1}{1 + \left(\frac{\omega - \omega_{\text{sin}}}{\omega_{\text{PLL}}} \right)^2} + \frac{1}{1 + \left(\frac{\omega + \omega_{\text{sin}}}{\omega_{\text{PLL}}} \right)^2} \right]. \quad (2.42)$$

As stated above, this approximation does not take the sampling process into account, nor the exact phase noise spectrum of a PLL-type clock generator [54–57]. Still, it is sufficiently accurate for many practical purposes. As an example, Fig. 2.20 shows

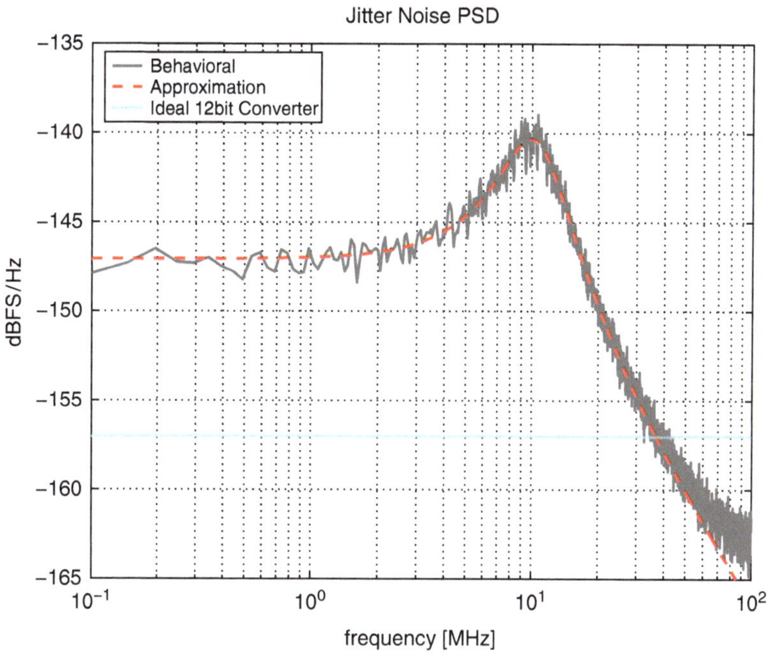

Fig. 2.20 Jitter noise PSD of sampled sine wave

the simulated jitter noise density of a 10 MHz sine wave with normalized amplitude $A_{\sin} = 1$. The signal is sampled with a 200 MHz clock having a 10ps long-term rms jitter and a PLL-bandwidth of 3.5 MHz.[7] The y-axis is normalized to the peak value of the sine wave. On this frequency scale, the approximate expression (2.42) rather accurately describes the noise peaking around the sine wave frequency, where most of the jitter noise power is concentrated. However, the signal frequency must be sufficiently small compared to the Nyquist-frequency, since the aliasing of the jitter noise due to the sampling process is not reflected in (2.42).

NRZ D/A-Converter

In NRZ D/A-converters the derivative in (2.37) is replaced by the difference of subsequent digital input codes, $d_k - d_{k-1}$. As shown in Fig. 2.21, the height of the output pulses does not change when the sampling instants jitter. Instead, the output pulses of the DAC get longer or shorter by ΔT_k, leading to an error pulse sequence $\Delta u(t)$:

[7] Typically, the PLL-bandwidth is around one tenth of the reference clock. We assume a 35 MHz quartz oscillator.

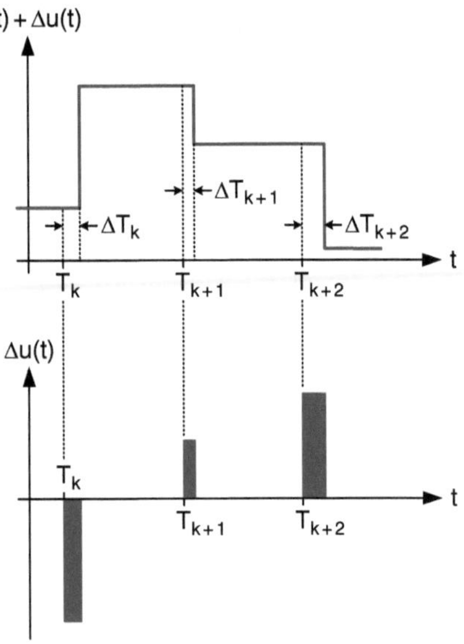

Fig. 2.21 DAC sampling clock error

$$\Delta u(t) = \sum_k (d_k - d_{k-1}) \cdot \mathrm{rect}\left(\frac{T - \frac{1}{2}\Delta T_k}{\Delta T_k} \right). \tag{2.43}$$

Since the error pulses of Eq. (2.43) are assumed to have a very short duration compared to the clock period, their spectral content is approximately constant within at least the first Nyquist band. The effect of the timing error sequence ΔT_k on the output signal of the D/A-converter, observed in the first Nyquist band, can therefore be modeled by an amplitude error sequence e_k. The amplitude error for each sample k corresponds to the area of the respective timing error pulse normalized to the sampling clock period [58]:

$$e_k = \sum_k (d_k - d_{k-1}) \frac{\Delta T_k}{T}. \tag{2.44}$$

This is a reasonable approximation, as long as we limit ourselves to frequencies much smaller than the bandwidth of the error pulses. In this case only the area of the error pulses is important, not their exact shape. A more detailed discussion can also be found in Appendix D.1.

When the NRZ-DAC processes a sine wave $d_k = A_{\sin} \sin(k\omega_{\sin} T)$, then the (one-sided) jitter noise PSD at the converter output becomes

$$S_{j,\mathrm{NRZ}}(\mathrm{e}^{j\omega T}) = A_{\sin}^2 \omega_{\sin}^2 \mathrm{sinc}^2\left(\frac{\omega_{\sin} T}{2}\right) \frac{S_T\left(\mathrm{e}^{j(\omega-\omega_{\sin})T}\right) + S_T\left(\mathrm{e}^{j(\omega+\omega_{\sin})T}\right)}{4}. \tag{2.45}$$

$S_T\left(e^{j\omega T}\right)$ is the (one-sided) PSD of the timing error sequence ΔT_k with power $\frac{1}{2}\sigma_{\mathrm{LT}}^2$ (see also Appendix B):

$$S_T(e^{j\omega T}) = 2T \cdot \frac{1 - e^{-2\omega_{\mathrm{PLL}} T}}{|1 - e^{-\omega_{\mathrm{PLL}} T} e^{-j\omega T}|^2} \cdot \frac{1}{2}\sigma_{\mathrm{LT}}^2$$

$$E\left(\Delta T_k^2\right) = \frac{1}{2\pi} \int_0^{\frac{\pi}{T}} S_T(e^{j\omega T}) \mathrm{d}\omega = \frac{\sigma_{\mathrm{LT}}^2}{2}. \tag{2.46}$$

Except for the sinc^2 term, Eq. (2.45) can be approximated again by (2.42). In contrast to an A/D-converter, where the differentiation of the signal occurs in the continuous-time domain, the jitter error in a D/A-converter depends on the difference of two consecutive samples. For this reason, the jitter noise PSD of an NRZ D/A-converter is additionally weighted with the sinc^2 term.

The total jitter noise power $N_{j,\mathrm{NRZ}}$ of the NRZ-DAC is found by integrating (2.45) over the first Nyquist band:

$$N_{j,\mathrm{NRZ}} = \frac{1}{2\pi} \int_0^{\frac{\pi}{T}} S_{\mathrm{jitter}}(e^{j\omega T}) \mathrm{d}\omega = \frac{1}{2} A_{\sin}^2 \omega_{\sin}^2 \mathrm{sinc}^2\left(\frac{\omega_{\sin} T}{2}\right) \frac{\sigma_{\mathrm{LT}}^2}{2}. \tag{2.47}$$

Also the signal power residing in the first Nyquist band is weighted with the same sinc^2 term:

$$P_{\sin} = \frac{1}{2} A_{\sin}^2 \mathrm{sinc}^2\left(\frac{\omega_{\sin} T}{2}\right). \tag{2.48}$$

Combining (2.47) and (2.48), again the generally valid expression for the SNR in Eq. (2.39) results.

Figure 2.22 shows the behavioral NRZ D/A-converter model including the normalized timing error of Eq. (2.44). To correctly describe the spectral shape of the jitter noise PSD, the input sequence d_k and the normalized timing error sequence e_k are applied to different pulse shapers. The input sequence is directly applied to the unit impulse response $p_{\mathrm{NRZ}}(t)$ with length T, whose output represents the ideal staircase signal. The normalized timing error sequence e_k, on the other hand, is converted to an analog error signal $e(t)$ by a pulse shaper $p_1(t)$ with a short impulse response $T_1 << T$.

Return-to-Zero DAC

In a return-to-zero (RZ) DAC both edges of the output pulses are affected by sampling jitter, see Fig. 2.23. We assume that the clock edges that control the current switching are generated in a PLL by an ideal frequency divider connected to a high-frequency VCO [59]. In principle, we only double the number of realizations of the

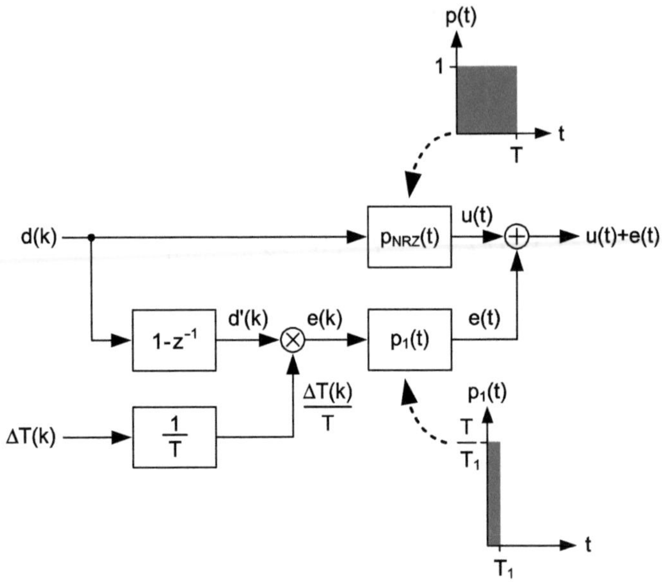

Fig. 2.22 DAC timing error model

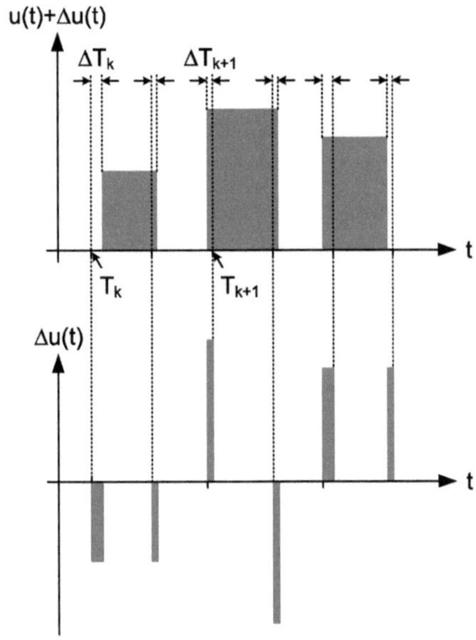

Fig. 2.23 Half-clock RZ-DAC with sampling jitter

random process ΔT_k, while the underlying jitter generation mechanism is the same. Therefore, the statistical properties of the clock jitter do not change when moving from an NRZ to a RZ-implementation.

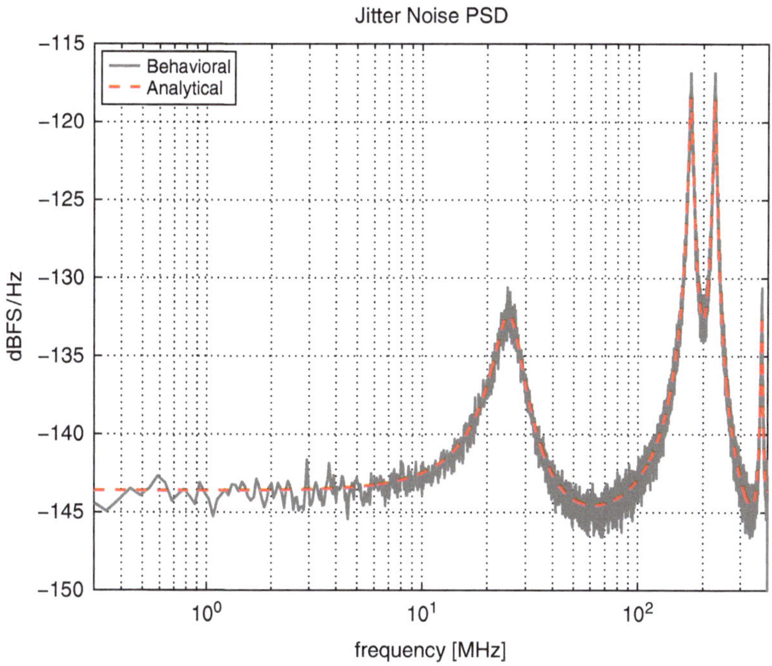

Fig. 2.24 Behavioral and analytical jitter noise PSD for RZ-DAC

In the following, we assume a half-clock RZ-DAC, i.e., the output pulses have a duty factor of $D = 0.5$. Because the PLL-bandwidth and the total power of the timing error sequence must remain the same, the RZ timing jitter sequence can be described by $S_T\left(e^{j\omega\frac{T}{2}}\right)$.

The (one-sided) phase noise PSD of a half-clock RZ-DAC processing a sine-wave signal is derived in Appendix B.3:

$$S_{j,\mathrm{RZ}}\left(e^{j\omega\frac{T}{2}}\right) = \frac{1}{T^2}A_{\sin}^2\sin^2\left(\frac{\omega_{\sin}T}{4}\right)\left[S_T\left(e^{j(\omega-\omega_{\sin})\frac{T}{2}}\right)+S_T\left(e^{j(\omega+\omega_{\sin})\frac{T}{2}}\right)\right]$$
$$+\frac{1}{T^2}A_{\sin}^2\cos^2\left(\frac{\omega_{\sin}T}{4}\right)\left[S_T\left(e^{j(\omega-\omega_0-\omega_{\sin})\frac{T}{2}}\right)+S_T\left(e^{j(\omega-\omega_0+\omega_{\sin})\frac{T}{2}}\right)\right]. \tag{2.49}$$

ω_0 is the sampling frequency of the converter. Figure 2.24 shows a comparison between the time-domain behavioral simulation result and the analytical expression according to Eq. (2.49) for a 200 MS/s half-clock RZ-DAC synthesizing a 25 MHz sine wave.

A unique property of the RZ-waveform is that jitter noise is also generated with a DC output signal. In an NRZ-DAC the switching activity is proportional to the signal frequency. At DC no current switching occurs and therefore no jitter noise is generated. In a RZ-DAC, on the other hand, the switching activity approaches a constant value for decreasing signal frequency. Jitter noise is therefore also generated at DC. For a very low-frequency sine wave and assuming $\omega_{PLL} << \omega_0$, we get in the limit $\omega_{\sin} \to 0$ for the total jitter noise power in the first Nyquist band (see Appendix B.3):

$$\lim_{\omega_{\sin}\to 0} N_{j,RZ} \approx N_{j,RZ,0} = \frac{4}{T^2} A_{\sin}^2 \frac{\sigma_{LT}^2}{2\pi} \cdot \frac{4\omega_{PLL}}{3\omega_0}. \tag{2.50}$$

According to Eq. (1.9), the baseband amplitude of a RZ-DAC is reduced by the duty factor $D = T_S/T$. Therefore, the signal power residing in the first Nyquist band with $D = 0.5$ is given by:

$$P_{S,RZ} = \frac{1}{2}\left(\frac{A_{\sin}}{2}\right)^2 = \frac{A_{\sin}^2}{8}. \tag{2.51}$$

In contrast to an NRZ-implementation, the jitter-limited SNR of a RZ-DAC, evaluated over the first Nyquist band, approaches a constant value for low signal frequencies:

$$\lim_{\omega_{\sin}\to 0} \mathrm{SNR}_{j,RZ} \approx \mathrm{SNR}_{j,RZ}(0) = -20\log_{10}\left(4\sigma_{LT}\sqrt{\frac{\omega_{PLL}\omega_0}{3\pi^3}}\right). \tag{2.52}$$

A comparison of the SNR as a function of the signal frequency for a 200 MS/s NRZ and RZ D/A-converter suffering from a 10ps rms long-term jitter with a PLL-bandwidth of 3.5 MHz is shown in Fig. 2.25. For high signal frequencies both DAC-types follow Eq. (2.39) with a −20 dB/decade drop in SNR vs. $f_{\sin}$. While this equation is generally valid for the NRZ-DAC, the SNR characteristic of the RZ-DAC flattens out for low signal frequencies and approaches the value given by (2.52). The approximation error is smaller than 0.5 dB. The RZ-architecture remains limited to an effective resolution of about 10 bits in this example.

By equating (2.39) and (2.52) it is possible to calculate the cross-over frequency, $\omega_{RZ,3dB}$, for which the SNR of the RZ-DAC experiences a 3 dB loss:

$$\omega_{RZ,3dB} \approx 4\sqrt{\frac{2}{3\pi^3}\omega_{PLL}\omega_0}. \tag{2.53}$$

Finally, the jitter noise SNR of a RZ-DAC can be approximated with a first-order behavior (also shown in Fig. 2.25):

$$\begin{aligned}\mathrm{SNR}_{j,RZ}(\omega_{\sin}) &\approx \mathrm{SNR}_{j,RZ}(0) - 10\log_{10}\frac{1}{1+\left(\frac{\omega_{\sin}}{\omega_{RZ,3dB}}\right)^2} \\ &= -10\log_{10}\left[\frac{\sigma_{LT}^2}{2}\omega_{\sin}^2\cdot\left(1+\frac{32}{3\pi^3}\frac{\omega_{PLL}\cdot\omega_0}{\omega_{\sin}^2}\right)\right]. \end{aligned} \tag{2.54}$$

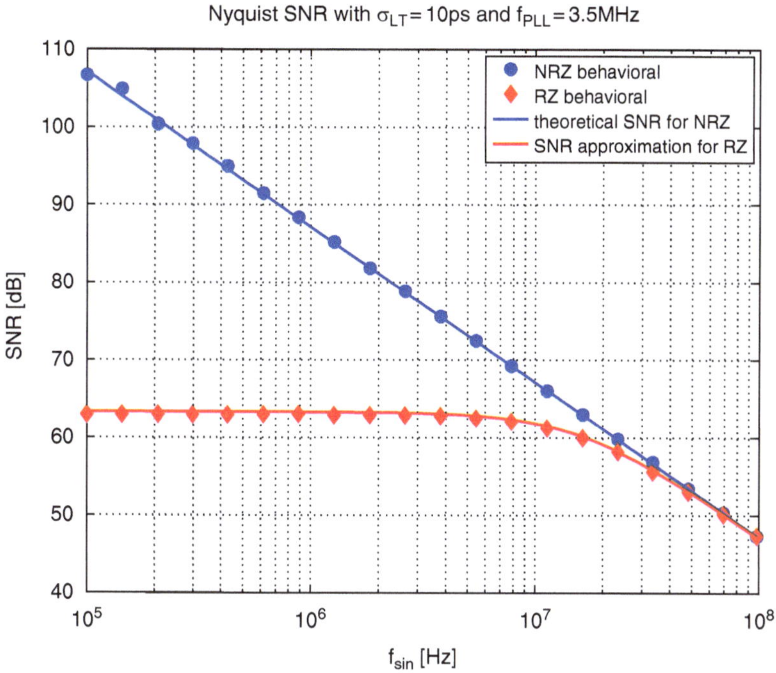

Fig. 2.25 Nyquist SNR with $\sigma_{\mathrm{LT}} = 10\mathrm{ps}$ and $f_{\mathrm{PLL}} = 3.5\,\mathrm{MHz}$

Equation (2.54) describes the full-Nyquist, jitter-limited SNR of a RZ-DAC as a function of the signal frequency and the sampling clock parameters. It is the equivalent of Eq. (2.39), the latter being valid for an NRZ-DAC, as well as for the sampling process of continuous-time signals in A/D-converters, i.e., the ideal sampler.

The main drawback of a RZ-DAC is the increased sensitivity to clock jitter, especially for low signal frequencies. The reason is that, in contrast to an NRZ-DAC, the switching activity approaches a constant value when the signal frequency is lowered toward DC. According to (2.52), the achievable resolution of a RZ-DAC decreases by 3 dB/octave with the clock frequency *and* the PLL-bandwidth, while the jitter-limited resolution of an NRZ-DAC for given rms jitter solely depends on the signal frequency.

Besides the inherent 6 dB signal loss, the pronounced jitter sensitivity is one of the reasons, why true RZ-implementations are seldomly used in embedded high-resolution D/A-converters for digital transceivers. Very low-jitter on-chip clock sources are always expensive in terms of silicon area and power dissipation.

Sampling Jitter in Multitone Systems

In a sampled multitone signal the individual carriers are independently affected by sampling jitter. Therefore, the jitter noise PSD's of the single carriers $S_j(\omega, \omega_k)$

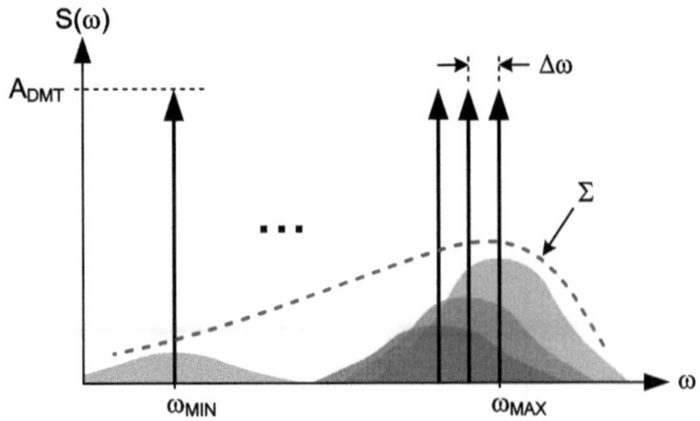

Fig. 2.26 Homogeneous multitone signal with Lorentzian sampling jitter

can be summed to yield the total jitter noise PSD of a sampled multitone signal (Fig. 2.26):

$$S_{j,\mathrm{DMT}}(\omega) = \sum_{k=0}^{N_C} S_j(\omega, \omega_k). \tag{2.55}$$

When the sampling clock for the data converters is generated by a PLL, then the phase noise contribution of each carrier can be approximated in its vicinity by Eq. (2.42). It is shown in Appendix B.4 that with sufficiently high carrier density and $\Delta\omega << \omega_{\mathrm{PLL}}$ the total jitter noise PSD of a homogeneous multitone signal synthesized by a NRZ-DAC can be approximated by:

$$\begin{aligned} S_{j,\mathrm{DMT}}(\omega) &\approx \frac{A_{\mathrm{peak}}^2}{\mathrm{CF}^2}\sigma_{\mathrm{LT}}^2 \cdot (B_j(\omega) + B_j(-\omega)) \\ B_j(\omega) &= \omega_{\mathrm{PLL}} + \frac{\omega \cdot \omega_{\mathrm{PLL}}}{\omega_{\max} - \omega_{\min}} \cdot \ln\left(\frac{\omega_{\mathrm{PLL}}^2 + (\omega - \omega_{\max})^2}{\omega_{\mathrm{PLL}}^2 + (\omega - \omega_{\min})^2}\right) \\ &\quad + \frac{\omega^2 - \omega_{\mathrm{PLL}}^2}{\omega_{\max} - \omega_{\min}}\left[\arctan\left(\frac{\omega_{\max} - \omega}{\omega_{\mathrm{PLL}}}\right) - \arctan\left(\frac{\omega_{\min} - \omega}{\omega_{\mathrm{PLL}}}\right)\right]. \end{aligned} \tag{2.56}$$

$B_j(\omega)$ characterizes the spectral shape of the jitter noise PSD. It depends only on the ratio of the maximum and minimum carrier frequency to the PLL-bandwidth. Figure 2.27 displays the calculated jitter noise PSD for a single tone and multitone signal with equal peak value corresponding to the converter full-scale. Also shown are the results of a time-domain behavioral simulation. Both signals are sampled at 200 MS/s. The sampling clock is affected by a colored timing jitter with a long-term rms value of $\sigma_{\mathrm{LT}} = 10\mathrm{ps}$ and a PLL-bandwidth of 3.5 MHz.

The multitone signal has the following parameters:

- Crest Factor CF = 4.5
- Minimum frequency $f_{\min} = 1\,\mathrm{MHz}$
- Maximum frequency $f_{\max} = 30\,\mathrm{MHz}$
- Number of carriers $N_C + 1 = 1345$

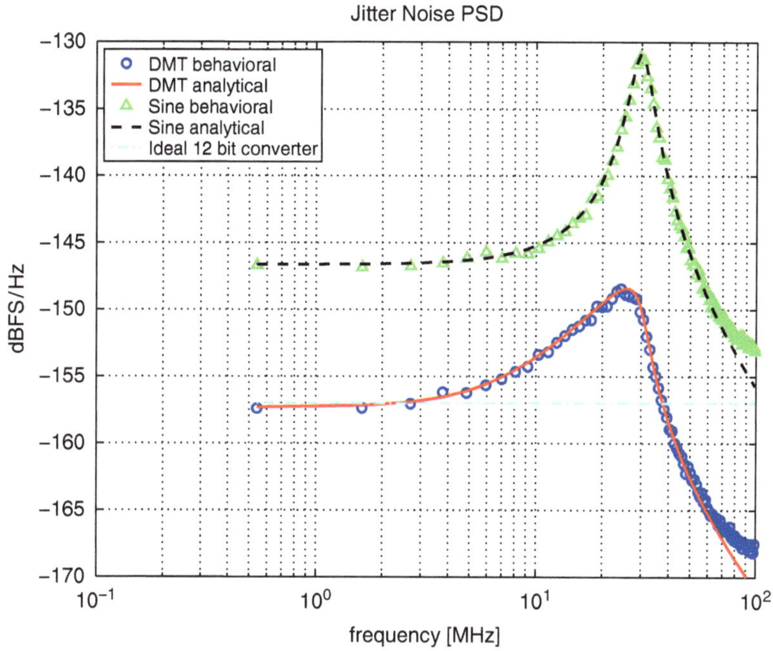

Fig. 2.27 Sine wave and DMT-signal with Lorentzian sampling jitter

The difference in the peak jitter noise density is 17.2 dB. Again, the noise is distributed over many carriers, making the multitone signal much less sensitive to clock jitter than a full-scale single-tone signal. Also visible in Fig. 2.27 is the inaccuracy of the analytical approximation close to the Nyquist frequency, since aliasing effects inherent to the sampling process have been neglected in the derivation of (2.56). The practical relevance of this limitation appears however marginal.

As with the correlated bias noise, the above results show that specifying the clock jitter for a broadband multitone system by using a full-scale single-tone criterion can lead to drastically overdesigned clock systems.

We can use (2.56) to calculate the jitter-limited and frequency dependent SNR for a single carrier embedded in a homogeneous multitone signal. Since, by definition, each carrier occupies a frequency band $\Delta\omega$, equal to the carrier spacing, we define the SNR/carrier as the ratio of the signal power to the total jitter noise located in that bandwidth. As shown in Fig. 2.28, in a sufficiently dense[8] multitone signal the variation of the jitter noise PSD $S_{j,\mathrm{DMT}}(\omega)$ within the carrier bandwidth $\Delta\omega$ is small, such that we can approximate it by the constant value $S_{j,\mathrm{DMT}}(\omega_k)$ obtained at the carrier frequency ω_k. With the carrier amplitude given by (2.7), the carrier

[8]The criterion is again $\Delta\omega << \omega_{\mathrm{PLL}}$.

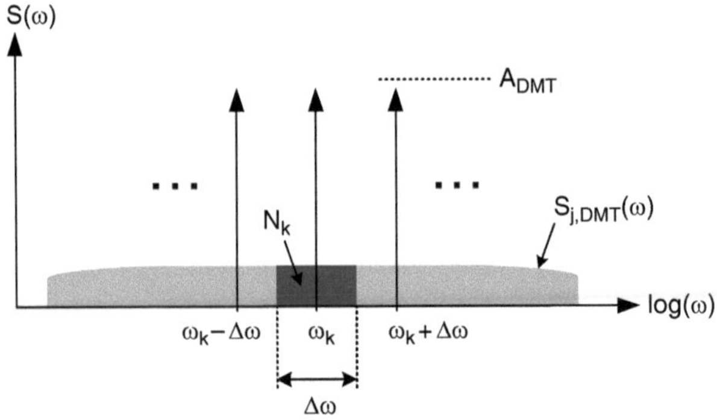

Fig. 2.28 SNR per carrier

power is $1/2 \cdot A^2_{\mathrm{DMT}}$. Following the derivation in Appendix B.4, the SNR/carrier for a homogeneous multitone signal results in [60]:

$$\mathrm{SNR}(\omega_k) = \frac{2\pi}{\sigma^2_{\mathrm{LT}}(\omega_{\max} - \omega_{\min})(B_j(\omega_k) + B_j(-\omega_k))}. \tag{2.57}$$

Equation (2.57) is the narrowband equivalent of the jitter-limited full-Nyquist SNR of a sampled sine wave given by (2.39). It is valid for a specific carrier at frequency ω_k, embedded in a sufficiently dense homogeneous multitone signal. Because both, the carrier power and the local jitter noise power are directly proportional to the tone spacing $\Delta\omega$, the carrier-specific jitter-limited SNR of a sampled multitone signal is actually independent of the carrier bandwidth.

In an analogous way, the jitter noise PSD and the carrier-specific SNR can also be derived for a homogeneous multiband multitone signal; see Appendix B.4.

Experimental Verification

To verify the developed formulas for the sampling jitter noise experimentally, the measurement setup of Fig. 2.29 is used. The sampling clock for the D/A-converter is generated with a sine wave coming from a low phase noise RF-generator (HP8644). This spectrally pure sine wave is converted into a square wave signal by an on-chip low-noise limiting amplifier. The initially low-jitter sampling clock is then fed to the converter and synchronizes the current switching in the DAC-elements.

The D/A-converters used for this experiment are described in Chap. 6. The on-chip limiting amplifier has a fully differential input and is optimized toward low-jitter performance for clock frequencies in the range 150–300 MHz [61].

Colored jitter with a first-order frequency roll-off is produced by filtering a wideband noise source (AWG420) with a first-order RC-filter. The filtered voltage noise is added to the spectrally pure sine wave at the input of the limiting amplifier

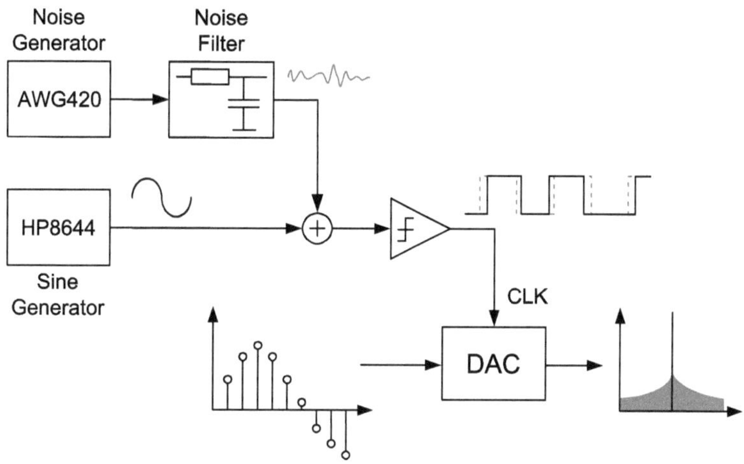

Fig. 2.29 Sampling jitter measurement setup

and randomly shifts the zero crossings. These random shifts are subsequently preserved in the edges of the square wave clock signal applied to the DAC. The sampling clock at the converter input thus exhibits a low-pass phase noise spectrum with the corner frequency set by the low-pass pole of the noise filter. With the sine-wave signal $v_{\mathrm{CLK}} = A_{\mathrm{CLK}} \sin(2\pi f_{\mathrm{CLK}} t)$, the (absolute) timing jitter at the input of the limiting amplifier is given by

$$\sigma_{tj} = \frac{\sigma_{\mathrm{LT}}}{\sqrt{2}} = \frac{v_{n,\mathrm{rms}}}{2\pi f_{\mathrm{CLK}} A_{\mathrm{CLK}}}. \tag{2.58}$$

We use a sampling clock of 210 MHz and a long-term jitter of 200ps, i.e. 4.2 % of the sampling period,[9] with a first-order "PLL-bandwidth" of 4 MHz. Figure 2.30 shows the spectrum and the measured output noise of a synthesized 30 MHz sine wave, as well as the calculated noise density according to Eq. (2.42).

The measured jitter noise for a homogeneous multitone signal with $f_{\mathrm{min}} =$ 300 kHz, $f_{\mathrm{max}} =$ 30 MHz and CF = 3.7 is shown in Fig. 2.31. The carrier spacing is $\Delta f = 70$ kHz, much smaller than the "PLL-bandwidth" f_{PLL}. The calculated noise according to Eq. (2.56) accurately predicts the measured jitter noise PSD. The maximum prediction error in this example is below 0.5 dB.

A multiband-DMT signal is shown in Fig. 2.32. The three bands are each 5 MHz wide with a 5 MHz gap in between. The carrier spacing is 35 kHz with CF = 3.45. Equation (B.35) accurately predicts the measured jitter noise PSD with a maximum error smaller than 1 dB.

[9]The sampling jitter is made intentionally large, such that no other effects limit the noise measurement.

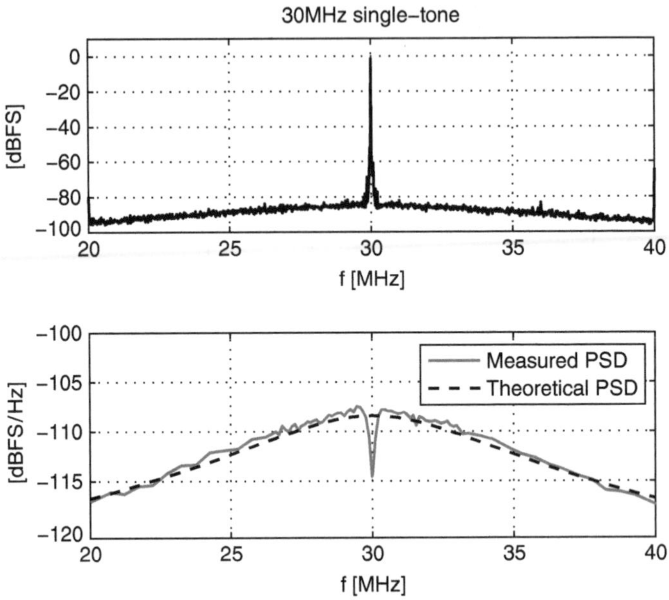

Fig. 2.30 Sine wave sampled with $\sigma_{LT} = 200\text{ps}$, $f_{PLL} = 4\,\text{MHz}$

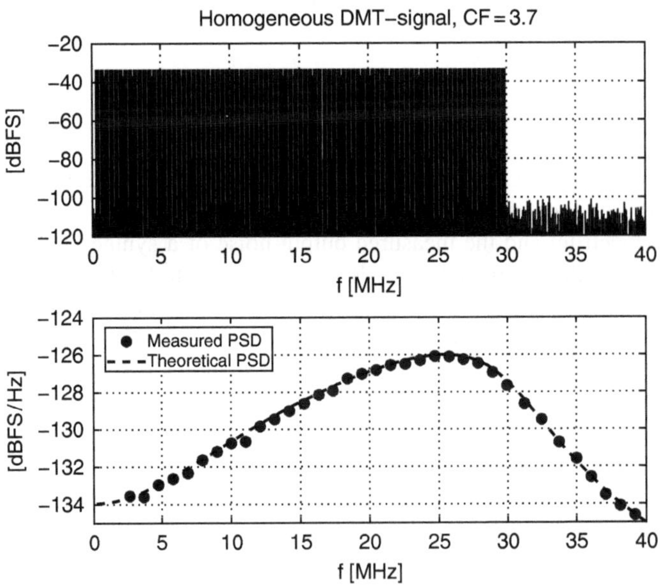

Fig. 2.31 Homogeneous DMT sampled with $\sigma_{LT} = 200\text{ps}$, $f_{PLL} = 4\,\text{MHz}$

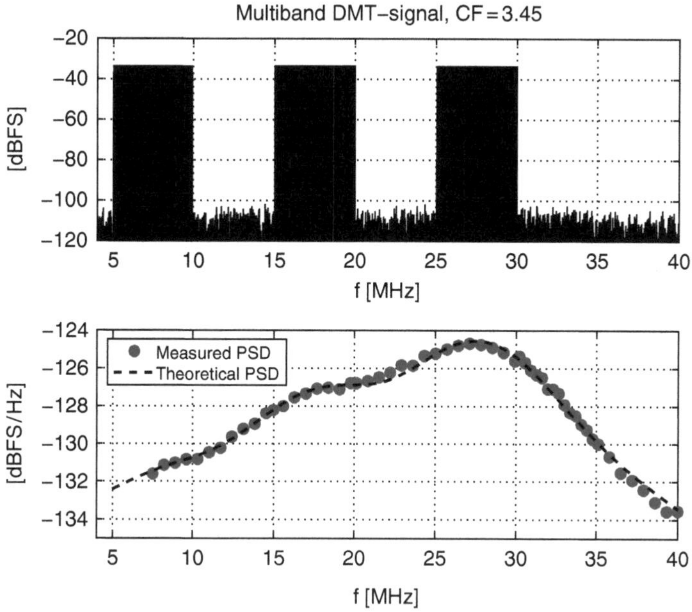

Fig. 2.32 Multiband DMT sampled with $\sigma_{LT} = 200\text{ps}$, $f_{PLL} = 4\,\text{MHz}$

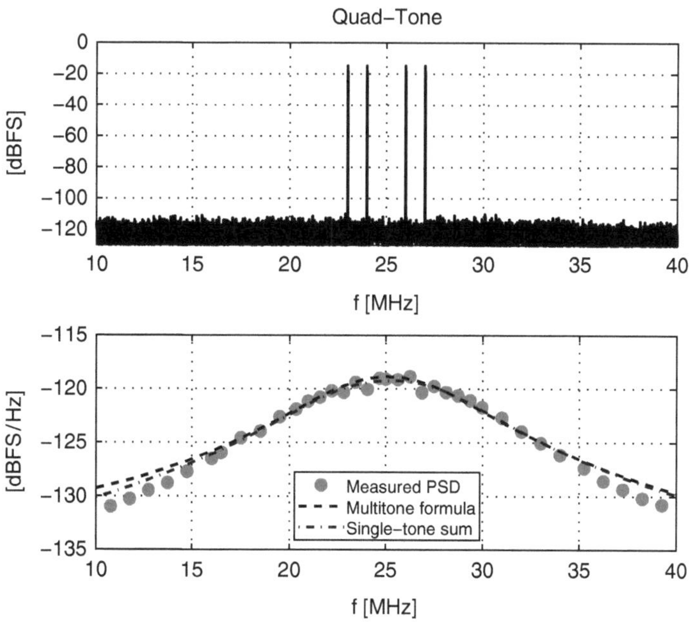

Fig. 2.33 Quad-tone sampled with $\sigma_{LT} = 200\text{ps}$, $f_{PLL} = 4\,\text{MHz}$

The multitone jitter formula has been derived for the limit case of an infinite number of carriers approaching zero frequency spacing. Figure 2.33 shows that it can be applied with reasonable accuracy also to a relatively small number of tones with considerably larger frequency spacing. In this example 4 carriers with 1–2 MHz spacing centered around 25 MHz are used. The signal has a Crest Factor of $CF = 2.78$. For practical purposes, the multitone formula, Eq. (2.56), is still nearly as accurate as the curve obtained by direct summation of the single carrier contributions described by Eq. (2.42). Only when the carrier spacing is made perceptibly larger than the "PLL-bandwidth," the single-tone summation becomes increasingly more accurate, because in this case the noise contributions of the single carriers overlap to an ever lesser extent.

Chapter 3
Static Linearity

3.1 Limitations for the Static Linearity

The static linearity of a D/A-converter is determined by the mismatch of the DAC-elements. "Matching" denotes the achievable degree of similarity of nominally identical devices and is ultimately limited by random fluctuations of process parameters during the fabrication of the integrated circuit. The resulting "random mismatch" solely depends on the specific process technology and is found to be inversely proportional to the device dimensions. In general, the statistical properties of the electrical parameters of closely spaced integrated circuit elements subject to random mismatch can be described by a normal distribution [39].

In practical implementations, systematic errors further deteriorate the achievable accuracy of the converter in addition to the random mismatch [35]. "Systematic mismatch" is introduced during fabrication by large-scale inhomogeneities in the layout patterns, process gradients, as well as packaging stress. During operation, temperature gradients can cause a difference in the device temperature, while unequal voltage drops along common supply rails displace the operating points of physically distant circuit elements sharing a common bias line [62]. As a consequence, systematic errors alone already cause deviations in the electrical behavior of otherwise identical devices.

In most cases, systematic mismatch effects that occur in current-steering D/A-converters can be efficiently suppressed by appropriate layout measures [33, 35, 40, 62–64]. Some effects, like packaging stress or temperature gradients are, however, difficult to predict beforehand. Furthermore, in a mixed-signal chip there are a variety of practical limitations that often prevent the optimum placement of a data converter to comply with all possible causes of systematic mismatch.

In this section we will first derive the expected static linearity error for different current-steering D/A-converter architectures, limited solely by random mismatch, under otherwise ideal conditions. The results can be interpreted as the "average" static nonlinearity for a large number of converter realizations, assuming an ideal layout and no other systematic errors due to fabrication, packaging and device

M. Clara, *High-Performance D/A-Converters*, Springer Series in Advanced Microelectronics 36, DOI 10.1007/978-3-642-31229-8_3,

operation, such as temperature gradients, for example. The knowledge of the underlying statistical properties then allows to establish a lower bound for the device sizes that comply with a given linearity specification and fabrication yield.

3.1.1 Matching of Current Sources

In a current-steering architecture the matching of the current sources constituting the DAC-array determines the static linearity of the converter. The relative variance of the drain current of a MOS-transistor in strong inversion is given by [37]:

$$\frac{\sigma^2(\Delta I_{\mathrm{DS}})}{I_{\mathrm{DS}}^2} = \frac{A_\beta^2}{WL} + \frac{A_{VT}^2}{WL} \cdot \left(\frac{gm}{I_{\mathrm{DS}}}\right)^2. \tag{3.1}$$

In the long-channel approximation for the strong inversion region, Eq. (3.1) becomes [65]

$$\frac{\sigma^2(\Delta I_{\mathrm{DS}})}{I_{\mathrm{DS}}^2} = \frac{A_\beta^2}{WL} + \frac{A_{VT}^2}{WL} \cdot \frac{4}{(V_{\mathrm{GS}} - V_T)^2}. \tag{3.2}$$

Note that in velocity saturation the factor 4 in (3.2) is replaced by 1, while in weak inversion the factor for the threshold voltage contribution to the drain current mismatch becomes $q^2/(nkT)^2$, with the subthreshold slope factor n [48]. A_{VT} and A_β are the technology specific matching constants for the threshold voltage and current factor mismatch of MOS-transistors.[1] The matching constants are the result of a variety of physical effects and heavily related to specific characteristics of the actual process technology [38].

Equations (3.1) and (3.2) suggest that the local drain current variation of a MOS-transistor, for a given technology and bias condition, only depends on its active gate area, irrespective of the actual W and L. However, if either of these geometrical dimensions gets very small, other effects, which tend to average only over one of them, will start to contribute significantly. In this case the matching behavior will deviate from the "universal" $1/\sqrt{WL}$ law [39]. Examples include the so-called "line edge roughness," which is related to the accuracy of the photolithographic process, as well as the influence of pocket implants in modern CMOS technologies [50].

In the following we assume that the current source transistors are sufficiently large, such that eventual matching effects that happen to depend only on one of the geometrical dimensions become small enough, such that they can be safely neglected. Then, following the relation (3.2), the matching of devices can be improved by increasing the device area $W \cdot L$. Furthermore, the influence of the

[1]Usually the matching constants are specified for transistor pairs. The single-device deviation from a (hypothetical) ideal device is then given by dividing the "pairwise" matching constants by $\sqrt{2}$.

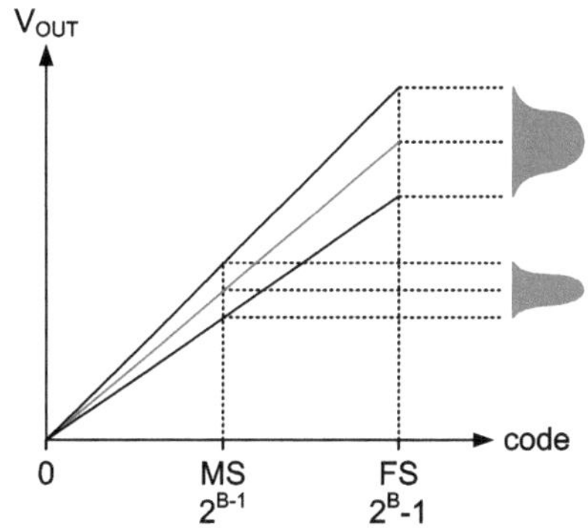

Fig. 3.1 INL-error accumulation

threshold voltage mismatch on the drain current variance can be minimized by choosing the gate overdrive voltage ($V_{GS} - V_T$) as large as possible.

To simplify the calculations, we introduce the combined current matching constant Γ, defined by the technological matching constants A_β and A_{VT}, as well as the gate overdrive voltage of the current source transistor:

$$\Gamma = A_\beta^2 + A_{VT}^2 \cdot \frac{4}{(V_{GS} - V_T)^2}. \tag{3.3}$$

As already stated, we assume the current-source transistors to be operating in strong inversion. With (3.3) Eq. (3.2) reduces to:

$$\frac{\sigma^2 (\Delta I_{DS})}{I_{DS}^2} = \frac{\Gamma}{WL}. \tag{3.4}$$

3.1.2 Statistical Description of the INL

The current sources of a current-steering DAC follow a normal distribution with a standard deviation depending on the matching constants, the transistor size, and the gate overdrive voltage; see Eq. (3.2). Moreover, the individual current sources constituting a D/A-converter are assumed to be statistically independent. Since the output signal of the DAC is the sum of an appropriate number of unit currents, also the static linearity error at any given code can be described by a normal distribution [66].

The output ramp of a B-bit D/A-converter is given by the cumulative sum of the individual unit currents. After eliminating the offset at code 0, the statistical deviation of the measured output voltage from the average ramp of a large ensemble of devices will be largest at the full-scale code $2^B - 1$. This is depicted schematically in Fig. 3.1.

The endpoint-INL definition (see Sect. 2.1.3), on the other hand, compares the measured output ramp with an ideal ramp having the *same* minimum and maximum output value. This effectively removes the variation at the full-scale code, since the endpoint of the ideal ramp is made intentionally equal to the measured full-scale voltage.

Therefore, the worst-case INL is expected at the midscale code 2^{B-1}, and it is given by summing the individual error contributions of all $2^B - 1$ DAC-elements and subsequent division by two [33]. The standard deviation of the midscale INL can thus be calculated as

$$\sigma\left(\mathrm{INL}\left(2^{B-1}\right)\right) = \frac{1}{2}\sqrt{\sum_{m=1}^{2^B-1} \frac{\sigma^2\left(\Delta I_{\mathrm{LSB}}\right)}{I_{\mathrm{LSB}}}} = \frac{1}{2}\sqrt{2^B-1}\cdot\frac{\sigma\left(\Delta I_{\mathrm{LSB}}\right)}{I_{\mathrm{LSB}}}. \tag{3.5}$$

With (3.10) we can show that the standard deviation of the INL-error at midscale for the binary array and also for the segmented array is equal to (3.5). However, the maximum INL does not necessarily occur at the midscale code. In [67] the analogy between the evolution of the INL-characteristic for increasing input codes and Brownian motion is described. Specifically, the endpoint-INL of a unary D/A-converter array with sufficiently many levels can be approximated with a so-called Brownian Bridge, i.e., a one-dimensional Wiener process with defined start and endpoint [68]. The standard deviation of the maximum INL is then given by [67, 68]:

$$\sigma\left(\max\left(|\mathrm{INL}|\right)\right) = \frac{\sigma\left(\Delta I_{\mathrm{LSB}}\right)}{I_{\mathrm{LSB}}}\sqrt{2^B-1}\cdot\max_{t\in[0,1]}\left(B_t\right). \tag{3.6}$$

B_t is the standard Brownian Bridge defined over the interval $t \in [0, 1]$

3.1.3 *Statistical Description of the DNL*

The DNL is the step-size error of the output characteristic, compared to the ideal ramp with equidistant steps of 1 LSB. Depending on the array coding, a different number of unit current sources contribute to the accuracy of the unit step size at the single code transitions. The expected DNL-error therefore strongly depends on the D/A-converter architecture.

Unary Array

In a unary architecture only one DAC-element is switched on or off when the input code changes by one digit. All current sources are identical and have a nominal current equal to the LSB-current, I_{LSB}, as well as the same standard deviation $\sigma(I_{\mathrm{LSB}})$, which can be estimated with (3.2). Because the ideal LSB-step is equal to the nominal current in one DAC-element, the standard deviation of the DNL-error in a B-bit unary array at every code transition is directly given by

$$E\left(\sqrt{\mathrm{DNL}(k)^2}\right) = \sigma(\mathrm{DNL}(k)) = \frac{\sigma\left(\Delta I_{\mathrm{LSB}}\right)}{I_{\mathrm{LSB}}} \qquad k \in [1; 2^B - 1]. \tag{3.7}$$

Binary Array

In a B-bit binary array the DAC-elements are "power-of-two" multiples of the smallest element, the so-called LSB-cell:

$$I_k = 2^k \cdot I_{\mathrm{LSB}} \qquad k \in [0; B-1]. \tag{3.8}$$

The largest element in a binary array is called MSB-cell with the MSB-current $I_{\mathrm{MSB}} = 2^{B-1} \cdot I_{\mathrm{LSB}}$. If the kth element in the binary-weighted array is a parallel combination of 2^k identical LSB-elements, each with area $W \cdot L$, then we can write with (3.2) and (3.8):

$$\frac{\sigma\left(\Delta I_k\right)}{2^k I_{\mathrm{LSB}}} = \sqrt{\frac{A_\beta^2}{2^k WL} + \frac{A_{VT}^2}{2^k WL} \cdot \frac{4}{\left(V_{\mathrm{GS}} - V_T\right)^2}} = \frac{1}{\sqrt{2^k}} \cdot \frac{\sigma\left(\Delta I_{\mathrm{LSB}}\right)}{I_{\mathrm{LSB}}}. \tag{3.9}$$

The standard deviation of the binary-weighted currents normalized to the LSB-current is therefore given by

$$\frac{\sigma\left(\Delta I_k\right)}{I_{\mathrm{LSB}}} = \sqrt{2^k} \cdot \frac{\sigma\left(\Delta I_{\mathrm{LSB}}\right)}{I_{\mathrm{LSB}}} \qquad k \in [0; B-1]. \tag{3.10}$$

The step-size error in a binary array is most critical at the MSB-transition, where all DAC-elements are switched. For this so-called "major-carry" transition at midscale, i.e., code 2^{B-1}, we can calculate the standard deviation of the DNL-error

$$\sigma\left(\mathrm{DNL}\left(2^{B-1}\right)\right) = \sqrt{\frac{\sigma^2\left(\Delta I_{\mathrm{MSB}}\right)}{I_{\mathrm{LSB}}} + \sum_{k=0}^{B-2} \frac{\sigma^2\left(\Delta I_k\right)}{I_{\mathrm{LSB}}}}. \tag{3.11}$$

Using (3.10) we get

$$\sigma\left(\mathrm{DNL}\left(2^{B-1}\right)\right) = \sqrt{2^B - 1} \cdot \frac{\sigma\left(\Delta I_{\mathrm{LSB}}\right)}{I_{\mathrm{LSB}}}. \tag{3.12}$$

Following the same calculation as above, we can also derive a general expression for the standard deviation of the DNL-error in a binary array at the code transitions 2^{B-k-1} with $k \in [0; B-1]$:

$$\sigma\left(\mathrm{DNL}\left(2^{B-k-1}\right)\right) = \sqrt{2^{B-k} - 1} \cdot \frac{\sigma\left(\Delta I_{\mathrm{LSB}}\right)}{I_{\mathrm{LSB}}} \qquad k \in [0; B-1]. \tag{3.13}$$

The worst-case DNL at the midscale code transition 2^{B-1}, given by (3.12), is a special case of the general Eq. (3.13) with $k = 0$.

Segmented Array

In a segmented array usually only the lowest segment is a binary-weighted array, while the higher segments are unary arrays. Again we assume, that all DAC-elements are multiples of the smallest unit, the LSB-element. In general, a segmented N-bit converter consists of L segments with resolutions $M_1, M_2, \ldots, M_L$ with $M_1 + M_2 + \cdots + M_L = B$. Let the first segment with resolution M_1 be a binary-weighted array with the largest element having a current of $2^{M_1-1} I_{\text{LSB}}$. Consequently, the unit current in the second, unary coded, segment is therefore $2^{M_1} I_{\text{LSB}}$. The kth segment has a unit current size of

$$I_k = 2^{\sum_{i=1}^{k-1} M_i} I_{\text{LSB}} \qquad k \in [0; L-1]. \tag{3.14}$$

The worst-case DNL is expected at the major-carry transitions where the next unit element of the highest segment is switched on, and all elements in the lower segments are switched off. These transitions occur at the codes $d_j = j \cdot 2^{M_1+M_2+\ldots M_{L-1}} = j \cdot 2^{B-M_L}$ with integer $j \in [1; 2^{M_L} - 1]$ and display a standard deviation for the DNL-error of

$$\sigma\left(\text{DNL}\left(d_j\right)\right) = \sqrt{\frac{\sigma^2\left(\Delta I_L\right)}{I_{\text{LSB}}^2} + \sum_{k=2}^{L-1} \sum_{m=1}^{2^{M_k}-1} \frac{\sigma^2\left(\Delta I_k\right)}{I_{\text{LSB}}^2} + \sum_{n=0}^{M_1-1} \frac{\sigma^2\left(\Delta I_n\right)}{I_{\text{LSB}}^2}}$$

$$d_j = j \cdot 2^{\sum_{i=1}^{L-1} M_i} = j \cdot 2^{B-M_L} \qquad j \in [1; 2^{M_L} - 1]. \tag{3.15}$$

If we assume homogeneous sizing, i.e., all DAC-elements are parallel combinations of an appropriate number of LSB-elements, then Eq. (3.10) holds and (3.15) reduces to

$$\sigma\left(\text{DNL}\left(d_j\right)\right) = \sqrt{2^{B-M_L+1} - 1} \cdot \frac{\sigma\left(\Delta I_{\text{LSB}}\right)}{I_{\text{LSB}}}. \tag{3.16}$$

3.1.4 *Minimum Area Requirements*

The assumed statistical properties of the static linearity errors allow us to estimate the minimum area required for the DAC-elements, such that the converter can satisfy a given linearity specification with a certain fabrication yield. In general, for a given performance parameter S, the yield is defined as the percentage of "good" devices that meet a given specification, e.g., $S \leq S_{\max}$. We therefore define the yield Y_S as the *probability* that a single realization out of a large ensemble of fabricated devices will meet this specification:

$$Y_S = P\left[S \leq S_{\max}\right]. \tag{3.17}$$

In practical designs usually a yield larger than 99 % is targeted, i.e., $Y_S > 0.99$. The standard procedure of 3σ design aims at 99.73 % fabrication yield.

INL-Limit

The asymptotic maximum INL-error, compliant with the endpoint definition (see Sect. 2.1.3), is given by Eq. (3.6), the scaled maximum of a standard Brownian bridge. For sufficiently many output levels, the INL-yield of a unary array can be approximated by [67]:

$$Y_{\mathrm{INL}} \approx 1 - 2\sum_{k=1}^{\infty}(-1)^{k-1}\mathrm{e}^{-2k^2\frac{\mathrm{INL}_{\max}^2}{(2^B-1)\cdot\left(\frac{\sigma(\Delta I_{\mathrm{LSB}})}{I_{\mathrm{LSB}}}\right)^2}}. \tag{3.18}$$

Equation (3.18) is the asymptotic limit of the INL-yield according to the endpoint definition when the number of output levels approaches infinity. In probability theory Eq. (3.18) is also known as the Kolmogorov-Smirnov formula, used in the asymptotic theory of empirical distribution functions [69, 70].

The formula given in [71], on the other hand, is a correct description of the yield, but for a slightly different INL-definition. In this case the ideal ramp is constructed by fixing only the startpoint and using the (ensemble) average of the slope of the measured output characteristics. This means that the gain error is only compensated for the ensemble average, but not for the actual device, such that the endpoint of the ramp does not always coincide with the ideal ramp. Applied to an endpoint-INL scenario this yield formula is too pessimistic [67].

An earlier result obtained in [65] for the yield with respect to endpoint-INL is still more pessimistic, while the modified formula, derived by the same authors in [72], appears to be too optimistic [67].

The simulated INL-yield of a 12-bit converter with different array coding and $\mathrm{INL}_{\max} = 0.5 \cdot \mathrm{LSB}$ is displayed in Fig. 3.2. The binary array has, on average, a smaller INL-error and thus a higher yield than a unary array with the same accuracy. However, in the practically relevant "high-yield" region the difference resulting from the actual segmentation level is at best marginal. Therefore, (3.18) can safely be used to describe the INL-yield, independent of the array coding [67].

For the calculation of the required current source area, we could make use of tabulated values for the Kolmogorov-Smirnov distribution [73] to find the inverse of Eq. (3.18). However, assuming a reasonably high INL-yield, we may also approximate the infinite sum in (3.18) by using only the first term with $k = 1$:

$$Y_{\mathrm{INL}} \approx 1 - 2\mathrm{e}^{-2\frac{\mathrm{INL}_{\max}^2}{(2^B-1)\cdot\left(\frac{\sigma(\Delta I_{\mathrm{LSB}})}{I_{\mathrm{LSB}}}\right)^2}}. \tag{3.19}$$

In Fig. 3.2 we also find the Kolmogorov-Smirnov formula (3.18) and its "high-yield" approximation (3.19). With (3.3), we can directly solve (3.19) for the minimum area

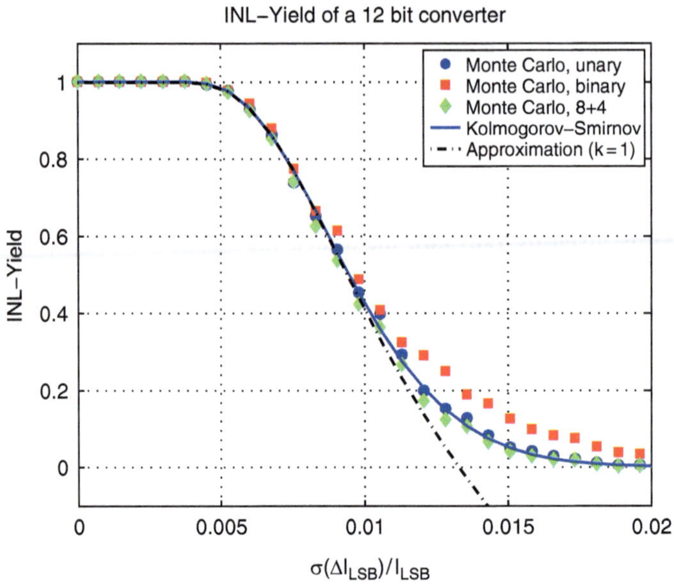

Fig. 3.2 Simulated INL-yield of a 12-bit converter

of the LSB-element that satisfies $\text{INL} \leq \text{INL}_{\text{max}}$ for a fraction $Y_{\text{INL}} \in [0, 1]$ of the total number of fabricated devices:

$$(WL)_{\text{LSB,INL}} = \frac{2^B - 1}{\text{INL}_{\text{max}}^2} \ln\left(\sqrt{\frac{2}{1 - Y_{\text{INL}}}}\right) \cdot \Gamma. \tag{3.20}$$

The relative error in terms of INL-yield that we make by approximating (3.18) with (3.19) is displayed in Fig. 3.3. For $Y_{\text{INL}} > 0.95$ the approximation error is smaller than 1 ppm and therefore largely irrelevant in practical situations.

DNL-Limit for the Unary Array

The DNL-error is a Gaussian vector with $2^B - 1$ elements and zero mean. In a unary array all elements also have the same standard deviation described by (3.7). Because the unit current sources are all statistically independent, the DNL-yield is given by:

$$Y_{\text{DNL},u} = P\left[\max\left(|\text{DNL}(k)|\right) \leq \text{DNL}_{\text{max}}\right] = \text{erf}\left[\frac{\text{DNL}_{\text{max}}}{\sqrt{2} \cdot \frac{\sigma(\Delta I_{\text{LSB}})}{I_{\text{LSB}}}}\right]^{2^B - 1}. \tag{3.21}$$

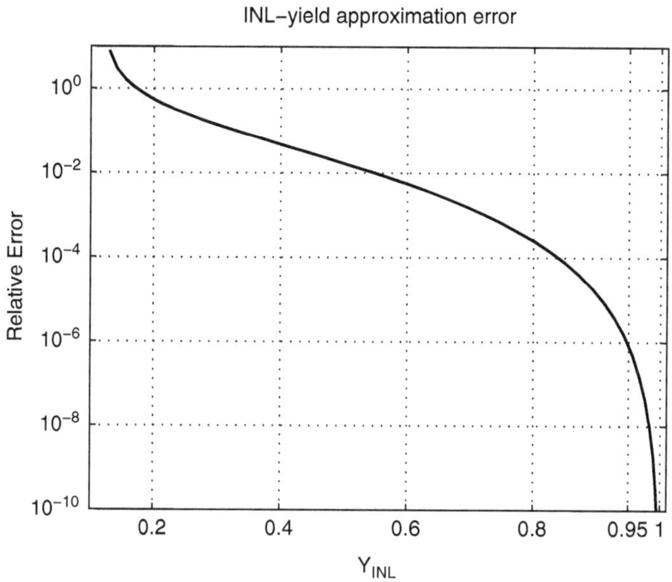

Fig. 3.3 Relative Error of INL-yield approximation

With (3.3) we can solve (3.21) for the minimum area of the current-source transistor in a unary array that satisfies DNL $\leq$ DNL$_{max}$ for a fraction $Y_{DNL,u} \in [0, 1]$ of the total number of fabricated devices:

$$(WL)_{DNL,u} = 2\left(\frac{\mathrm{erf}^{-1}\left[(Y_{DNL})^{\frac{1}{2^B-1}}\right]}{\mathrm{DNL}_{max}}\right)^2 \cdot \Gamma. \tag{3.22}$$

$\mathrm{erf}^{-1}(x)$ is the inverse of the error function, defined by $x = \mathrm{erf}^{-1}(\mathrm{erf}(x))$.

DNL-Limit for the Binary Array

In a binary array the DNL-error is statistically described by the standard deviation given in (3.13). If we assume negligible correlation of the DNL-errors at different code transitions, the DNL-yield of the binary array can be written as

$$Y_{DNL,b} = \prod_{k=1}^{B} \mathrm{erf}\left[\frac{\mathrm{DNL}_{max}}{\sqrt{2}\cdot\sqrt{2^k-1}\cdot\frac{\sigma(\Delta I_{LSB})}{I_{LSB}}}\right]. \tag{3.23}$$

Since the midscale transition is expected to display the worst-case DNL-error, we can approximate (3.23) by

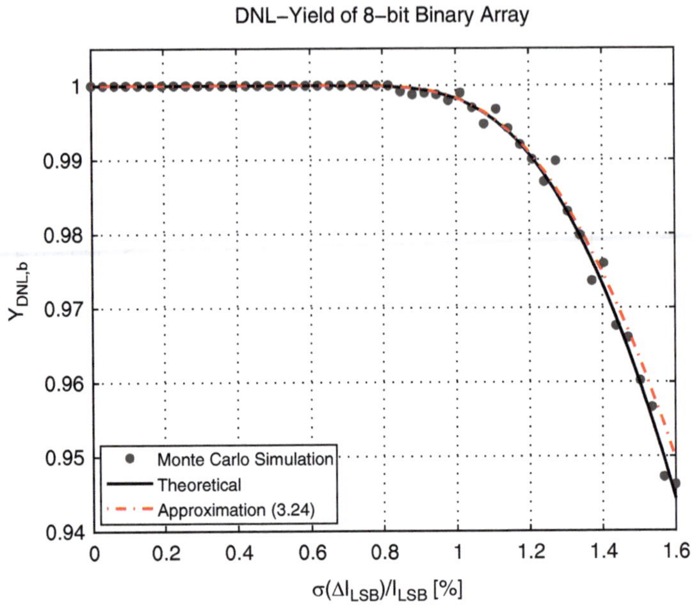

Fig. 3.4 DNL-yield of 8-bit binary array

$$Y_{\mathrm{DNL},b} \approx \mathrm{erf}\left[\frac{\mathrm{DNL_{max}}}{\sqrt{2}\cdot\sqrt{2^B-1}\cdot\dfrac{\sigma\left(\Delta I_{\mathrm{LSB}}\right)}{I_{\mathrm{LSB}}}}\right]. \tag{3.24}$$

Since Eq. (3.24) only considers the major-carry transition at midscale, it is an upper bound to the actual DNL-yield characteristic. In the "high-yield" region, however, the approximation error is practically negligible. This is also shown in Fig. 3.4 for a hypothetical 8-bit converter and $\mathrm{DNL_{max}} = 0.5$ LSB.

From (3.24) the required area of the current-source transistor in the LSB-element can be calculated as

$$(WL)_{\mathrm{DNL},b} = 2\left(2^B-1\right)\cdot\left(\frac{\mathrm{erf}^{-1}\left[Y_{\mathrm{DNL}}\right]}{\mathrm{DNL_{max}}}\right)^2\cdot\Gamma. \tag{3.25}$$

DNL-Limit for the Segmented Array

In a segmented array with $2^{M_L}-1$ MSB-sources we encounter as many major-carry transitions, where we would expect the DNL-error to assume its worst-case values. Because all elements in the lower segments are reused in every of these code transitions, the single DNL-errors described by (3.16) are not statistically independent.

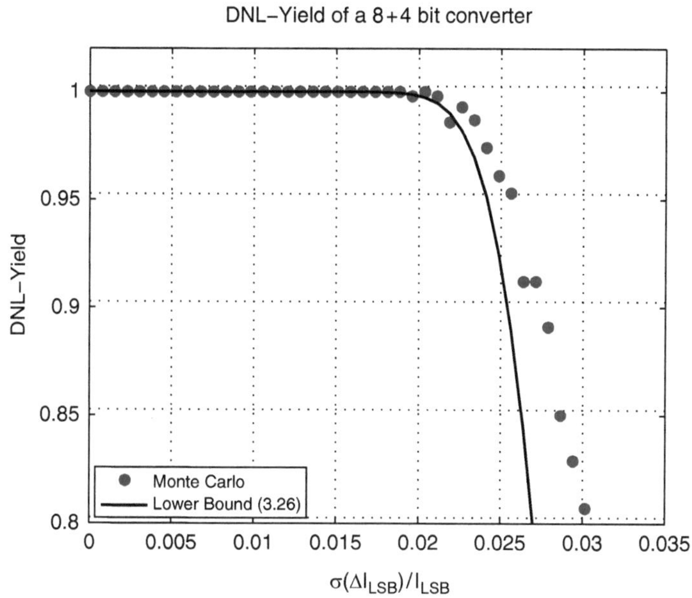

Fig. 3.5 DNL-yield of 8+4 bit segmented array

A lower bound on the DNL-yield can be derived by assuming statistical independence of all major-carry transitions in the segmented array.[2] The actual DNL-yield is then approximated by

$$Y_{\mathrm{DNL},s} \approx \operatorname{erf}\left[\frac{\mathrm{DNL}_{\max}}{\sqrt{2}\cdot\sqrt{2^{B-M_L+1}-1}\cdot\dfrac{\sigma\left(\Delta I_{\mathrm{LSB}}\right)}{I_{\mathrm{LSB}}}}\right]^{2^{M_L}-1}. \tag{3.26}$$

For all practical purposes, the approximation error is sufficiently small in the "high-yield" region. This is also shown in Fig. 3.5 for a segmentation of 8 + 4 and $\mathrm{DNL}_{\max} = 0.5$ LSB. From (3.26) the minimum area of the current-source transistor in the LSB-array can be calculated:

$$(WL)_{\mathrm{DNL},s} = 2\left(2^{N-M_L+1}-1\right)\left(\frac{\operatorname{erf}^{-1}\left[\left(Y_{\mathrm{DNL}}\right)^{\frac{1}{2^{M_L}-1}}\right]}{\mathrm{DNL}_{\max}}\right)^2\cdot\Gamma. \tag{3.27}$$

[2]This is equivalent to having $2^{M_L}-1$ independent versions of the lower segments, such that no element thereof is used twice in any major-carry transition.

Table 3.1 3-sigma yield LSB-area for a hypothetical 12-bit converter

Array coding	$(WL)_{\mathrm{INL}}$	$(WL)_{\mathrm{DNL}}$
Unary array	36.6 μm²	0.066 μm²
Binary array	36.6 μm²	99.7 μm²
Segmented 10+2	36.6 μm²	0.42 μm²
Segmented 8+4	36.6 μm²	1.63 μm²
Segmented 6+6	36.6 μm²	5.75 μm²
Segmented 4+8	36.6 μm²	19.4 μm²

As a numerical example, Table 3.1 shows the minimum area of the LSB-element for a hypothetical, mismatch-limited 12-bit converter in a 3-sigma design, i.e., for a yield of 99.73 %. The following parameters were used:

- $\mathrm{INL}_{\max} = 0.5\,\mathrm{LSB}$
- $\mathrm{DNL}_{\max} = 0.5\,\mathrm{LSB}$
- $A_\beta = 1\,\% \cdot \mu\mathrm{m}$
- $A_{VT} = 3\,\mathrm{mV} \cdot \mu\mathrm{m}$
- $V_{\mathrm{GS}} - V_T = 0.25\,\mathrm{V}$

Even for a 4+8 segmentation, the LSB-element area is still determined by the INL-specification. Only for $M_L < 3$ the area of the LSB-element is determined by the DNL-specification and approaches the value for the binary array, which is almost 3 times larger than the INL-limited area.

3.1.5 *Code-Dependent Output Impedance*

The current sources forming a current-steering D/A-converter display a finite output impedance and thus form a code-dependent current divider with the load resistor. Even at DC, this current division results in a small, signal-dependent error of the output current delivered to the load resistor, and therefore distorts the INL-characteristic of the converter. The general situation is depicted in Fig. 3.6. Since, without mismatch, the unit current cells are indistinguishable, we can assume, without loss of generality, that the converter consists of N identical unit current cells, each with a unit current I_{unit} and output impedance Z_{unit}. The full-scale current of the converter core is thus $N \cdot I_{\mathrm{unit}}$ with a minimum output impedance–seen into one of the output terminals–of Z_{unit}/N.

Single-Ended Converter

With the input code $k \in [0; N]$, the output currents delivered to the single-ended load impedances Z_L are given by

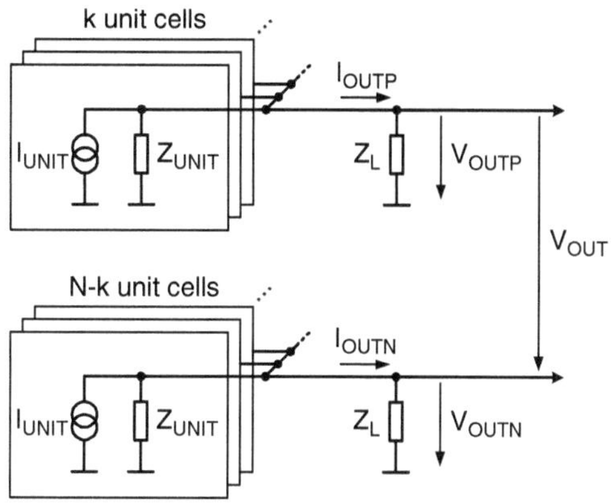

Fig. 3.6 Current-steering DAC with finite output impedance

$$I_{\mathrm{OUTP}}(k) = kI_{\mathrm{unit}} \cdot \frac{Z_{\mathrm{unit}}}{Z_{\mathrm{unit}} + kZ_L}$$
$$I_{\mathrm{OUTN}}(k) = (N-k)I_{\mathrm{unit}} \cdot \frac{Z_{\mathrm{unit}}}{Z_{\mathrm{unit}} + (N-k)Z_L}. \tag{3.28}$$

For the static linearity only the real part of the impedances is relevant. With $R_L = Re(Z_L)$ and $R_{\mathrm{unit}} = Re(Z_{\mathrm{unit}})$ the voltage observed at the positive single-ended output is given by:

$$V_{\mathrm{OUTP}}(k) = kI_{\mathrm{unit}} \cdot \frac{R_L R_{\mathrm{unit}}}{R_{\mathrm{unit}} + kR_L}. \tag{3.29}$$

Assuming $R_{\mathrm{unit}} >> N \cdot R_L$, the maximum deviation from the ideal output ramp with the same endpoints as (3.29) can be derived as [74]

$$\Delta V_{\max} \approx \frac{I_{\mathrm{unit}} R_L^2 N^2}{4 R_{\mathrm{unit}}}. \tag{3.30}$$

As shown in Appendix C.1, the maximum deviation of the converter output from the ideal ramp occurs at about midscale. Dividing (3.30) by the single-ended LSB-voltage $V_{\mathrm{LSB}} = I_{\mathrm{unit}} R_L$, the achievable INL for a single-ended current-steering D/A-converter is given by:

$$\mathrm{INL}_{\max,s} \approx \frac{\Delta V_{\max}}{I_{\mathrm{unit}} R_L} = \frac{N^2}{4} \cdot \frac{R_L}{R_{\mathrm{unit}}}. \tag{3.31}$$

For a single-ended DAC the INL-error resulting from the finite output resistance of the current sources is an even function with respect to the midscale code.

The DNL-error of a single-ended converter as a function of the input code is derived in [75]. The authors also derive the resulting third-order intermodulation for a fully differential converter.

Fully Differential Converter

In Fig. 3.6 the differential output current delivered to the load for input code $k \in [0; N]$ is given by

$$\begin{aligned} I_{\mathrm{OUT}}(k) &= I_{\mathrm{OUTP}}(k) - I_{\mathrm{OUTN}}(k) \\ &= kI_{\mathrm{unit}} \cdot \frac{Z_{\mathrm{unit}}}{Z_{\mathrm{unit}} + kZ_L} - (N-k)I_{\mathrm{unit}} \cdot \frac{Z_{\mathrm{unit}}}{Z_{\mathrm{unit}} + (N-k)Z_L}. \end{aligned} \quad (3.32)$$

The differential output voltage at DC is therefore

$$V_{\mathrm{OUT}}(k) = \frac{kR_L R_{\mathrm{unit}} I_{\mathrm{unit}}}{R_{\mathrm{unit}} + kZ_L} - \frac{(N-k)R_L R_{\mathrm{unit}} I_{\mathrm{unit}}}{R_{\mathrm{unit}} + (N-k)R_L}. \quad (3.33)$$

The digital input codes giving the maximum deviation from the ideal output ramp (endpoint-fit) are derived in Appendix C.2:

$$k_{\mathrm{INL,max}} \approx N \cdot \frac{3 \pm \sqrt{3}}{6} = \frac{N}{2} \pm \frac{\sqrt{3}}{6} N. \quad (3.34)$$

Assuming once again $R_{\mathrm{unit}} >> N \cdot R_L$, we obtain the maximum deviation from the ideal ramp in a fully differential implementation with perfectly matched load resistors:

$$\Delta V_{\max} \approx \frac{\sqrt{3} I_{\mathrm{unit}} R_L^3 N^3}{18 R_{\mathrm{unit}}^2}. \quad (3.35)$$

With the differential LSB-voltage $V_{\mathrm{LSB}} = 2I_{\mathrm{unit}} R_L$ the achievable INL for a fully differential current-steering D/A-converter with finite current-source output resistance becomes

$$\mathrm{INL}_{\max,d} \approx \frac{\Delta V_{\max}}{2I_{\mathrm{unit}} R_L} = \frac{\sqrt{3} N^3}{36} \cdot \frac{R_L^2}{R_{\mathrm{unit}}^2}. \quad (3.36)$$

In a fully differential implementation the finite output resistance of the current sources results in a purely odd-order INL-characteristic.

Dividing (3.31) by (3.36) we get the ratio of the achievable INL for a single-ended and fully differential version of the same current-steering D/A-converter:

$$\frac{\mathrm{INL}_{\max,s}}{\mathrm{INL}_{\max,d}} = \frac{9}{\sqrt{3} N} \cdot \frac{R_{\mathrm{unit}}}{R_L} > 1 \Leftrightarrow R_{\mathrm{unit}} > \frac{N}{\sqrt{3^3}} \cdot R_L. \quad (3.37)$$

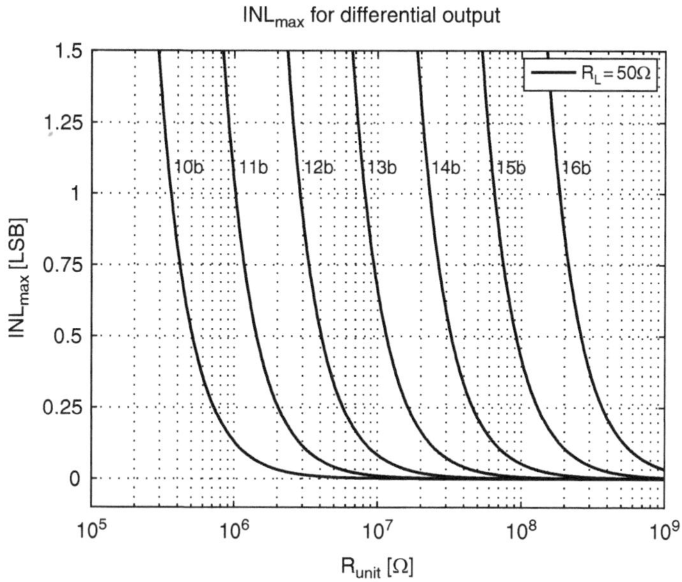

Fig. 3.7 $\mathrm{INL}_{\max} = f(R_{\text{unit}})$ of differential converter

Table 3.2 Minimum R_{unit} for $\mathrm{INL}_{\max} \leq 0.5\,\mathrm{LSB}$

Resolution	Differential	Single-ended
10 bit	508 kΩ	26.2 MΩ
11 bit	1.44 MΩ	105 MΩ
12 bit	4.1 MΩ	420 MΩ
13 bit	11.5 MΩ	1.68 GΩ
14 bit	32.5 MΩ	6.7 GΩ
15 bit	92 MΩ	27 GΩ
16 bit	260 MΩ	107 GΩ

In practical situations we always have $R_{\text{unit}} >> N \cdot R_L$. This means that the differential architecture is way less sensitive to the finite output resistance of the current sources than the single-ended implementation, although built with the *same* set of current sources [75].

Figure 3.7 shows the maximum achievable INL of a differential converter as a function of the DC output impedance R_{unit} of the LSB-cell for 50 Ω load resistors. Table 3.2 reports the minimum impedance required to achieve $\mathrm{INL}_{\max} \leq 0.5\,\mathrm{LSB}$ for a differential and single-ended output. For a resolution up to 14 bits a single cascode transistor will be sufficient in a differential implementation, if the current switch is kept in saturation. For 16 bits a current source with double cascode transistor will become necessary, as described in [76].

Note that the finite output impedance of the unit current cells at DC fundamentally limits the static linearity of the D/A-converter, since it cannot be corrected by element shuffling or by calibration methods based on the comparison with a single reference element.[3]

Even if we can implement the current sources with sufficiently large output impedance, the static linearity of the converter is still limited by the matching of its unit elements. Once all systematic errors are eliminated, only the random fluctuations of the current source transistors' electrical properties will distort the output characteristic of the converter. As discussed in Sect. 3.1.4, a given static linearity specification is directly linked with a minimum gate area required for the current sources. In the ideal case, once the converter architecture and the bias point of the current source transistors have been chosen, the required silicon area only depends on the matching constants of the used technology.

Although a static linearity of 14 bits based on intrinsic matching has been shown in [64], the converter area seems to become prohibitively large for SoC-integration, once a resolution of more than 11 bits is required.[4]

As an alternative, special circuit techniques to increase the static linearity beyond the limits imposed by random mismatch can be employed. These allow the design of compact high-resolution converter cores suitable for embedded integration into complex mixed-signal chips. Two basic methods are available to increase the static linearity of data converters:

- Time-domain averaging of unit elements
- Calibration of unit elements

Section 3.2 gives an overview on mismatch averaging achieved by element shuffling, while Sect. 3.3 describes the principles and available circuit techniques to perform unit element calibration.

3.2 Dynamic Element Matching Techniques

In unary arrays, the mapping function that determines which DAC-elements to use for a given digital input code is not unique. Because the members of the array are, at least in principle, not distinguishable, the digital input code can be represented by different combinations of DAC-elements. In a unary array with N elements only the full-scale codes have a unique representation, i.e., all elements switched

---

[3]Only a code-by-code calibration of the overall converter characteristic is able to correct also the static error introduced by the finite output impedance of the current cells. This can, e.g., be achieved with the method described in Sect. 3.3.4.

[4]An exception are perhaps $\Sigma\Delta$-DACs with low physical resolution. With a small number of DAC unit cells the converter can indeed be designed for an accuracy approaching 14 bits, while still maintaining a reasonable silicon area. Of course, perfect layout is mandatory to even approach such a high degree of intrinsic matching.

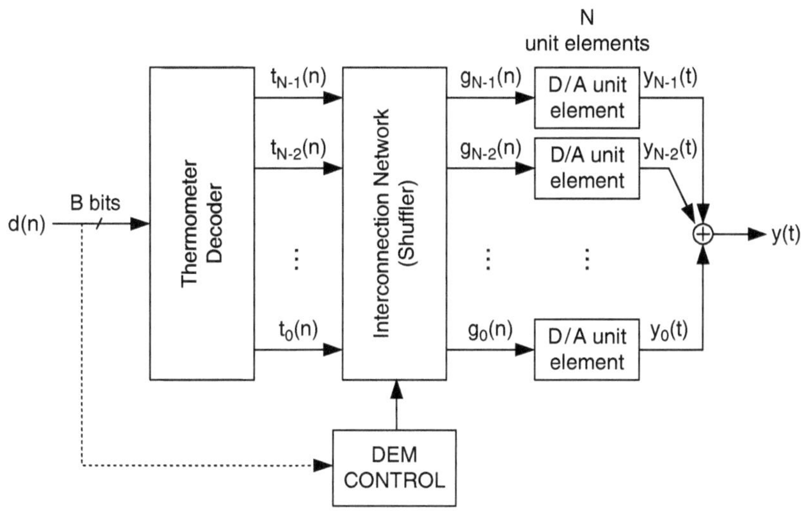

Fig. 3.8 Principle of DEM in unary DAC

on or off. All other codes can be represented by more than one combination of DAC-elements.[5]

Dynamic element matching (DEM) techniques take advantage of the redundancy in the code representation of unary arrays by continuously representing each output level with a different set of DAC-elements. If all members of the array are appropriately included in the interchanging process, then the errors of the individual components are averaged out over time. In this way, the *percepted* static accuracy of the single DAC-levels increases considerably. The minimum averaging time is given by the total interchanging time period [31].

The principle of dynamic element matching (DEM) in a D/A-converter is shown in Fig. 3.8 [77]. The thermometer-decoder converts the binary input code $d(n)$ into a thermometer coded data vector $t(n)$. Instead of addressing the $N = 2^B - 1$ unit DAC-elements directly with $t(n)$, a controlled interconnection network, or shuffler, generates a modified data vector $g(n) = S[t(n)]$ that contains the same number of 1's and 0's as $t(n)$, but in a different order for each consecutive sampling interval. The DAC-elements are then selected according to the modified data vector g.

Depending on the shuffling algorithm S used, the static error resulting from the mismatching DAC-elements is converted into wideband noise, or shifted in frequency. The effect of DEM is thus first of all perceived as an apparent increase of the static converter linearity. In conjunction with oversampling, a major portion of

[5]In an N-element unary array the number of possible combinations that represent a digital input code k is given by the binomial factor $r_k = \binom{N}{k} = \frac{(N)!}{k!(N-k)!}$, with $k \in [0; N-1]$.

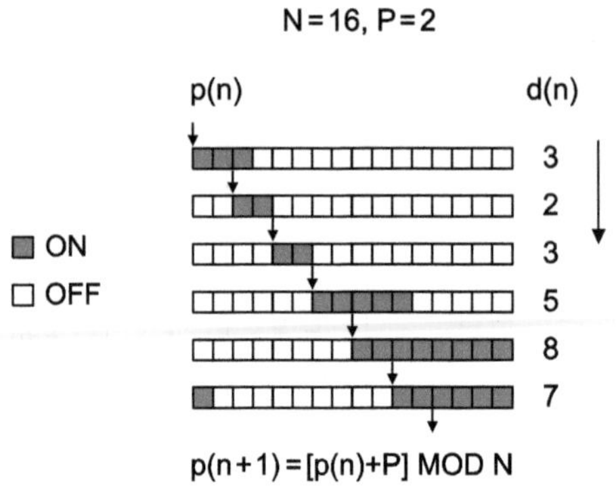

Fig. 3.9 Clocked level averaging

the mismatch noise, now placed outside the bandwidth of interest, can be eliminated by filtering. Subsequently, the effective in-band resolution is increased as well.

3.2.1 Clocked Level Averaging

A barrel-shift algorithm using a constant pointer increment rotates the DAC-elements in a periodic fashion and at a constant rate. In the literature, this method is referred to as clocked level averaging (CLA) [78]. The interconnection network in Fig. 3.8 is basically controlled by a counter that advances a pointer by a fixed value P.

As shown in Fig. 3.9, in each sample, the DAC-elements are switched on starting from the current pointer position, indicated by the arrow. When the end of the array is reached, the process revolves around to the first element of the array. The pointer position for the next sample $n+1$ can thus be calculated as

$$p(n+1) = (p(n) + P) \bmod N. \tag{3.38}$$

With CLA the static nonlinearity of the transfer characteristic is modulated to subharmonics of the clock frequency. Although the in-band resolution increases, a potential problem is the generation of tones at multiples of the element repetition frequency[6]:

$$f_{\text{rep}} = f_{\text{CLK}} \cdot \frac{\gcd(N, P)}{N}. \tag{3.39}$$

$\gcd(a, b)$ is the greatest common divisor of integers a and b. Additionally, due to the modulation of the static nonlinearity with the element rotation process, the input signal and its harmonic distortion products appear around these frequencies.

[6]After $1/f_{\text{rep}}$ s, the index pointer reaches again its starting point. Therefore, f_{rep} is the fundamental frequency of the CLA-cycle.

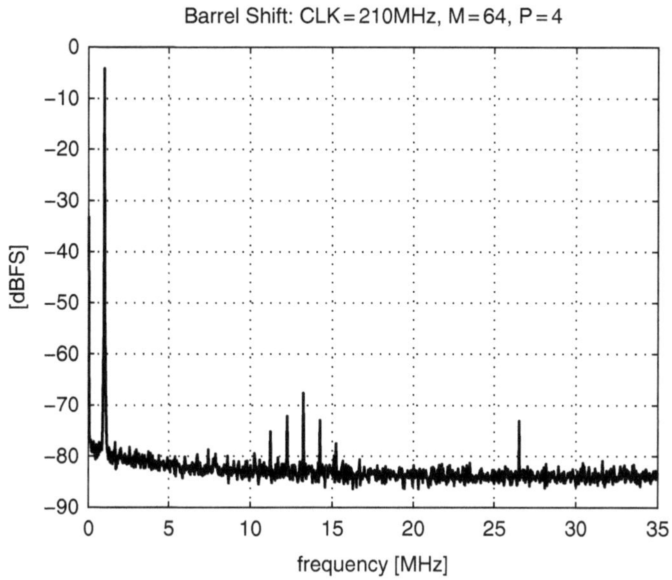

Fig. 3.10 Output spectrum with CLA ($P = 4$)

With a low oversampling ratio, the generated tones can even fall into the passband and degrade the converter performance [79].

Figure 3.10 shows the measured output spectrum of a segmented 13-bit D/A-converter using a CLA element rotation in the 6-bit MSB-segment. With a clock frequency of 210 MHz, $N = 64$ elements, and a pointer step size of $P = 4$, tones are generated at multiples of $\frac{4}{64} \cdot 210\,\text{MHz} = 13.125\,\text{MHz}$. Also the signal and its harmonics appear around these frequencies.

The advantage of CLA consists in a very simple hardware realization for the digital circuitry. It is therefore often used if the tones around multiples of f_{rep} do not fall inside the signal bandwidth, and can thus be adequately suppressed by analog filtering. This algorithm is used in Sect. 5.2 to realize a 14-bit D/A-converter for ADSL-applications with only 9 mW of total power consumption.

3.2.2 *Data-Weighted Averaging*

Data weighted averaging (DWA) is a rotation algorithm that cycles through all DAC-elements at the maximum possible rate [80]. As shown in Fig. 3.11 the pointer that indicates the first element to be switched on in the current sampling period is pointing to the first switched-off element of the previous cycle:

$$p(n+1) = [p(n) + d(n)] \bmod N. \tag{3.40}$$

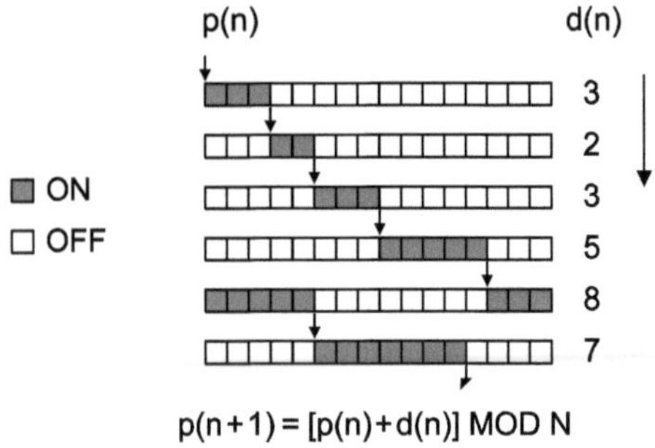

Fig. 3.11 Principle of DWA

DWA can be implemented as a barrel-shift algorithm that advances the element pointer according to the actual data sample. Besides the interconnection network, it requires an adder and a register to store the current pointer position.

A desirable property of the DWA algorithm is the first-order mismatch shaping, by which the mismatch noise is shaped by a first order highpass filter [44]. For sufficiently high oversampling, the in-band excess noise introduced by the element mismatch is thus very small, since the majority is pushed to higher frequencies.

As a drawback, also DWA is subject to tone generation, especially when processing a DC or low-frequency periodic waveform [44]. Variations of the basic DWA algorithm have been proposed to reduce the tonal behavior, usually at the cost of an increased in-band noise floor. For example, a bidirectional DWA [81] maintains two independent, data controlled rotation states in opposite directions and switches between them after every sampling period. These variations generally tend to increase the digital hardware complexity.

However, with a sufficiently busy and random input signal, as is the case for the transmitter of a digital transceiver, even a standard DWA algorithm usually does not generate noticeable tones and could be used with good results. A specific problem of DWA is the strongly data correlated switching activity. Asymmetries in the current switching transitions generate even-order harmonic distortion [82], while the charge sharing effect leads to strong odd-order harmonic distortion products. These effects are discussed in Sect. 4.1.

In Sect. 5.2 a hardware realization of a current-steering D/A-converter with DWA is described. The switching error problem is solved by introducing an interleaved current-cell architecture that results in an effective return-to-zero (RZ) behavior. In this way the data correlated switching errors are eliminated. The converter achieves almost 12-bit resolution in a 30 MHz bandwidth with a low oversampling ratio.

3.2.3 Other DEM Techniques

The simple element rotation algorithms described above have the advantage of a very inexpensive hardware implementation. To circumvent the inherent disadvantages of these methods, or to improve their mismatch-shaping properties, more sophisticated element interchange strategies can be used. Since the resulting increase in hardware complexity is restricted to the digital domain, the migration to process technologies with smaller minimum gate lengths should make the usage of more complex DEM-algorithms increasingly attractive.

Individual Level Averaging

To reduce the tonal behavior of the barrel-shift method, but still avoid the in-band noise generation of the randomized element selection, the individual level averaging (ILA) algorithm was proposed in [83]. It tries to use all unit elements equally often for each individual digital input code. This is performed by maintaining a separate rotation state for each output level [77]. Therefore the current pointer position for each code must be stored to decide which elements to use for the next data sample. ILA thus requires more digital hardware than barrel-shift algorithms like CLA and DWA, especially if the number of unit elements becomes large.

ILA is less affected by a possible correlation between the mismatch shaping and the D/A input signal. As a result, the generation of tones is not very likely, even when processing DC or slowly changing input signals. As a drawback, the in-band noise floor is higher when compared to the previously described data-driven element rotation (DWA), because the ILA algorithm converges more slowly to the averaged condition [44].

A hardware realization implementing the ILA algorithm is described in [78].

Random Level Averaging

A random selection of the unit elements to represent a given digital input code leads to a conversion of the static mismatch errors into wideband noise. With perfect randomization, i.e., no correlation of the resulting static errors during any two sampling periods, no tones will be generated in the output spectrum and the mismatch error is converted into white noise [79]. With sufficient oversampling, a large portion of the mismatch noise can then be filtered out and the in-band resolution is increased considerably.

A hardware realization of the Random Level Averaging is described in [84]. In this case the randomization of the element selection is achieved by a pseudo random number generator controlling the interconnection network. To reduce the hardware complexity, only a subset of all possible element combinations can be selected. This approach is also called partial randomization dynamic element matching (PRDEM) [85].

Higher-Order Mismatch Shaping

By implementing a noise-shaping loop for each unit element of the D/A-converter, also called a vector quantizer, higher-order mismatch shaping can be implemented [44]. The same is possible with the tree-structured mismatch shaper described in [86]. At the expense of an increasingly complex digital hardware, these generalized methods can implement mismatch shaping of arbitrary order with increasingly better in-band suppression of the mismatch noise.

Because in practice higher-order mismatch-shaping loops are only effective with sufficiently high oversampling and also susceptible to instability, hardware implementations mostly employ second-order mismatch shaping [87–89].

Modified Mismatch Shaping

With an appropriate modification to the vector quantizer, dynamic errors associated with the switching of unit elements can be decoupled from the input signal, while still maintaining the mismatch-shaping property [90]. The modified algorithm essentially tries to make the total number of switching events in each sampling period approximately constant and thus largely avoids the generation of signal-correlated energy, manifesting itself as harmonics. At the same time, mismatch shaping is performed by appropriately selecting the elements to be switched. The required amount of digital hardware appears to restrict this method to low-resolution D/A-converters. It can however be used to advantage in the MSB-array of a segmented D/A-converter to make the number of switching events per sample approximately constant [91].

Unless redundant elements are included, modified mismatch shaping (MMS) and any other element shuffling technique cannot completely eliminate asymmetrical transition errors, nor the charge-sharing effect (see Sect. 4.1), because the net number of cells that switch must at least follow the differentiated digital input code.

3.3 Current Source Calibration

Instead of trying to average out the static linearity error over time during normal operation, the goal of calibration is the minimization of the mismatch *before* the DAC-elements are used for signal processing. Depending on the calibration strategy, the unit currents are either trimmed only once at the end of the manufacturing process, in a dedicated power-on calibration phase, between regular operation intervals, or even continuously in the background.

Common to all calibration scenarios is the requirement to measure the single DAC-elements and to trim their output current to the desired value. Usually the current sources are compared one by one with a constant reference device. During the calibration process the difference between the output current of the DAC-element under calibration and the reference current is then minimized by appropriately changing the value of the DAC-element. The principle of this negative feedback loop is shown in Fig. 3.12. The trimming information for the single DAC-elements must be stored, such that the DAC-element can remember the correct value during normal operation.

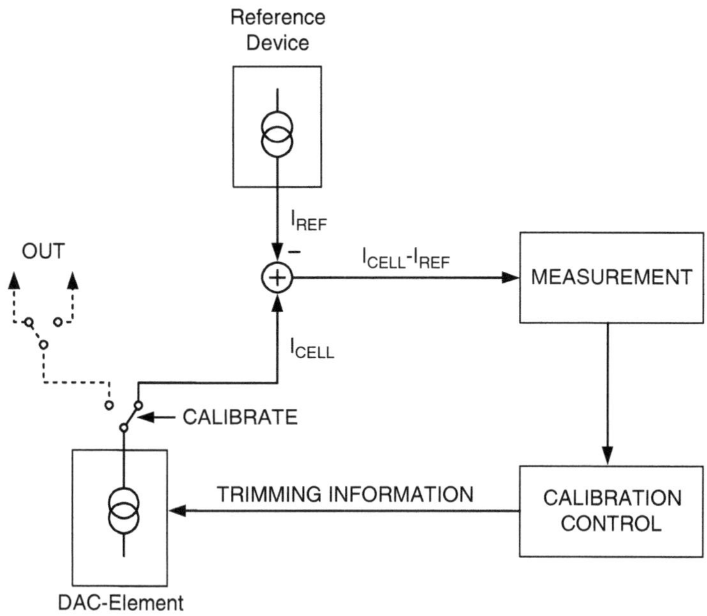

Fig. 3.12 Principle of current-source calibration

In segmented D/A-converters the MSB-array is the most critical segment in terms of matching. Its elements must match with the accuracy of the entire D/A-converter resolution. For the lower segments the required accuracy is subsequently relaxed. For example, in a 14-bit converter with a 6-bit MSB-array, the intermediate segment must only be accurate to 8 bits. For this reason, in most implementations only the matching critical elements in the uppermost segment are trimmed. The required matching of the lower segments must then be guaranteed by design, i.e., by proper sizing.

However, the sum current of the lower segments must be matched to the calibrated unit current of the MSB-array. This can either be achieved by accurate division of a calibrated MSB-current [76], biasing of the lower segments with a current mirror driven by a calibrated MSB-current [92], or by a dedicated trimming of the segment boundaries [42], see also Sect. 6.1. Direct trimming of different segments is possible with a variable reference cell [93], as described in Sect. 6.2.3.

3.3.1 Factory Trimming

Data converters can be calibrated at the end of the manufacturing process to achieve high-accuracy. In this case the measurement and calibration control equipment is generally off-chip.

Early bipolar data converter designs used laser trimming of integrated thin-film resistors [22,23,94]. Usually the degeneration resistors of the bipolar current sources are fine-trimmed for near-perfect matching. Since the calibration must be performed at wafer level, systematic errors induced by the packaging cannot be compensated. If the packaging stress can be handled appropriately, laser trimming yields the highest possible accuracy [13].

A different approach is "Zener-zapping." By forcing a special base-emitter junction into sustained avalanche breakdown, a stable metallic short circuit is formed by localized melting of the silicon [95]. A Zener-zap is a (one-time) programmable switch that can be used to trim analog circuit elements, like the input stage of a precision operational amplifier [96]. In a current-steering D/A-converter small correction currents can be added to the current sources by shorting the corresponding Zener diodes [97]. A big advantage of this method is that it does not need special process options, and it can be performed after packaging, provided that at least one extra pin for applying the zapping current is made available. Compared to laser trimming, this method is not only cheaper, but also potentially less accurate, because Zener-trimming is inherently discrete.

Another example of one-time programmable switches are metal fuses [98] and poly fuses [99]. These are mainly used to trim the central bandgap circuit or the D/A-converter reference, e.g., to achieve a very accurate transmitter gain [100]. Because fuses are relatively expensive in terms of silicon area and testing time, trimming an entire data converter by "fusing" is not practical in an SoC-environment.

In general, factory trimming is rather expensive and eventually requires special process technology options. It is therefore only used for "stand-alone" building blocks requiring the highest precision. If high-accuracy converters are needed in inexpensive digital transceiver products, usually self-calibration concepts must be implemented.

3.3.2 *Self-calibration*

Self-calibration enables an electronic system, e.g. a D/A-converter, to trim its performance critical elements autonomously, without depending on external resources. Specifically, the module must contain a reference device, the measurement equipment, as well as the entire calibration control logic.

The value of each DAC-element, upon comparison with the reference element, must be trimmed within a certain range to correct for its mismatch. In this case the mismatch is defined as the deviation from the value of the reference element, while the latter is assumed to be ideal. If the full-scale range (gain error) of the converter is of importance, the reference element itself must be guaranteed to have an accurate value. This can be achieved by appropriately trimming the value of the reference element in an independent calibration step, e.g., by fusing.

In contrast to factory trimming, self-calibration can be performed repeatedly. In the extreme case a background calibration provides continuous retrimming of the DAC-elements during normal operation. Usually, self-calibration requires neither special process technology options nor extra trimming equipment during production and test, which makes it a very attractive method for integrated data converters used in mixed-signal systems. Several circuit techniques are available to perform the DAC-element trimming, the most important ones are described in the following sections.

Foreground Calibration

Foreground calibration trims all DAC-elements *before* the converter can be used for signal synthesis. This means that calibration and data processing are two distinct operating modes. In most cases a foreground calibration is performed after the power supply is switched on. In this case it is also called start-up or power-on calibration. In some applications the converter can be re-trimmed, either periodically after a certain time span, or when the system detects changes of critical environmental parameters that are known to influence the converter accuracy.

The primary characteristic of a foreground calibration is that the converter cannot be used for data processing during the calibration phase. Although most applications allow a power-on calibration during system start-up, a re-calibration of the data converters during normal operation is generally not possible in digital transceivers.

In multi-channel solutions, independent channels integrated on the same chip can switch on or off arbitrarily. In this case the temperature distribution on the common silicon die can change considerably during normal operation, but also the supply voltage and other biasing parameters may vary. In such systems a power-on calibration usually cannot guarantee the required converter accuracy, at least not over a prolonged operation period.

Background Calibration

Background calibration trims the DAC-elements *during* normal operation without interrupting the data processing. To accomplish this, a certain degree of redundancy must be included in the data converter. A DAC-element that is switched into calibration mode must be replaced on the fly by a redundant, previously calibrated element that takes over its data processing task. When the calibration is finished, the trimmed element is switched back into normal operation. Then the redundant element either replaces another DAC-element that is being trimmed, or it is getting calibrated itself. The calibration control logic must guarantee that the converter remains transparent for the data processing, although the single DAC-elements are periodically switched into calibration mode and back.

The main advantage of background calibration is that relatively slowly changing "environmental" conditions can be tracked. Examples are power supply drifts and

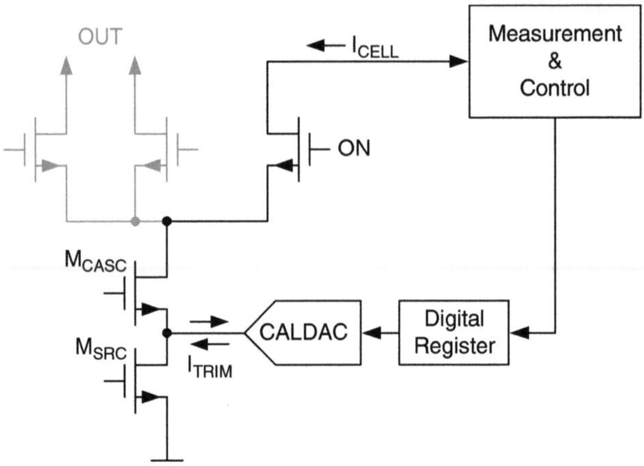

Fig. 3.13 Current-source calibration with local CALDAC

temperature changes, which may occur during prolonged operation, or even happen randomly in multi-channel applications.

A disadvantage is that background calibration periodically interchanges DAC-elements on the fly during normal operation. This may lead to the generation of spurious tones that appear at multiples of the refresh frequency inherent to the calibration cycle.

3.3.3 Local Calibration DAC

Current cells can be trimmed with a small calibration DAC (CALDAC) added to each unit element [76, 101, 102]. The local CALDAC allows to trim the value of the DAC-element in discrete steps by adding a small correction current to the main current source.

During calibration, the current difference between the DAC-element and the reference device must be digitized. A comparator decides whether the element current is smaller or larger than the reference current. This digital information is then used by the control logic to set the CALDAC, e.g., with a SAR-algorithm [76]. A digital register is required to store the trimming information.

The local CALDAC provides a static correction current I_{TRIM}, which is added at the source node of the cascode transistor; see Fig. 3.13. In this way the CALDAC is decoupled from the switching of the output current during normal operation.

Although the trimming with local CALDAC is mostly used for start-up calibration, e.g., as described in [76], also background calibration is possible, provided that redundant DAC-elements are implemented.

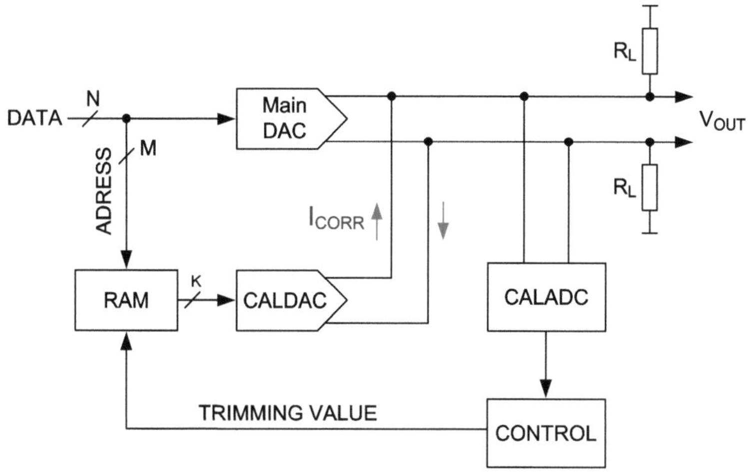

Fig. 3.14 Main DAC calibration with global CALDAC

3.3.4 Global Calibration DAC

A different approach is the use of a global CALDAC integrated together with the main converter [103]. Whereas a local CALDAC situated in the unit cell delivers a static current that is added before the current switches, the global CALDAC adds the appropriate correction current at the output of the main DAC for each input code. It is thus not only a static correction device, but it also has an influence on the dynamic performance of the converter (Fig. 3.14).

During calibration, the output levels of the main DAC must be measured with an accuracy greater than the final resolution to determine the exact trimming value for each input code. This requires the integration of a very accurate A/D-converter (CALADC) together with the D/A-converter. Typically, a single-bit $\Sigma\Delta$ ADC would be used to measure the DC output signal of the converter with high resolution. Again a control logic minimizes the deviation from the ideal output value in a feedback loop. The final digital trimming values for each input code are stored in a random access memory (RAM). During operation, the RAM is addressed by the input data sample and feeds the corresponding digital correction value to the CALDAC.

In [103], a 14-bit main DAC with a 6 bit MSB-segment is implemented. The CALDAC requires 8 bit resolution to trim the MSB-sources. A 16-bit CALADC is needed to measure the MSB-levels with sufficient accuracy.

With this method the complete static characteristic of the converter is measured code-by-code and calibrated. Note, that in this way also, the static error due to the finite output resistance of the current cells is corrected. On the other hand, only foreground calibration is possible.

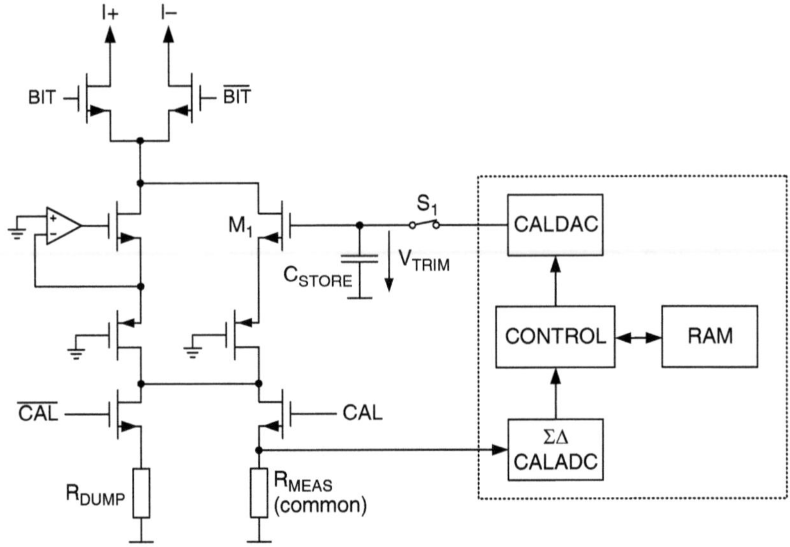

Fig. 3.15 Trimmable floating current source

3.3.5 Trimmable Floating Current Source

A floating current source with backside measurement access allows to calibrate the DAC-element in the background, while at the same time the cell current is used for signal synthesis in the foreground [92, 104].

Figure 3.15 shows schematically the implementation of the trimmable floating current source with backside measurement, described in [104]. It uses a 0.35 μm CMOS process with a 3.3 V supply. During calibration the current of the DAC-element is switched to the resistor R_{MEAS}, which is common for all cells. The voltage drop across R_{MEAS} is measured with a ΣΔ-CALADC and an appropriate trimming voltage is applied via the 12-bit CALDAC to the gate of the trimming device M_1. The capacitor C_{STORE} stores the analog trimming voltage until the cell is recalibrated.

Because the floating current sources are trimmed during normal operation, this method does not require any redundant DAC-element that replaces the current cell under calibration. As a consequence, calibration spurs, normally generated by the periodic switching of DAC-elements into calibration mode and back into normal operation, are greatly attenuated [104].

The main drawback of the floating current source is that it needs considerable voltage headroom due to the stacking of several transistors. It can therefore only be used in conjunction with a current folding output stage, along with an increase of the static power consumption by at least a factor of three.[7]

The direct measurement of the voltage across R_{MEAS} with a dedicated A/D-converter can also be replaced by a comparison of the cell current with a reference current. Instead of the CALDAC and the digital feedback control we can then use an analog feedback loop to set the trimming voltage to the correct value. A 1.8 V implementation with such a fully analog calibration circuit in 0.18 μm CMOS is described in [92]. At this supply voltage level, however, the biasing of the current cell and the calibration circuit already become quite complex because of the limited headroom. A migration of this concept to even lower supply voltages will therefore become increasingly difficult, if not impractical.

3.3.6 Dynamic Current Calibration

Dynamic current calibration is based on the principle of dynamic current copying. This method is described in [105, 106] for the construction of very accurate current mirrors that are independent of device matching. The basic concept can be extended to the calibration of entire arrays of current sources forming a current-steering D/A-converter [42, 92, 104, 107].

In principle, the trimming element is a single transistor controlled by an analog voltage stored on a capacitor. This voltage is set to the correct value during the calibration of the unit current source by comparison with a reference current in a negative feedback loop. However, due to leakage the calibration information on the storage capacitor is gradually lost and the dynamically calibrated DAC-element requires periodic retrimming. For this reason, dynamic current calibration can only be used for background calibration of current-steering D/A-converters.

Principle of Operation

The principle of dynamic current calibration is shown in Fig. 3.16 [31, 107, 108]. During calibration, switch S_1 connects drain and gate of transistor M_1 and switch S_2 connects the reference current I_{REF} to the drain node of M_1. Because M_1 is diode connected, its drain current is equal to I_{REF} and the corresponding gate-source voltage V_{GS} appears across the effective gate-source capacitance.

After the gate-source voltage has settled, S_1 can be opened and the drain current of M_1 is switched to the output node. Now M_1 is a current source that delivers an exact copy[8] of the current I_{REF}, e.g., to serve as a unit-element current in a

[7] Each output branch of the current folder must carry the full-scale current plus some extra current to prevent the cascode transistors from turning off near full scale.

[8] Hence the name dynamic current copying.

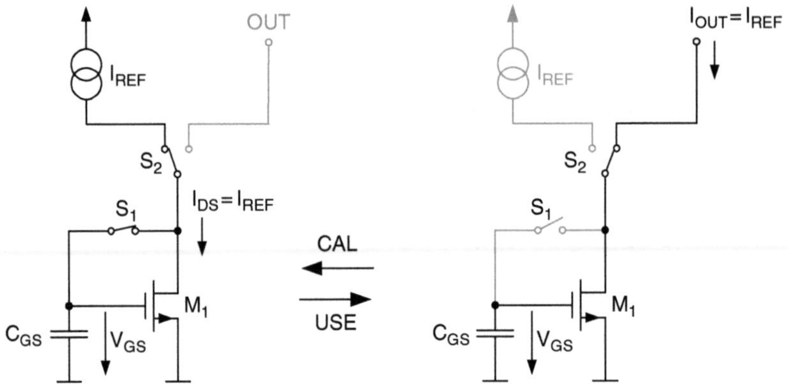

Fig. 3.16 Principle of dynamic current copying

D/A-converter. As the calibration algorithm proceeds, I_{REF} is subsequently copied to all unit current cells and the initial mismatch of the DAC-elements is greatly reduced.

The analog trimming information is stored as charge on the total capacitance C_{GS} connected between the gate and source terminal of M_1. In the simplest case, C_{GS} is given by the intrinsic gate-source capacitance of M_1 [107].

Nonideal Effects and Limitations

As shown in Fig. 3.17, when S_1 is switched off at the end of the calibration period at time $t = t_0$, a portion of the mobile channel charge of S_1 is released onto the gate node of M_1 and stored on C_{GS}. Additional charge couples via the gate-drain overlap capacitance of S_1 when its gate is switched to ground. This effect is called charge injection [109]. The total injected charge Δq_0 at the end of the calibration slot changes the gate-source voltage and the drain current of M_1 by [107]:

$$\Delta V_{\mathrm{GS}}(0) = \frac{\Delta q_0}{C_{\mathrm{GS}}} \longrightarrow \Delta I_{\mathrm{DS}}(0) = gm\frac{\Delta q_0}{C_{\mathrm{GS}}}. \tag{3.41}$$

Charge injection introduces an initial error $\Delta I_{\mathrm{DS}}(0)$ in the output current of each calibrated DAC-element. Because the switch S_1 is included in the current cell, the injected error charge is subject to mismatch, meaning that the achievable accuracy has a lower bound given by the matching of Δq_0 and the ratio gm/C_{GS}.

During normal operation leakage currents gradually change the value of the stored gate-source voltage and the drain current of M_1 between repetitive calibration slots. Assuming a constant total leakage current I_{LEAK}, the time-varying drain current of M_1 is given by [107]

$$\Delta I_{\mathrm{DS}}(t) = gm\frac{I_{\mathrm{LEAK}}}{C_{\mathrm{GS}}}\,(t - t_0)\,. \tag{3.42}$$

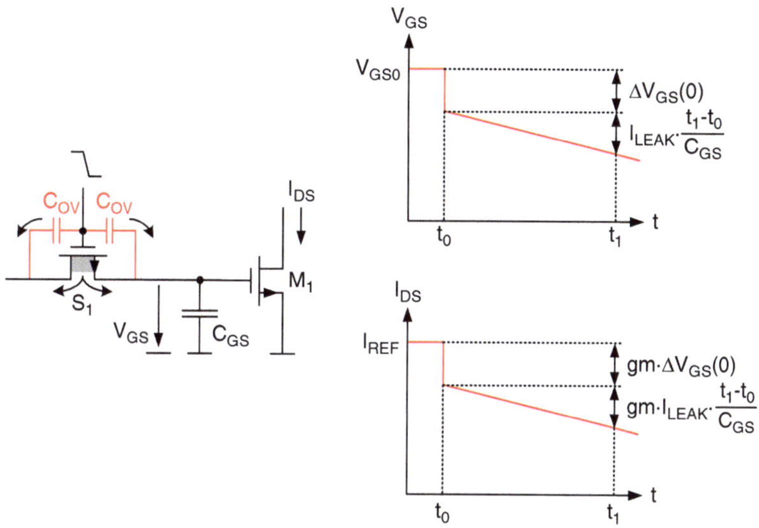

Fig. 3.17 Charge injection at the end of the calibration slot

Leakage currents in the calibrated state are due to junction leakage of the reverse-biased source-bulk junction belonging to switch transistor S_1, the non-zero off-current of S_1, and, especially in newer CMOS technologies, also due to the gate leakage of the calibrated current-source transistor M_1 [110–112].

Before the deviation of the drain current becomes excessively large, the unit cell must be recalibrated. Equation (3.42) thus puts an upper limit on T_{refresh}, the time span one unit cell can spend in normal operation mode between two successive calibration slots. Assuming an acceptable deviation in the drain current of $\delta_1 \cdot I_{\text{LSB}}$, the upper limit for the refresh cycle time is given by

$$T_{\text{refresh}} < T_{\text{refresh,max}} = \frac{C_{\text{GS}}}{gm} \cdot \frac{\delta_1 \cdot I_{\text{LSB}}}{I_{\text{LEAK}}}. \tag{3.43}$$

At the same time, with N unit cells in the calibration loop, the average time span $\overline{T_{\text{CAL}}}$ used for the calibration of one unit cell, i.e., one calibration slot, must not exceed:

$$\overline{T_{\text{CAL}}} < T_{\text{CAL,max}} = \frac{C_{\text{GS}}}{N \cdot gm} \cdot \frac{\delta_1 \cdot I_{\text{LSB}}}{I_{\text{LEAK}}}. \tag{3.44}$$

During calibration the gate of M_1 is controlled in a negative feedback loop to force the drain current equal to the reference current. The required settling time of this loop determines the minimum time span $T_{\text{CAL,min}}$ that must be assigned for the calibration of one unit cell.

We assume that the closed loop system has a dominant pole at $-\omega_{p1}$. Moreover, the negative feedback loop requires P time constants for complete settling. Usually, P is in the range of 3…5. The minimum calibration slot length is therefore

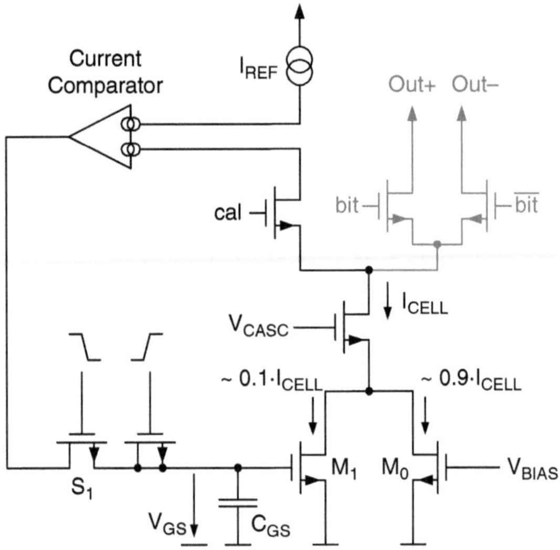

Fig. 3.18 Practical implementation of dynamic current copying

$$T_{\mathrm{CAL}} > T_{\mathrm{CAL,min}} = \frac{P}{\omega_{p1}} = 3\ldots5 \cdot \frac{1}{\omega_{p1}}. \tag{3.45}$$

With (3.44) we can also express the required minimum bandwidth for the calibration loop as:

$$\omega_{p1} > \omega_{p1,\mathrm{min}} = N \cdot P \cdot \frac{gm}{C_{\mathrm{GS}}} \cdot \frac{I_{\mathrm{LEAK}}}{\delta_1 \cdot I_{\mathrm{LSB}}}. \tag{3.46}$$

Practical Implementation

To reduce the sensitivity of the total cell current to the described nonideal effects, in practical implementations, only a small fraction in the order of 5–10 % of the total cell current is provided by M_1, whereas the major part of the current is flowing in transistor M_0 with fixed gate bias, see Fig. 3.18 [31, 107]. Since M_1 carries only a fraction $1/M$ of the total cell current, the sensitivity of the output current to voltage changes at the gate of M_1 is decreased, to first order, by a factor M, of course assuming that the gate overdrive of M_1 is approximately kept constant.

The effect of charge injection can be reduced by introducing a charge cancellation scheme [106, 107, 113, 114]. With a properly sized dummy transistor switched in the opposite direction, a part of the error charge Δq_0 can be compensated and the initial current error $\Delta I_{\mathrm{DS}}(0)$ minimized. It is advisable, however, to minimize also the size of switch S_1 and thereby Δq_0 in the first place, because the charge compensation will never be perfect.

Equation (3.41) suggests that the ratio gm/C_{GS} must be made as small as possible. Besides trying to minimize the transconductance of M_1 by choosing a large nominal gate overdrive, also increasing C_{GS} by adding extra capacitance to the gate node of M_1 further decreases the effective gate-source voltage change due to charge injection.

Two hardware realizations of 13-bit current-steering D/A-converters with dynamic current calibration are described in Chap. 6. In these designs the refresh tone problem is solved by randomizing the calibration slot length within the limits described by Eqs. (3.44) and (3.45). In this way the energy residing in the calibration tones is spectrally spread out and pushed toward the noise floor. The basic element-wise calibration concept is also extended to dual-polarity current cells, as well as to segmented architectures. Section 6.2 introduces a multilevel reference element, which is itself calibrated using the same principle and allows direct calibration of DAC-elements in differently weighted segments [93].

In the used 130 nm CMOS technology the gate leakage current of the regular, thin-oxide transistors seems to start dominating the overall leakage in the dynamic current calibration scheme. For the converter described in Sect. 6.1 the estimated worst-case gate leakage current is in the range of 10 pA. The maximum simulated off-current in the switch is about 5 pA, while the transconductance of the trim transistor M_1 (see Fig. 3.18) is $80\,\mu A/V$ and $C_{GS} \approx 2\,pF$. Using (3.44) with $\delta_1 = 0.25$ and $N = 66$ we can calculate $T_{CAL,max} = 3.8\,\mu s$. Although just an estimate, this value complies with the measurement result in Fig. 6.15, which shows the second harmonic distortion as a function of the average calibration time assigned to a single DAC-element.

Chapter 4
Dynamic Linearity

4.1 Limitations for the Dynamic Linearity

The static linearity of a current-steering D/A-converter is ultimately limited by the low-frequency output impedance of the unit current-sources and the mismatch in their output current, provided that other systematic errors can be adequately suppressed [35]. For the synthesis of higher signal frequencies, though, a number of dynamic effects become important. These are linked with the frequency-dependent output impedance of the current sources, as well as with transition errors related to the imperfect switching of the currents in the single DAC-elements. Limited dynamic linearity is perceived as increased distortion at the output of the D/A-converter. Dynamic errors tend to dominate at higher signal frequencies, not only because the impedance of internal circuit nodes becomes smaller, but also because the switching activity of the converter increases. With data-driven scrambling even the linearity at very low signal frequencies can be heavily compromised by dynamic errors (see Sects. 4.1.3 and 4.1.4).

4.1.1 Frequency-Dependent Output Impedance

When the D/A-converter of Fig. 3.6 processes a sine-wave with normalized amplitude $A_{\sin} \in [0; 1]$, the time-varying output admittance seen into the single-ended output can be written as [115]:

$$Y_{\mathrm{OUT},P}(t) = Y_L + Y_{\mathrm{unit}} \cdot N \cdot \left(\frac{1 + A_{\sin} \sin(\omega_{\sin} t)}{2} \right). \tag{4.1}$$

M. Clara, *High-Performance D/A-Converters*, Springer Series in Advanced Microelectronics 36, DOI 10.1007/978-3-642-31229-8_4,

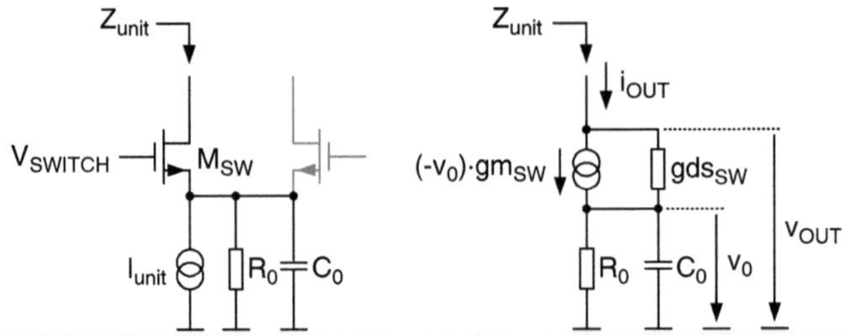

Fig. 4.1 Output impedance of single-polarity current cell

The single-ended output voltage is therefore:

$$V_{\mathrm{OUT},P}(t) = \frac{N\left(A_{\mathrm{sin}}\sin(\omega_{\mathrm{sin}}t) + 1\right) I_{\mathrm{unit}}}{2Y_L + Y_{\mathrm{unit}} N\left(A_{\mathrm{sin}}\sin(\omega_{\mathrm{sin}}t) + 1\right)}. \tag{4.2}$$

Applying a Taylor series expansion to (4.2), the second and third harmonic distortion products can be derived [35, 115]:

$$\begin{aligned} \mathrm{HD}_2 &= \left| \frac{N A_{\mathrm{sin}} Z_L}{2Z_{\mathrm{unit}} + N Z_L} \cdot \frac{8Z_{\mathrm{unit}}^2 + 8N Z_L Z_{\mathrm{unit}} + N^2 Z_L^2\left(2 + 2A_{\mathrm{sin}}^2\right)}{16 Z_{\mathrm{unit}}^2 + 16 N Z_L Z_{\mathrm{unit}} + N^2 Z_L^2 \left(4 + 3A_{\mathrm{sin}}^2\right)} \right| \\ \mathrm{HD}_3 &= \left| \frac{N^2 A_{\mathrm{sin}}^2 Z_L^2}{4\left(2Z_{\mathrm{unit}} + N Z_L\right)^2} \cdot \frac{16 Z_{\mathrm{unit}}^2 + 16 N Z_L Z_{\mathrm{unit}} + N^2 Z_L^2\left(4 + 5A_{\mathrm{sin}}^2\right)}{16 Z_{\mathrm{unit}}^2 + 16 N Z_L Z_{\mathrm{unit}} + N^2 Z_L^2\left(4 + 3A_{\mathrm{sin}}^2\right)} \right|. \end{aligned} \tag{4.3}$$

If we assume $|Z_{\mathrm{unit}}| >> N \cdot |Z_L|$, we can approximate (4.3) by:

$$\begin{aligned} \mathrm{HD}_2 &\approx \frac{N A_{\mathrm{sin}}}{2} \cdot \left| \frac{Z_L}{2Z_{\mathrm{unit}} + N Z_L} \right| \approx \frac{N A_{\mathrm{sin}}}{4} \cdot \left| \frac{Z_L}{Z_{\mathrm{unit}}} \right| \\ \mathrm{HD}_3 &\approx \frac{N^2 A_{\mathrm{sin}}^2}{4} \cdot \left| \frac{Z_L^2}{\left(2Z_{\mathrm{unit}} + N Z_L\right)^2} \right| \approx \frac{N^2 A_{\mathrm{sin}}^2}{16} \cdot \left| \frac{Z_L^2}{Z_{\mathrm{unit}}^2} \right|. \end{aligned} \tag{4.4}$$

In a fully differential architecture, the even harmonics will be suppressed, while the odd harmonics do not cancel in the differential output signal. In practice, the attenuation of even harmonics is limited by the matching of the two single-ended output loads. With on-chip loads and careful layout, a matching of 1 % and better can relatively easily be achieved [116], resulting in a theoretical even-order harmonic suppression of at least 40 dB.

Figure 4.1 shows the simplified model of a unit current cell for a single-polarity DAC and the corresponding small-signal model. We assume that the current switch transistor M_{SW} is biased in the strong inversion region, when switched on. It therefore works as an additional cascode transistor. The current source itself is

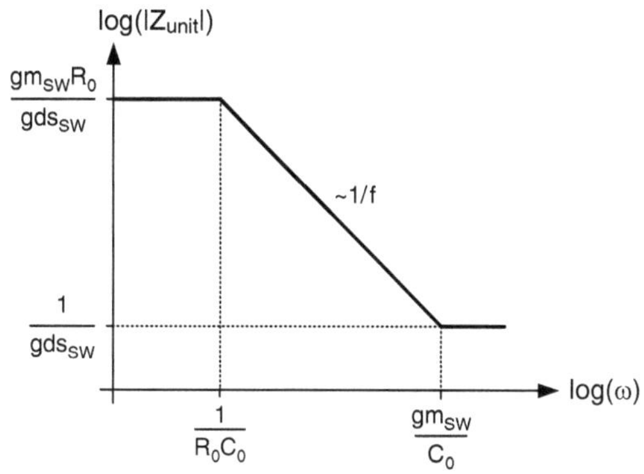

Fig. 4.2 Output impedance vs. frequency

modeled with a parasitic impedance to ground, $Z_0 = R_0||C_0$. The output impedance of this current cell is easily calculated as [115]

$$Z_{\text{unit}}(j\omega) = \frac{1 + gm_{\text{SW}}R_0}{gds_{\text{SW}}} \cdot \frac{1 + j\omega\dfrac{C_0}{gm_{\text{SW}}}}{1 + j\omega R_0 C_0}. \tag{4.5}$$

Figure 4.2 shows the magnitude of the output impedance of the unit current cell as a function of frequency. It has a pole determined by the parasitic impedance $R_0||C_0$ and a zero given by gm_{SW}/C_0. A more detailed discussion of different single-polarity current cell architectures and possible optimization strategies can be found in [35].

Inserting (4.5) into (4.4) and assuming $Z_L = R_L$ yields:

$$\begin{aligned} \text{HD}_2 &\approx \frac{NA_{\text{sin}}}{4} \cdot \frac{gds_{\text{SW}}R_L}{gm_{\text{SW}}R_0} \cdot \left| \frac{1 + j\omega R_0 C_0}{1 + j\omega\dfrac{C_0}{gm_{\text{SW}}}} \right| \\ \text{HD}_3 &\approx \frac{N^2A_{\text{sin}}^2}{16} \cdot \frac{gds_{\text{SW}}^2R_L^2}{gm_{\text{SW}}^2R_0^2} \cdot \left| \frac{1 + j\omega R_0 C_0}{1 + j\omega\dfrac{C_0}{gm_{\text{SW}}}} \right|^2. \end{aligned} \tag{4.6}$$

For low frequencies the second and third harmonic distortion products due to the finite output impedance of the current sources are determined by the ratio of the load resistance to the output resistance of the unit current cell. The pole of the unit current source impedance given by $R_0||C_0$ introduces a zero in the frequency dependence of HD_2 and HD_3, beyond which they increase with 20 dB/decade and 40 dB/decade, respectively. At gm_{SW}/C_0 the second and third harmonic distortion products level off and eventually decrease again beyond the pole frequency of the

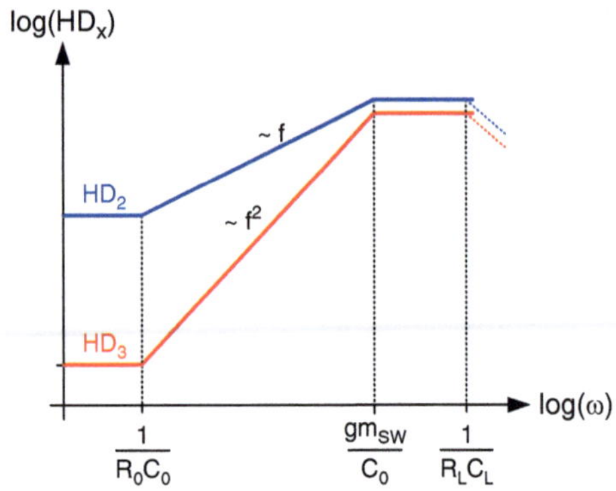

Fig. 4.3 Frequency dependence of harmonic distortion

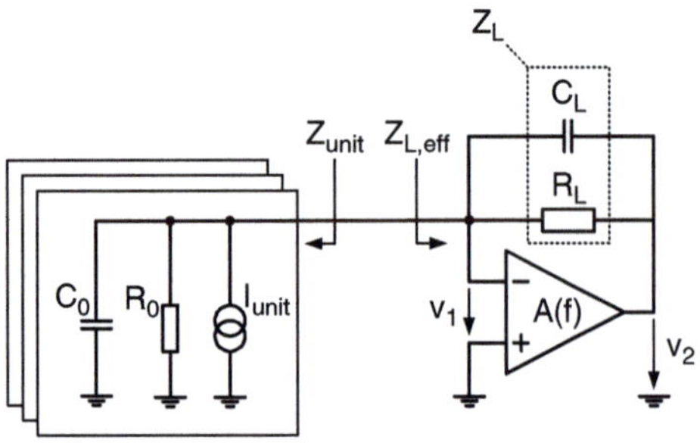

Fig. 4.4 Impedance levels in active output stage

output impedance $R_L||C_L$. Also note, that in a fully differential implementation the second harmonic distortion will be further attenuated by a matching-limited factor, as discussed above (Fig. 4.3).

The situation is somewhat different with an active output stage. Figure 4.4 shows the respective single-ended model. In this case the current switches are usually working in the linear region, such that the output impedance of the unit current cell is directly given by $Z_{\text{unit}} = Z_0 = R_0||C_0$. The effective load seen by the current source array, though, is the input impedance of the active output stage when looking into the virtual ground node:

$$Z_{L,\text{eff}}(j\omega) = \frac{Z_L}{1 + A(j\omega)}. \tag{4.7}$$

With a single-pole opamp model having a DC-gain A_0 and a gain-bandwidth product GBW the open-loop transfer function of the amplifier is given by

$$A(j\omega) = \frac{A_0}{1 + j\omega \dfrac{A_0}{2\pi \text{GBW}}}. \tag{4.8}$$

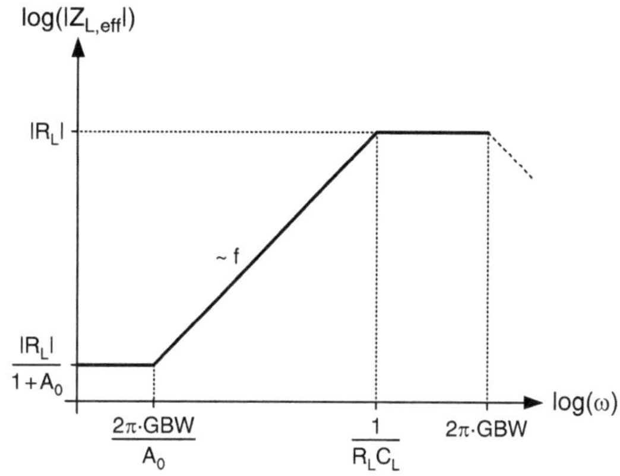

Fig. 4.5 Input impedance of active output stage

With (4.8), Eq. (4.7) becomes:

$$Z_{L,\text{eff}}(j\omega) = \frac{Z_L}{1+A_0} \cdot \frac{1 + j\omega \dfrac{A_0}{2\pi \text{GBW}}}{1 + j\omega \dfrac{A_0}{1+A_0} \cdot \dfrac{1}{2\pi \text{GBW}}} \approx \frac{Z_L}{1+A_0} \cdot \frac{1 + \dfrac{A_0}{2\pi \text{GBW}}}{1 + \dfrac{1}{2\pi \text{GBW}}}. \tag{4.9}$$

As seen in Eq. (4.9) and in Fig. 4.5, the input impedance of the transimpedance stage has a zero at the dominant pole of the operational amplifier, a pole given by the feedback network $Z_L = R_L || C_L$, as well as a pole at the gain-bandwidth (GBW) of the amplifier.

Below the unity-gain frequency of the operational amplifier, the feedback around the active element reduces the effective load impedance seen by the current-cell array. Inserting (4.9) into (4.4) we have with $1 + A_0 \approx A_0$:

$$\begin{aligned} \text{HD}_2 &\approx \frac{NA_{\text{sin}}}{4} \cdot \frac{R_L}{A_0 R_0} \cdot \left| \frac{(1 + j\omega R_0 C_0)\left(1 + j\omega \dfrac{A_0}{2\pi \text{GBW}}\right)}{(1 + j\omega R_L C_L)\left(1 + j\omega \dfrac{1}{2\pi \text{GBW}}\right)} \right| \\ \text{HD}_3 &\approx \frac{N^2 A_{\text{sin}}^2}{16} \cdot \frac{R_L^2}{A_0^2 R_0^2} \cdot \left| \frac{(1 + j\omega R_0 C_0)\left(1 + j\omega \dfrac{A_0}{2\pi \text{GBW}}\right)}{(1 + j\omega R_L C_L)\left(1 + j\omega \dfrac{1}{2\pi \text{GBW}}\right)} \right|^2 . \end{aligned} \tag{4.10}$$

The frequency dependency of the second and third harmonic distortion products is more complex, as shown in Fig. 4.6. The first zero occurs at the pole frequency of the unit current source, $1/(R_0 C_0)$. A second zero is determined by the dominant pole of the operational amplifier $2\pi \text{GBW}/A_0$, because from there on the effective

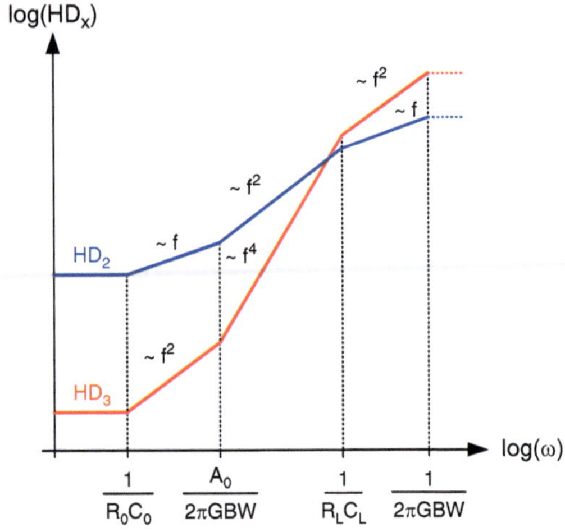

Fig. 4.6 Harmonic distortion with active output stage

load impedance seen into the virtual ground node starts decreasing. Beyond this frequency HD_2 increases with 40 dB/decade and HD_3 even with 80 dB/decade, until the first pole is met at $1/(R_L C_L)$, the pole frequency of the feedback impedance. A second pole is at the unity-gain frequency of the operational amplifier. However, the maximum signal frequency of interest is usually located below $1/(R_L C_L)$.

4.1.2 A Generalized Switching Error Model

In a current-steering D/A-converter each change of the digital input code triggers a number of switching transitions, which must occur within a very short time span at the beginning of each sampling period. Thereby the current of the corresponding DAC-cells is switched from the positive output to the negative output or vice versa, depending on the value of the controlling digital input signal. Any deviation from the ideal switching transition in the single DAC-cells introduces a temporary error in the output current that can have a negative impact on the dynamic performance of the converter.

Figure 4.7 shows a single polarity current cell with the associated parasitic elements that potentially affect the switching transition. The two transistors M_1 and M_2 are the current switches that steer the tail current I_{unit} either to the positive or the negative output, depending on the input bit. The synchronization logic is necessary to align the instant of the switching transition of all current cells in the DAC-array to the sampling clock. Additionally, dedicated switch drivers control the gates of the current switches M_1 and M_2 with an optimized crossing point in an attempt to minimize the impact of the switching transition on the current source [117].

We define the ON-transition of a current cell, when I_{unit} is switched from the negative output (OUTN) to the positive output (OUTP). Likewise, in the

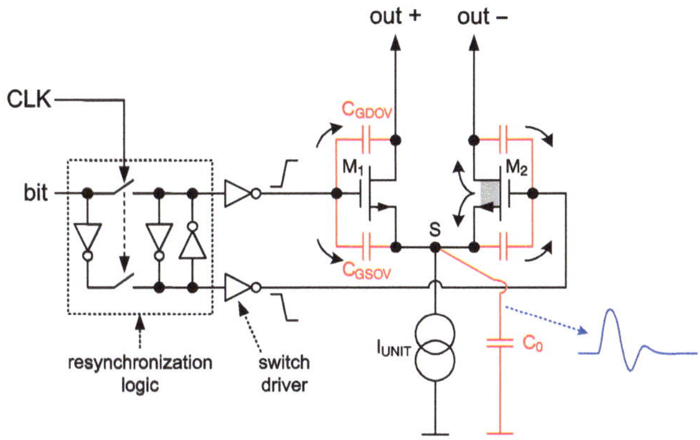

Fig. 4.7 Switching transition of current cell

OFF-transition I_{unit} is switched from the positive output (OUTP) to the negative output (OUTN). The unit current pulse observed at the differential output is defined by a switch-ON transition followed by a switch-OFF transition with a duration of one sampling period. The area under the unit current pulse is the unit charge, i.e., the amount of charge delivered by the current source in one sampling period, $Q_{\text{unit}} = I_{\text{unit}}T$.

In practical implementations, a number of nonideal effects are associated with the switching transition:

- Switch control signal feedthrough–The overlap capacitances C_{GDov} and C_{GSov} couple the voltage transitions of the switch control signals at the gates of M_1 and M_2 directly to the respective output node and also to the source node (S). The injected extra charge appears as a net charge error added to the unit charge.
- Channel charge mismatch–During the switch-ON transition channel charge must flow out from M_2 and into the forming channel of M_1. A mismatch between the switch transistors generates a net charge error that must be delivered by the current source and is subtracted from the unit charge.
- Non-optimum switch control–If switch M_2 turns off too early during the ON-transition, then both switches are off for a short time span. Then the tail current source immediately starts to discharge the tail capacitance C_0. On the other hand, if M_2 turns off too late, then the tail current source is shortly connected to both output nodes simultaneously. In both cases an error in the unit charge results.
- Local switch-ON/switch-OFF timing skew–A timing skew between the switch control lines generates a different delay for the switch-ON and switch-OFF transition with respect to the sampling clock edge.
- Rise and fall time mismatch–A different rise and fall time of the unit current pulse is generated by a mismatch in the current gain of M_1 and M_2, by a different rise and fall time of the switch control signals, as well as by a mismatch in the driving strength of the switch drivers.

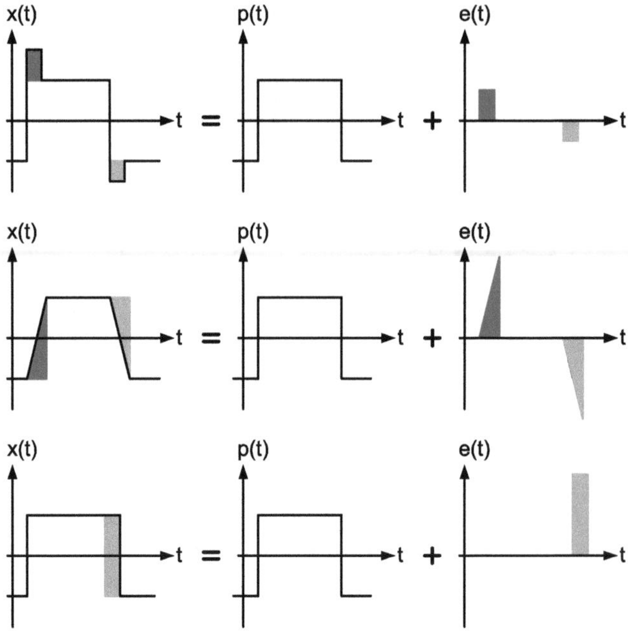

Fig. 4.8 Nonideal unit current pulses

- Charge sharing–The tail–node S sees an attenuated version of the voltage at the output node to which the tail current flows momentarily. The capacitor C_0 stores a corresponding amount of charge. During the switching transition, C_0 is switched to the complementary output. The voltage at this output node is in general different, such that a different amount of charge must be accumulated on C_0. This charge difference must be delivered, within a very short time span, by the tail current source and again introduces an error into the unit current pulse.

All described switching imperfections generate extra charge packets that are added or subtracted to the unit current pulse during the short transition phase. Figure 4.8 shows examples of nonideal switching transitions and the corresponding error pulses, which we get by subtraction of the ideal current pulse from the nonideal one.

In practical situations, all described effects will be present concurrently. Also, there will be dynamic mismatch between the different current cells in the D/A-array that considerably complicates the analysis [118]. However, the in-depth treatment of nonlinear effects caused by the dynamic interaction of switched current cells in an array, i.e. differences in the pulse transitions of single DAC-cells with respect to other DAC-cells and to different code changes, is beyond the scope of this text. For completeness, some known dynamic effects belonging to this class of "holistic" nonlinear error sources in current-steering D/A-converters are briefly discussed in Sect. 4.1.6. In the following we will concentrate on the distortion related to error charges injected by identical, time-invariant current cells, whose behavior is not influenced by the activity of the rest of the array.

Because of the finite impedance seen at the output nodes, the exact settling behavior of the DAC-network depends on the number of current cells that are connected to the positive and negative output nodes and on the settling of the tail nodes and other internal nodes during previous samples. A current-steering D/A-converter is thus a time-variant network with memory, which makes an exact analysis quite impossible [46]. Nevertheless, it would be advantageous to extract simple formulas to estimate the fundamental dynamic performance limitations of current-steering DACs.

To that aim, we try to simplify the analysis by considering an idealized NRZ unary array with the following properties:

- The unit cells are perfectly matched, statically and dynamically. This means that the behavior of all unit cells receiving the same input bit is identical. All unit cells receiving the same input bit are indistinguishable among themselves.
- The current source in each cell is ideal. Therefore, its DC-value does not depend on the voltage at the tail-node S.
- The behavior and the properties of the current switches do not depend on the voltage seen at the output nodes, i.e., the current switch transistors' drain nodes.
- The switch transistors do not have a capacitive bypass path, i.e., $C_{\mathrm{DS}} = 0$.
- The switch drivers are effective voltage sources, i.e., displaying zero output impedance.
- The settling behavior of the output nodes and of the internal tail-node S is independent of the input code.
- The network is memoryless in-between samples. Any settling effects due to injected or redistributed charges occur within a very short time span at the beginning of each sampling period and do not influence the next sample. Especially does the voltage of the tail-node S at the end of the sampling period only depend on the actual voltage of the output node the current source is being steered to.

The last property is equivalent to assuming that the internal time constants of the DAC-network are very small compared to the sampling period, i.e., complete settling of the tail nodes of all unit cells within each sample. The unit current cells in this simple model are effectively decoupled and memoryless, except for a very short transition phase. As a consequence, in this model, each DAC-element undergoing the same transition generates exactly the same error during that transition, independent of the activity of all other DAC-elements and of all previous transitions. This effectively means that we also assume a perfect power and clock distribution within the current-cell array. Nevertheless, we will show that even in this very idealized case nonlinear errors occur, which fundamentally limit the dynamic performance of current-steering D/A-converters.

To be able to analyze a broad class of switching errors we introduce a general model for a unary NRZ current-steering DAC, shown in Fig. 4.9. The upper branch constitutes the ideal DAC. The sampled data with infinite resolution $d(k)$ plus the quantization error $q(k)$ are applied to the ideal unit pulse $p_{\mathrm{NRZ}}(t)$, see Eq. (1.4). The ideal output signal of the NRZ-DAC is thus given by

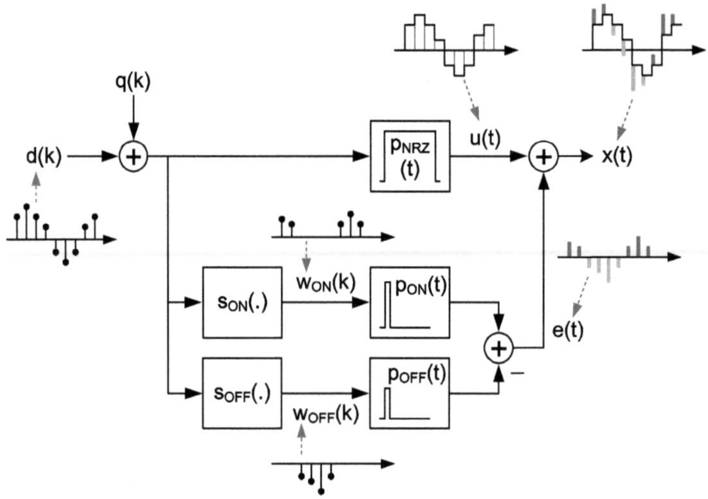

Fig. 4.9 Switching error model for NRZ-DAC

$$u(t) = \sum_k (d(k) + q(k)) \cdot p_{\text{NRZ}}(t - kT). \tag{4.11}$$

The quantized input data $d(k) + q(k)$ is also applied to the switching functions $s_{\text{ON}}(\cdot)$ and $s_{\text{OFF}}(\cdot)$. The two resulting switching sequences $w_{\text{ON}}(k)$ and $w_{\text{OFF}}(k)$ represent the number of unit cells that are being switched on and switched-off at the beginning of each sample k. The switching sequences are subsequently filtered by the pulse shapers $p_{\text{ON}}(t)$ and $p_{\text{OFF}}(t)$. Their combined output $e(t)$ represents the error signal that is finally added to the ideal DAC output $u(t)$:

$$e(t) = \sum_k \left[w_{\text{ON}}(k)p_{\text{ON}}(t - kT) - w_{\text{OFF}}(k)p_{\text{OFF}}(t - kT)\right]. \tag{4.12}$$

As discussed in Appendix D.1, if the switching transitions are very fast compared to the clock period T, then the unit error pulses $p_{\text{ON}}(t)$ and $p_{\text{OFF}}(t)$ are completely characterized by their area, A_{ON} and A_{OFF}. This approximation is valid at least over the first Nyquist band $0 \ldots \pi/T$. In this case the baseband spectrum of the error signal $e(t)$ can be obtained by reconstruction with an ideal brickwall-filter of bandwidth π/T:

$$\begin{aligned} E(j\omega) &= E\left(e^{j\omega T}\right) \cdot \text{rect}\left(\frac{\omega T}{\pi}\right) \\ &\approx \frac{A_{\text{ON}}}{T} \cdot W_{\text{ON}}(j\omega) - \frac{A_{\text{OFF}}}{T} \cdot W_{\text{OFF}}(j\omega). \end{aligned} \tag{4.13}$$

According to (4.13), the spectral components of the error signal $e(t)$ in the first Nyquist band can be obtained by evaluating the spectra of a scaled version of the continuous-time representation of the switching sequences, $w_{ON}(t)$ and $w_{OFF}(t)$. The scaling factors are given by the respective error pulse area divided by the clock period. This is equivalent to the averaging of the switching errors over the full sampling period.

The continuous-time functions $w_{ON}(t)$ and $w_{OFF}(t)$ are obtained from the switching sequences $w_{ON}(k)$ and $w_{OFF}(k)$ by ideal analog reconstruction according to the sampling theorem. Reversely, the number of switching transitions in ON-and OFF-direction for each data sample k are directly obtained by sampling $w_{ON}(t)$ and $w_{OFF}(t)$ at integer multiples of the sampling period T.

As long as we are only interested in the baseband spectrum, the unit error pulses can be completely characterized by their area, which makes a unified treatment for different shapes of the switching error possible.

4.1.3 *Switching Transition Mismatch*

According to [33], unary arrays with thermometer coding should be insensitive to switching effects, because the introduced error is always linearly dependent on the input code change. This is true for symmetrically injected charge errors, e.g., gate control feedthrough via the switches' overlap capacitances. However, we will show that a difference between the switch-ON and switch-OFF transition introduces even-order distortion, also in fully differential implementations. This effect is especially detrimental in an NRZ D/A-converter when strongly data correlated switching in the form of data weighted averaging (DWA) is used, in which case the increase in static accuracy obtained by the mismatch shaping is completely ruined by the introduced dynamic nonlinearity [82].

Thermometer Coding

In a thermometer-coded unary array, a number of identical DAC-elements corresponding to the input code are switched on sequentially, always starting from the first element in the array, see Fig. 4.10. Let the signal radian frequency be normalized to 1. Then the clock period T can be expressed as

$$2\pi f_{\text{sin}} = 1 \rightarrow T = \frac{1}{f_{\text{CLK}}} = \frac{2\pi f_{\text{sin}}}{f_{\text{CLK}}} = \omega_{\text{sin}} T = \frac{\pi}{\text{OSR}}. \tag{4.14}$$

In a unary array with N elements, the decoded input data sequence for a single tone with normalized amplitude $A_{\text{sin}} \in [0, 1]$ is given by

$$d(k) = \frac{N}{2} \left(1 + A_{\text{sin}} \cos(kT)\right). \tag{4.15}$$

Fig. 4.10 Thermometer-coded unary array

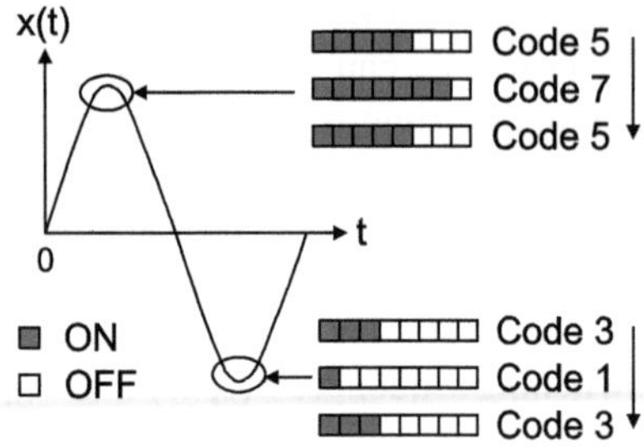

Using (4.11), the output signal of the DAC becomes:

$$x(t) = \sum_k \left(\frac{N}{2} (1 + A_{\sin} \cos(kT)) + q(kT) \right) \cdot p_{\text{NRZ}}(t - kT) + e(t). \quad (4.16)$$

If the signal gradient $d(k) - d(k-1)$ is positive, then DAC-cells are only switched ON, whereas for negative signal gradients DAC-cells are only switched OFF. In Appendix D.2 the harmonic distortion products generated by the imperfect switching transitions with thermometer coding are derived:

$$\text{for } n = 1, 2 \ldots :$$

$$\text{HD}_{2n} = 2 \cdot \frac{|A_{\text{ON}} - A_{\text{OFF}}|}{\Delta \cdot T} \cdot \frac{\sqrt{2 - 2\cos\left(\frac{\pi}{\text{OSR}}\right)}}{\pi(2n+1)(2n-1)} \cdot \frac{\frac{\pi}{2\text{OSR}}}{\sin\left(\frac{\pi}{2\text{OSR}}\right)}$$

$$\text{HD}_{2n+1} = 0 \quad (4.17)$$

Δ is the LSB-size, i.e., the value of the unit DAC-elements. If the switching transitions are asymmetric, $|A_{\text{ON}} - A_{\text{OFF}}| \neq 0$, then even-order harmonic distortion is generated, *also in fully differential implementations*. The reason is that, obviously, asymmetric switching transitions are not differentially balanced, since both outputs are affected differently.

A special case of switching asymmetry is the rise and fall time mismatch of the unit current pulse. This is often caused by unequal driving strength of the inverter buffers that drive the current switches, together with a slight imbalance in the generation of the switch timing. In the case of rise and fall time mismatch equation (4.17) becomes

$$\text{for } n = 1, 2 \ldots :$$

$$\text{HD}_{2n} = \frac{|T_{\text{rise}} - T_{\text{fall}}|}{T} \cdot \frac{\sqrt{2 - 2\cos\left(\frac{\pi}{\text{OSR}}\right)}}{\pi(2n+1)(2n-1)} \cdot \frac{\frac{\pi}{2\text{OSR}}}{\sin\left(\frac{\pi}{2\text{OSR}}\right)}$$

$$\text{HD}_{2n+1} = 0. \quad (4.18)$$

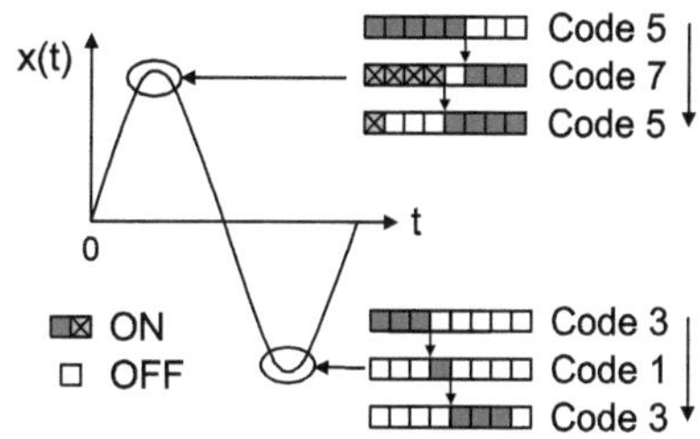

Fig. 4.11 Unary array with DWA

As long as quantization effects are negligible, the distortion does not depend on the number of current cells and is proportional to the ratio of the differential glitch area $|A_{\mathrm{ON}} - A_{\mathrm{OFF}}|$ and the unit pulse area $\Delta \cdot T$. This in turn means that for constant glitch area and oversampling ratio the harmonic distortion will rise linearly with the clock frequency at 20 dB/decade. Note, that the generated harmonics are also subject to aliasing due to the sampling process.

Data Weighted Averaging

DWA is a widely used DEM-algorithm (see Sect. 3.2.2) to improve the effective resolution of a unary array. As shown in Fig. 4.11, the DAC-elements are switched on left to right beginning from the first switched-off element of the last sample period. If the end of the array is reached, the process wraps around to the first element at the left. In this way the energy residing in the mismatch error is spectrally shaped with a first-order transfer function, offering the possibility for a huge gain in resolution, especially with large oversampling ratios [44, 80].

Figure 4.11 also shows the directional asymmetry in the switching activity of the DWA-algorithm in relation to a sine wave. In the negative half period every "turned-on" DAC-element (switched to OUTP) is replaced by a previously "turned-off" element (switched to OUTN) in the next output sample. In the positive half period instead, a code-dependent number of "turned-on" DAC-elements do not change their status (crossed boxes).

The resulting harmonic distortion (see Appendix D.3) is given by

$$\text{for } n = 1, 2 \ldots :$$

$$\mathrm{HD}_{2n} = 2 \cdot \frac{|A_{\mathrm{ON}} - A_{\mathrm{OFF}}|}{\Delta \cdot T} \cdot \frac{\sqrt{2 + 2\cos\left(\frac{\pi}{\mathrm{OSR}}\right)}}{\pi(2n+1)(2n-1)} \cdot \frac{\frac{\pi}{2\mathrm{OSR}}}{\sin\left(\frac{\pi}{2\mathrm{OSR}}\right)}$$

$$\mathrm{HD}_{2n+1} = 0. \tag{4.19}$$

Again, if the switching error is asymmetric, $|A_{\text{ON}} - A_{\text{OFF}}| \neq 0$, even-order harmonic distortion products are introduced in the differential output signal. Also the linear relation with the clock frequency for constant glitch area is maintained.

In the case of a rise and fall time mismatch equation (4.19) becomes

$$\begin{aligned}
&\text{for } n = 1, 2 \ldots : \\
&\text{HD}_{2n} = \frac{|T_{\text{rise}} - T_{\text{fall}}|}{T} \cdot \frac{\sqrt{2 + 2\cos\left(\frac{\pi}{\text{OSR}}\right)}}{\pi(2n+1)(2n-1)} \cdot \frac{\frac{\pi}{2\text{OSR}}}{\sin\left(\frac{\pi}{2\text{OSR}}\right)} \\
&\text{HD}_{2n+1} = 0.
\end{aligned} \tag{4.20}$$

A property of DWA is that the switching activity is only weakly dependent on the signal frequency. In contrast to thermometer coding, with DWA the aggregate switching error is maximum at very low signal frequencies. In the limit $\omega_{\text{sin}} \to 0$, we get for the second harmonic with rise and fall-time mismatch:

$$\lim_{\omega_{\text{sin}} \to 0} \text{HD}_2 = \frac{2}{3\pi} \cdot \frac{|T_{\text{rise}} - T_{\text{fall}}|}{T}. \tag{4.21}$$

As a numerical example, a linearity of 80 dB with a sampling frequency of 100 MHz requires a maximum rise and fall-time difference of only 4.7 ps for the unit current pulses of a perfectly matched current cell array.

The normalized second harmonic distortion product for thermometer coding and DWA, obtained by dividing (4.17) and (4.18) through the normalized differential glitch area $2 \cdot |A_{\text{ON}} - A_{\text{OFF}}| / (\Delta \cdot T)$, is given by:

$$\begin{aligned}
\text{hd}_{2,\text{therm}} &= \frac{\sqrt{2 - 2\cos\left(\frac{\pi}{\text{OSR}}\right)}}{3\pi} \cdot \frac{\frac{\pi}{2\text{OSR}}}{\sin\left(\frac{\pi}{2\text{OSR}}\right)} \\
\text{hd}_{2,\text{DWA}} &= \frac{\sqrt{2 + 2\cos\left(\frac{\pi}{\text{OSR}}\right)}}{3\pi} \cdot \frac{\frac{\pi}{2\text{OSR}}}{\sin\left(\frac{\pi}{2\text{OSR}}\right)}.
\end{aligned} \tag{4.22}$$

Figure 4.12 shows the normalized second harmonic distortion for asymmetrical switching transitions. With thermometer coding the data correlation of the switching errors increases with the signal frequency and has a maximum at the Nyquist frequency (OSR $= 1$). With DWA the maximum level of harmonic distortion is achieved for very low signal frequencies, where the data correlation of the switching errors is strongest. Interestingly, the two curves intersect at OSR $= 2$ ($f_{\text{sin}} = f_{\text{CLK}}/4$), and for still higher signal frequencies, in the range $1 < \text{OSR} < 2$,

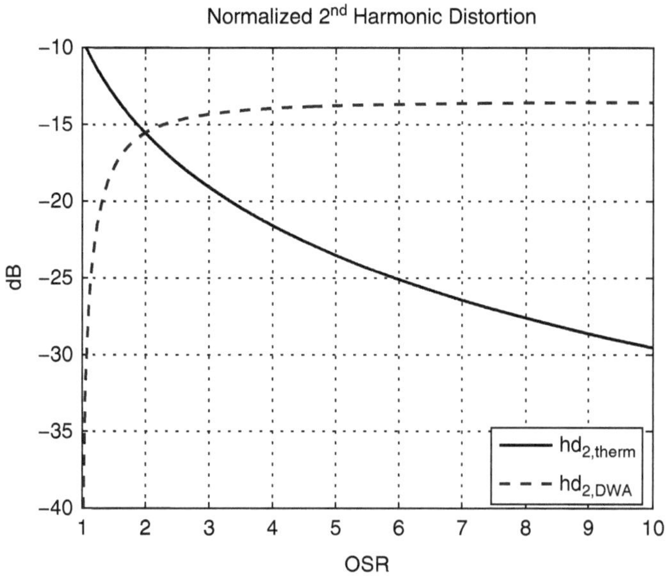

Fig. 4.12 Normalized hd_2 with asymmetrical switching error

with DWA, the data correlation of the switching errors is gradually lost. Note, however, that DWA is normally used with a decent amount of oversampling in order to achieve first-order mismatch shaping.

4.1.4 Charge Sharing at the Switching Node

The structure of the DAC-network and its initial conditions for each sample are not only influenced by the input code, which determines the way in which the DAC-elements are connected together, but also by the settling of the tail node voltages during the previous samples. More specifically, some of the tail-node capacitors C_0 of the unit current-cells together with their associated charges are connected to a different node at the beginning of the sampling period, as dictated by the change in input code. The charges on the capacitors, accumulated by the end of the previous sample, constitute the initial conditions for the settling of the DAC-network during the actual sampling period.

The DAC-network is thus time variant and has memory. The exact analysis of the charge-sharing effect in a current-steering DAC requires to solve the Kirchhoff equations for each sample, taking into account that the initial conditions depend on the settled end values of the last sample. This behavior can be modeled quite efficiently in the state-space domain and simulated with the aid of a linear time-domain solver [119]. However, it is not easily possible to extract simple equations from the complex state-space model [46].

The problem can be simplified by assuming that the internal time constants are small compared to the clock period. In this case the DAC-network is approximately memoryless between different samples; therefore, the settled end value does not depend on the tail-node capacitances. The tail-node capacitors C_0 store a fraction ρ of the voltage present at the output node they are indirectly connected to via the ON current switch. When the cell switches, the tail-node capacitor will be connected to a different output node with a different voltage. The net charge difference stored on C_0 between samples is injected into the output. Again, we assume that the charge redistribution takes place within a very short time span, much smaller than the clock period T. In this case the exact shape of the error pulse is not relevant for the resulting error spectrum in the first Nyquist band. Only the amount of injected charge relative to the unit charge–the unit current pulse area–is of importance.

In practical implementations, a nonzero feedthrough of the output voltage to the tail-node S, i.e. $\rho > 0$, also means a finite output impedance of the current cell. Due to the code-dependent current division, this effect already generates nonlinear distortion by itself, as described in Sect. 4.1.1. In Sect. 4.1.5 we will show that in most practical implementations the charge-sharing effect remains dominant compared to the finite output impedance with respect to HD_3 generation.

For the single-polarity current cell of Fig. 4.1 with $R_0 \rightarrow \infty$ and the switch transistor operating in strong inversion, we have by inspection

$$\rho(\omega) = \left|\frac{V_S}{V_{\text{out}}}\right| = \frac{gds}{|gm + gds + j\omega C_0|} \approx \frac{gds}{gm} \cdot \frac{1}{\left|1 + j\omega \dfrac{C_0}{gm}\right|}. \tag{4.23}$$

For frequencies below gm/C_0 the output signal feedthrough is constant and given by gds/gm. Above gm/C_0 the feedthrough decreases with a first-order roll-off. Since in practical implementations C_0 is small, this pole will be at a very high frequency.

For the dual-polarity current cell with active output stage, assuming that the switches operate in the linear region with small R_{ON}, we get from Fig. 4.4:

$$\rho(\omega) = \frac{1}{|A(j\omega)|} \approx \frac{\left|1 + j\omega \dfrac{A_0}{2\pi \text{GBW}}\right|}{A_0}. \tag{4.24}$$

The dominant pole of the operational amplifier at $-2\pi\text{GBW}/A0$ introduces a zero into the frequency dependence of the feedthrough factor ρ. In contrast to the single-polarity architecture with passive output stage, the first pole of the operational amplifier is at a relatively low frequency.

The feedthrough factor given in (4.23) and (4.24) is only valid for sufficiently low signal frequencies. A calculation of the exact expression of ρ must take the sampled-data nature of the voltage across C_0 into account. Only the settled value of this voltage at the end of the clock period determines the amount of error charge injected into the output during the following sample. In case the characteristic frequencies of the circuit are not much larger than the signal and clock frequency, the actual value of ρ can differ from the simple approximations given above. In Appendix D.6 we derive the expression based on a sampled-data model for the single-polarity D/A-converter with passive output stage:

$$\rho(\omega) = \frac{gds}{gm} \cdot \left| \frac{1 - \mathrm{e}^{-\frac{T}{R_L C_L}}}{\mathrm{e}^{j\omega T} - \mathrm{e}^{-\frac{T}{R_L C_L}}} \right|, \tag{4.25}$$

where we assume $C_0/gm << R_L C_L, T$, as well as a reasonably flat transimpedance transfer function over the first Nyquist band. In case the fundamental tone and the harmonic product experience a different attenuation up to the converter output, we have to include also this factor. The complete expression is found in Eq. (D.36).

For the dual-polarity DAC-model of Fig. 4.4 a similar derivation (see Appendix D.6.2) with $R_0 \to \infty$ leads to:

$$\rho(\omega) = \left| 1 + \frac{1 - \frac{1}{A_0}}{1 - K_\tau} \cdot \frac{\mathrm{e}^{j\omega T} - 1}{\mathrm{e}^{j\omega T} - \mathrm{e}^{-2\pi \mathrm{GBW} \cdot T}} + \frac{\frac{K_\tau}{A_0} - 1}{1 - K_\tau} \cdot \frac{\mathrm{e}^{j\omega T} - 1}{\mathrm{e}^{j\omega T} - \mathrm{e}^{-\frac{T}{R_L C_L}}} \right|, \tag{4.26}$$

where we have introduced the dynamic circuit constant K_τ, defined as:

$$K_\tau = R_L C_L \cdot 2\pi \mathrm{GBW}. \tag{4.27}$$

The complete expression including the transimpedance transfer function of the output stage is found in Eq. (D.39).

Thermometer Coding

The tail nodes of the current cells follow the respective output voltages, attenuated by the feedthrough factor ρ. The charges of the tail-node capacitors C_0 switched to the opposite output at the beginning of each sample generate a code-dependent charge error that affects the output signal.

In Appendix D.4 the resulting harmonic distortion for a single-tone signal is derived:

$$\begin{aligned} &\text{for } n = 3, 5, 7 \ldots: \\ &\mathrm{HD}_n = \frac{4\rho \cdot C_0 R_L N A_{\sin}}{\pi T n (n+2)(n-2)} \cdot \\ &\quad \cdot \sqrt{4 - (n^2 - 4)\cos^4\left(\frac{\pi}{2\mathrm{OSR}}\right) + (n^2 - 8)\cos^2\left(\frac{\pi}{2\mathrm{OSR}}\right)} \cdot \frac{\frac{\pi}{2\mathrm{OSR}}}{\sin\left(\frac{\pi}{2\mathrm{OSR}}\right)} \\ &\text{for } n = 2, 4, 6 \ldots: \\ &\mathrm{HD}_n = 0. \end{aligned} \tag{4.28}$$

In a single-polarity current-steering DAC terminated with a pair of load resistors R_L the output voltage feedthrough is given by gds/gm for signal frequencies below $gm/(2\pi C_0)$. It can be shown that for $f < \frac{1}{5} f_{CLK}$ a reasonable linear approximation of 4.28 for $n = 3$ is given by[1]:

$$\mathrm{HD}_3(\omega) \approx \frac{2}{5\pi} R_L N A_{\sin} C_0 \frac{gds}{gm} \omega. \tag{4.29}$$

According to (4.6) the third-order distortion due to the finite output impedance of the current cells for sufficiently high signal frequencies below $gm/(2\pi C_0)$ can be approximated by

$$\mathrm{HD}_3(\omega) \approx \frac{N^2 A_{\sin}^2}{16} \cdot \frac{gds^2}{gm^2} R_L^2 C_0^2 \omega^2. \tag{4.30}$$

The upper "intercept"-frequency ω_U, above which the output impedance of the current cells becomes the limiting factor for the third-order nonlinearity in the linear approximation, can be calculated as

$$\omega_U = \frac{32}{5\pi} \cdot \frac{1}{N A_{\sin}} \cdot \frac{gm}{gds R_L C_0}. \tag{4.31}$$

Likewise, we can calculate the lower "intercept"-frequency ω_L, below which the output impedance dominates the third-order nonlinearity:

$$\omega_L = \frac{5\pi}{32} \cdot N A_{\sin} \cdot \frac{gds R_L}{gm C_0 R_0^2}. \tag{4.32}$$

Thus, for frequencies in the range $\omega_L < \omega < \omega_U$ the charge-sharing effect will dominate the generation of odd-order harmonic distortion in a fully differential single-polarity current-steering DAC with current cells according to Fig. 4.1, at least as long as the model assumptions stated in Sect. 4.1.2 remain approximately valid.

Data Weighted Averaging

With DWA the harmonic distortion due to the charge-sharing effect becomes (see also Appendix D.5)

[1] By comparison with (4.6), we find that the HD_2 at the *single-ended* output due to the output impedance is a factor $5\pi/8 \approx 2$ larger than the HD_3 due to the charge-sharing effect.

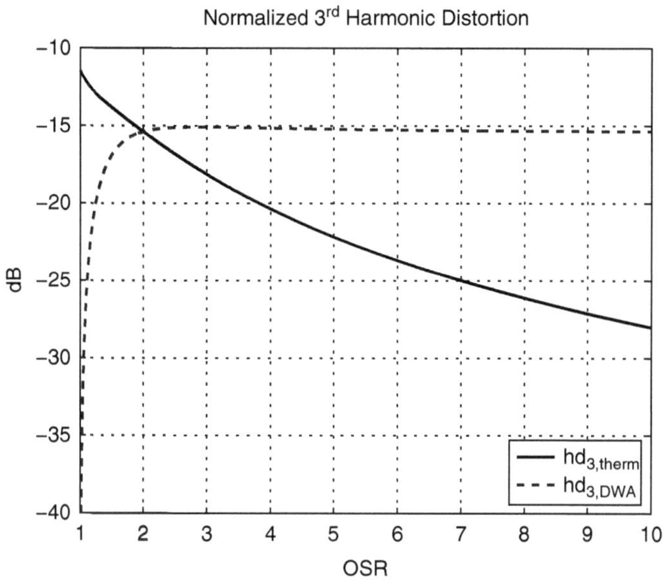

Fig. 4.13 Normalized hd_3 with charge sharing

$$
\begin{aligned}
&\text{for } n = 3, 5, 7 \ldots: \\
&\mathrm{HD}_n = \frac{4\rho \cdot C_0 R_L N A_{\sin}}{\pi T n(n+2)(n-2)} \sqrt{n^2 \cos^2\left(\frac{\pi}{2\mathrm{OSR}}\right) - (n^2-4)\cos^4\left(\frac{\pi}{2\mathrm{OSR}}\right)} \cdot \\
&\qquad \cdot \frac{\frac{\pi}{2\mathrm{OSR}}}{\sin\left(\frac{\pi}{2\mathrm{OSR}}\right)} \\
&\text{for } n = 2, 4, 6 \ldots: \\
&\mathrm{HD}_n = 0.
\end{aligned}
\tag{4.33}
$$

Again, with DWA, the switching activity is only weakly dependent on the signal frequency. For $\omega_{\sin} \to 0$ the third harmonic is limited to

$$
\lim_{\omega_{\sin} \to 0} \mathrm{HD}_3 = \frac{8}{15\pi T} \cdot \rho C_0 R_L N A_{\sin}. \tag{4.34}
$$

The expressions for the normalized third-order harmonic distortion for thermometer coding and DWA, shown in Fig. 4.13, are given by:

$$\begin{aligned}
\mathrm{hd}_{3,\mathrm{therm}} &= \frac{4}{15\pi}\sqrt{4-5\cos^4\left(\frac{\pi}{2\mathrm{OSR}}\right)+\cos^2\left(\frac{\pi}{2\mathrm{OSR}}\right)}\cdot\frac{\frac{\pi}{2\mathrm{OSR}}}{\sin\left(\frac{\pi}{2\mathrm{OSR}}\right)} \\
\mathrm{hd}_{3,\mathrm{DWA}} &= \frac{4}{15\pi}\sqrt{9\cos^2\left(\frac{\pi}{2\mathrm{OSR}}\right)-5\cos^4\left(\frac{\pi}{2\mathrm{OSR}}\right)}\cdot\frac{\frac{\pi}{2\mathrm{OSR}}}{\sin\left(\frac{\pi}{2\mathrm{OSR}}\right)}.
\end{aligned} \tag{4.35}$$

Again, the two curves intersect at OSR $= 2$. Approaching the Nyquist frequency, the third harmonic distortion is maximum with thermometer coding, suggesting maximum correlation of the injected charge error with the input data. For DWA the correlation between input data and charge error is lost at OSR $= 1$ and the odd harmonics vanish near the Nyquist frequency.

4.1.5 A SPICE-Simulation Example

As an example, a series of SPICE-simulations with an idealized single-polarity NRZ DAC-model is performed. An ideal tail current source and two MOS-switches, as shown in Fig. 4.1, form the unit current cell. The tail capacitance is deliberately increased to enhance the charge-sharing effect. The following circuit parameters are used:

- $N = 127$ (7 bit)
- Ideal LSB-section
- $f_{\mathrm{CLK}} = 100\,\mathrm{MHz}$
- $|T_{\mathrm{rise}} - T_{\mathrm{fall}}| = 20\,\mathrm{ps}$ (2 % of T)
- $C_0 = 200\,\mathrm{fF}$
- $R_0 = \infty$
- $\rho = 0.03 \approx \frac{gds}{gm}$
- $R_L = 50\,\Omega$
- $gm/C_0 >> \pi/T$

Figure 4.14 shows the simulated and calculated HD_2 and HD_3 for thermometer coding and DWA with the circuit parameters given above. Except for signals very close to the Nyquist frequency, a good correspondence between the calculated and the simulated nonlinearity is achieved.

In this fully differential example the dominating nonlinearity is the charge-sharing effect (HD_3), followed by the rise and fall time mismatch (HD_2). With DWA the NRZ-DAC is limited to $\mathrm{HD}_2 = -67.5\,\mathrm{dBc}$ and $\mathrm{HD}_3 = -63.8\,\mathrm{dBc}$ at low signal frequencies.

Also shown in Fig. 4.14 is the effect of the finite output impedance of the current cell, given by $|Z_{\mathrm{unit}}|^{-1} \approx \rho \cdot \omega C_0$. Using 4.31, the upper "intercept" frequency in the case of thermometer coding can be calculated to $\omega_U/(2\pi) = 8.5\,\mathrm{GHz}$, almost

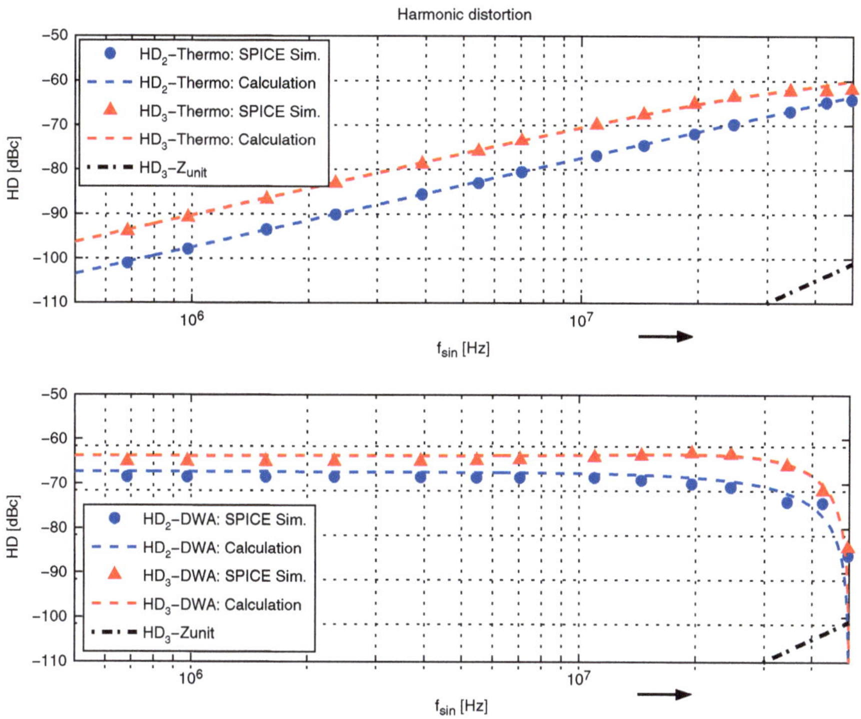

Fig. 4.14 Comparison of SPICE-simulation with calculated distortion

two decades above the sampling frequency. Furthermore, assuming a moderate output resistance of $R_0 = 100\,\text{k}\Omega$ for the MSB current source, Eq. (4.32) yields $\omega_L/(2\pi) = 7.4\,\text{kHz}$. Thus, as long as the output impedance can be made reasonably large, its effect on the dynamic linearity of a fully differential converter with a current cell architecture according to Fig. 4.1 will remain mostly negligible.

Note, however, that the output impedance of the unit current cells is a true fundamental limit for the achievable dynamic linearity of a current-steering D/A-converter. This is because the signal dependency of the switching errors can, at least in principle, be eliminated, as discussed in the following sections. The error energy related to the switching events is then no longer concentrated in harmonic distortion components, but is, e.g., shifted to the clock rate.

The output impedance of the current sources, on the other hand, is by definition always switched in parallel to the effective load impedance. It is therefore a persistent source of error in a current-steering D/A-converter. At the moment the output impedance of the current cells should become the limiting factor for DAC-linearity, only a suitable predistortion in the digital domain–with all the problems related to the identification of the nonlinear model coefficients–can improve the situation further.

4.1.6 *Other Nonlinear Effects*

As described above, the dynamic performance of current-steering D/A-converters is fundamentally limited, even with perfectly matched DAC-elements, by the finite output impedance of the current sources (Sect. 4.1.1), switching transition asymmetries (Sect. 4.1.3), as well as code-dependent charges involved in the settling of the DAC-network (Sect. 4.1.4). In practical implementations there are also other effects that cause harmonic distortion in the output signal. Generally, these depend on the amount of switching activity that takes place in the converter[2]:

- Dynamic nonlinearity is caused by a timing mismatch of the DAC-elements. If some of the cells are updated with slightly different timing as compared to others, harmonic distortion products are generated [120]. Systematic timing offsets must therefore be minimized by special layout measures.
- Timing differences can also depend on the signal activity [121]. The clock buffer that drives the sampling clock to the resynchronizing flip–flops in the DAC-elements may be influenced by the previously stored data in these flip–flops. This data-dependent loading of the clock buffer causes a timing deviation that is directly related to the signal and therefore generates harmonic distortion.
- A similar effect is power-rail interference. Due to the finite impedance of the power supply network, the drivers for the current switches are influenced by the number of cells that switch. If more unit cells are switching at the beginning of the sampling period, the voltage drop on the power rails will be larger. The delay of the switch drivers is in turn influenced by the supply voltage at the moment of the switching. Power-rail interference thus causes a signal-dependent timing deviation and, as a consequence, harmonic distortion.
- The switching activity in the unit cells is also directly correlated with the digital activity in the data decoders needed for the unary segments. Although the decoders are generally connected to a separate digital supply, an increased digital switching activity can produce more substrate noise and influence the resynchronization latch and the switch drivers in the unit cell [122].

4.2 Methods to Improve the Dynamic Performance

In the following sections we review the available circuit techniques to improve the dynamic linearity of current-steering D/A-converters, given the above imperfections. We can, in general, distinguish three strategies. First, the dynamic nonlinearity

---

[2]This must be clearly distinguished from digital interference effects caused by parasitic coupling of digital signals into the analog signal path. In many cases such effects also depend on the switching activity in the digital aggressor circuit that may be correlated with the converter activity. This makes a clear identification sometimes difficult in practice.

can be reduced by trying to minimize its source quantitatively. Examples are the minimization of the charge injection by the current switches, see Sect. 4.2.1, and the active reduction of the effective tail node capacitance by bootstrapping, Sect. 4.2.2.

The second strategy consists in making the nonideal effect "spectrally invisible" by introducing a redundant effect of the same type that makes the total error independent of the signal. In this case the energy of the nonideal effect will be concentrated, e.g., at the clock rate, while the in-band portion of the output signal will at most experience a linear error. Examples are described in Sects. 4.2.4 and 4.2.5.

Note that the modified mismatch shaping (MMS) algorithm described in 3.2.3 also falls in this category because besides its mismatch-shaping property, it is designed to provide an approximately constant switching activity. It is thus also effective against power-rail interference and clock-loading effects.

The third strategy consists in trying to make the signal-dependent portion of the transition phase of the current cells "invisible" to the output. This general concept is known as return-to-zero (RZ) and described in Sects. 4.2.6–4.2.8.

4.2.1 Current Switch with Reduced Gate Voltage Swing

The general rule to minimize the charge injection of the current switches consists in making the size of the switch transistors as small as possible. While the channel length of the switches is usually chosen minimum, their width will be limited by the cell current and the available voltage headroom.

The amount of charge that during every transition directly couples onto the output via the gate-drain overlap capacitances is also called the gate control feedthrough charge. In a differential implementation the injected charge per switching transition is given by

$$Q_{ov} = 2WC'_{\mathrm{GDov}}\Delta V_G \tag{4.36}$$

C'_{GDov} is the overlap capacitance per unit gate width. As shown in Eq. (4.36), besides minimizing the switch transistors' width, also the amplitude of the gate drive can be chosen smaller than rail-to-rail to minimize the gate control feedthrough. The theoretical minimum is given by the gate voltage difference required to steer the cell current I_{cell} completely to one side of the differential switch pair [20]:

$$\Delta V_G \geq \Delta V_{G,\mathrm{min}} = \sqrt{2}\,(V_{\mathrm{GS}} - V_T) = \sqrt{2}\cdot\sqrt{\frac{2I_{\mathrm{cell}}}{\mu C_{\mathrm{ox}}\frac{W}{L}}}. \tag{4.37}$$

There are several possibilities to reduce the voltage swing at the gate of the switch transistor. A switch driver with reduced output swing is described in [123] and shown in Fig. 4.15. The gate control signals for the current switches are generated with an inverter loaded with MOS-diodes, M_{DP} and M_{DN}. With transistor M_B the driving stage is operated in class-AB for faster transitions. The low and the high voltage for the switches can be tailored by optimizing the aspect ratio of

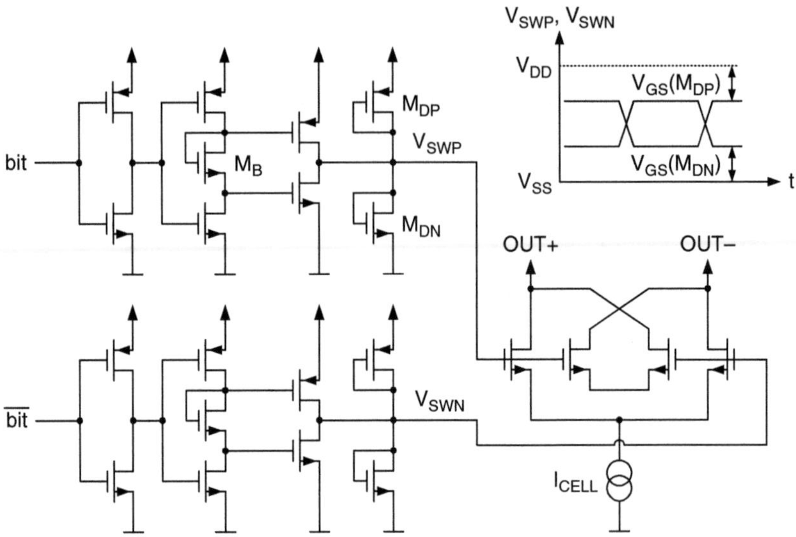

Fig. 4.15 Reduced swing switch driver

the driver transistors and the diode loads. However, for fast switching, the static current in the driver circuit can become quite large. Also, it will be subject to considerable variations over supply voltage and temperature changes, as well as transistor parameter fluctuations. For large current-cell arrays this approach is therefore not used.

Figure 4.15 also shows the switching pair with dummy transistors, as described in [123]. The idea is to cancel the clock feedthrough with a similar transistor, whose drain is connected to the other output. Although the charge injected via $C_{GD,ov}$ can in theory be canceled with perfectly matched transistors, the voltage at the source nodes of the dummy transistors will follow the respective output voltage. Since the channel charges of switch transistor and dummy transistor do not match, and neither will the impedances at the source nodes, the overall effect is not easily predictable.

A better control over the power dissipation of the switch driver is possible with the circuit described in [124], see Fig. 4.16. The high and low voltage is generated using a current source with diode load and a cascode transistor M_C. The current switches are operated by selecting either the high or the low voltage with a passive multiplexer controlled by the resynchronization latch. The R_{ON} of the MOS-switches in the passive multiplexer limits the rise and fall time of the switch control signals. Additionally, the transient recovery of the voltage generator depends on the impedance at the source and drain of M_C. A faster recovery after a switching event requires a higher bias current in the voltage generator and increases the overall power consumption.

In [125] the switch driver and the resynchronization latches are implemented as current mode logic (CML) circuits. This automatically ensures a reduced voltage swing at the current switches, while allowing precise and very fast switching. Because CML-logic is based on a differential pair, the current drain is approximately

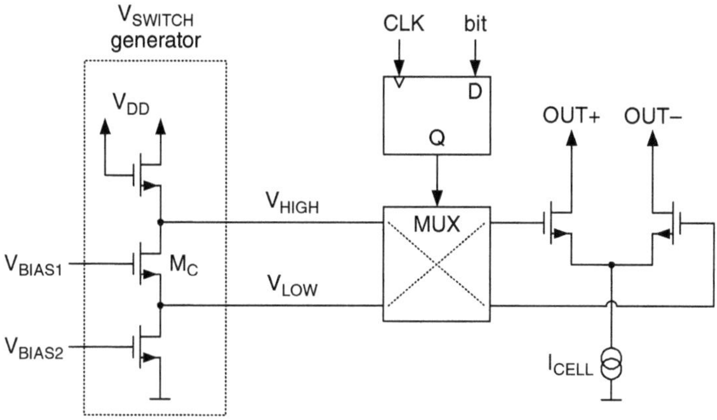

Fig. 4.16 Constant current switch driver

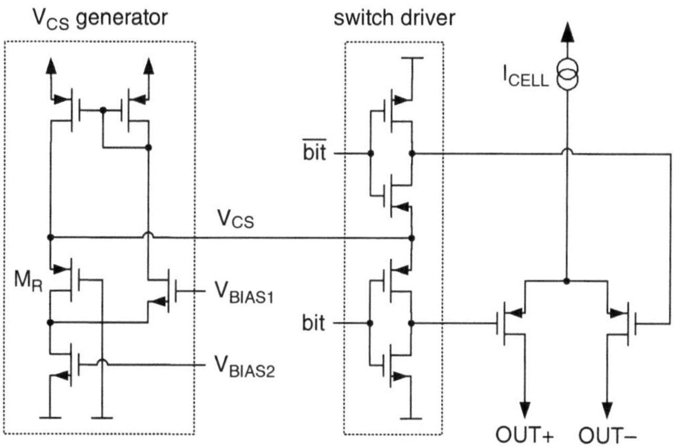

Fig. 4.17 Switch driver with VCS generator

constant and independent of the switching activity. This property practically eliminates power-rail interference. The inherent symmetry of CML-latches also avoids data-dependent clock loading. However, the use of current-mode digital logic results in a considerable increase in power consumption.

Other implementations reduce only the gate voltage for the switch transistor in the off-condition. In [114] a common bias line sets the off-voltage for the current switches. It is not mentioned, however, whether this bias voltage is generated on-chip.

The off-voltage for the (PMOS) current switches can be generated locally in the unit cell by replicating the source voltage of the "on" switch to the gate of the "off" switch with a so-called VCS-generator, shown schematically in Fig. 4.17 [76]. The source voltage V_{CS} of the replica switch M_R is used as the power rail for the switch drivers. This arrangement ensures that the off-voltage tracks the transistor parameters. A local feedback loop helps to recover faster from the transient disturbances caused by the switching transitions. Additional dummy switch drivers connected to V_{CS} produce dummy transitions when the input data does not change and thus ensure constant switching activity and data-independent settling of V_{CS}.

In [117] the resynchronization latches and switch drivers are connected to a dedicated supply voltage with a reduced value compared to the load supply. This allows to reduce the on-voltage of the NMOS current switches. The advantage of this approach is that the switch drivers are connected between a low impedance supply and ground, such that their behavior is not influenced by a locally generated bias voltage with finite output impedance. The complementary approach is described in [126, 127]. In these implementations, the switch drivers are connected to a lower supply voltage than the current sources, and the off-voltage for the PMOS current switches is consequently reduced. A significant advantage of such a split-supply architecture is that, besides the switch drivers, also the resynchronization latches and the digital decoders can run from the lower supply voltage. This brings a significant advantage in overall power consumption.

In general, the implementation of a switch driver that delivers a reduced voltage swing to the gates of the current switch transistors requires extra power consumption, unless a dedicated supply voltage with the correct voltage level is available.

4.2.2 Source Node Bootstrapping

The output voltage feedthrough to the source node of the current switches gives rise to dynamic nonlinearity, see Sect. 4.1.4. The reason is the corresponding charge stored on the parasitic capacitance connected to the source node. When the current cell is switched, the source node must settle to a different voltage, and the resulting charge difference on the capacitor is injected into the output.

Figure 4.18 shows the solution described in [76]. The main portion of the source node capacitor consists of the source-bulk capacitances of the current switches and the drain-bulk capacitance of the cascode transistor. The switches and the cascode transistor share the same n-well. A unity-gain amplifier with levelshifter copies the variations of the source node potential to the n-well contact. The effective source node capacitance is, to first order, eliminated and the nonlinearity due to the charge-sharing effect drastically reduced. The levelshift is required to safely reverse bias the source-bulk diode of the cascode transistor.

The bandwidth of the unity-gain amplifier puts a limit to the bootstrapping effect. The amplifier bandwidth must be larger than the maximum signal frequency to be synthesized. The unity-gain amplifier is mainly loaded by the well to substrate capacitance $C_{n-\mathrm{well}}$. To keep this capacitor small, the cascode transistor and the current switches must be placed closely together. Furthermore, care must be taken that the capacitance of the wiring that connects the cascode transistor with

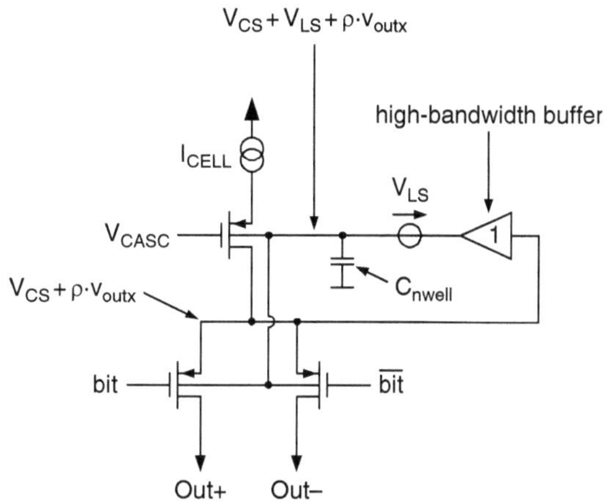

Fig. 4.18 Bootstrapping of the switch-node capacitance

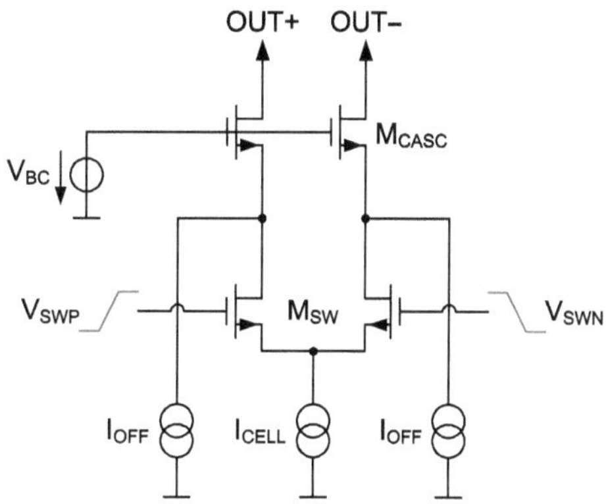

Fig. 4.19 Cascode isolation of the switch-node capacitance

the current switches is also referenced to the same *n*-well; otherwise, it is not bootstrapped out.

4.2.3 Source Node Isolation

Superior isolation of the parasitic tail capacitance connected to the source node of the current switches can be achieved by introducing additional cascode transistors between the current switch pair and the output nodes [128], shown schematically in Fig. 4.19. With the cascode transistors M_{CASC} biased in the active region, the voltage swing at their sources is already attenuated by the gain of M_{CASC}. Provided that the current switch in its on-state is also working in the active region, the voltage swing seen by the parasitic tail capacitance C_0 is then roughly attenuated by the product of the voltage gain of M_{CASC} and M_{SW} [128].

This current-cell architecture requires more voltage headroom in order to accommodate the additional cascode transistor M_{CASC}. Auxiliary currents I_{OFF} must be introduced to keep the cascode transistor in the "off-branch" sufficiently turned

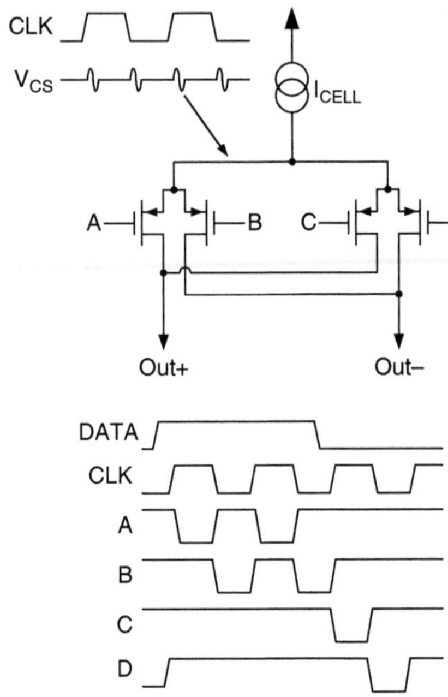

Fig. 4.20 Differential quad switching

on; otherwise, the gate-source capacitance variation of the cascode transistors will be too large during switching events, and this would seriously affect the dynamic linearity [128]. Obviously, to keep M_{SW} in the active region, the gate of M_{CASC} must be biased at a higher voltage than the switch driver's supply, assumed equal to the "high" voltage at the gate control nodes, V_{SWP}, V_{SWP}.

4.2.4 *Differential Quad Switching*

The charge injection of the current switches, including also the clock feedthrough, generates harmonic distortion, because the number of switching events and thus the injected charge depends on the input data activity. If the switching rate could be kept constant, the amount of injected charge is the same in each sampling period and thus does not produce a nonlinear signal component.

Differential quad switching makes the charge injection independent from the data pattern [129]. Instead of using two current switches that toggle only when the input data bit undergoes a change, four switches are connected to the cascode transistor of the current source, see Fig. 4.20 [126]. In each sampling period exactly two switches are toggling and thus the error energy related to the switching events will be concentrated at the sampling rate. The input data determines whether the current is steered to the other output by this switching or routed to the same output, but via a different transistor [126].

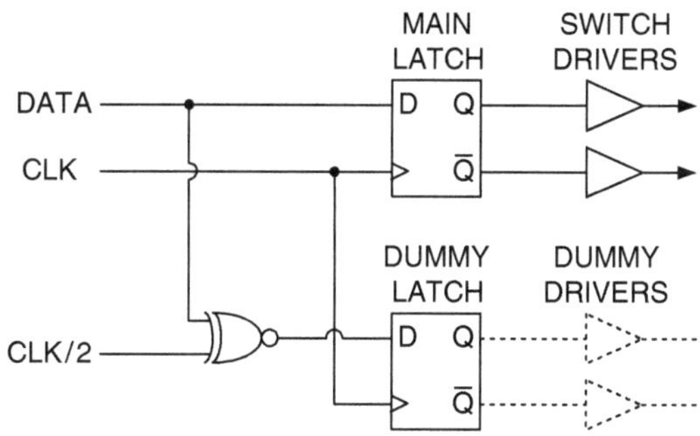

Fig. 4.21 Dummy latch for constant clock loading

Differential quad switching eliminates the nonlinear distortion due to charge injection, source node disturbance and asymmetric transition errors, because the cell current is rerouted in each sampling period. Effectively, differential quad switching implements two interleaved RZ switches [130].

However, the charge-sharing effect described in Sect. 4.1.4 is not eliminated by differential quad switching, because the signal-dependent charge related to the output signal feedthrough, which is stored on the source node capacitor, is still injected into the complementary output node when the input data changes.

Another drawback of differential quad switching is increased power consumption in the switching part of the converter. Because two switching events take place in each sampling period, the dynamic power is at least twice as high as with a normal switching pair with the converter operated near the Nyquist frequency.

4.2.5 *Constant Digital Activity*

Dynamic effects associated with the digital switching activity can be mitigated by ensuring that the number of digital transitions in each sampling period is constant and therefore independent of the input data pattern. In this way, the power of the disturbance will be concentrated at the clock rate and will not be visible in the output spectrum of the converter. Since the normal data switching must be complemented by a given amount of "dummy switching", constant activity architectures require a certain degree of redundancy. The price for improved dynamic performance is paid for in terms of silicon area and power consumption.

Problems with data-dependent clock buffer loading and power-rail interference can be eliminated by making the number of latches that toggle at the beginning of each sampling period independent of the input data activity. A possible implementation includes a dummy latch in each DAC-element which is forced to toggle whenever the input data bit does not change, see Fig. 4.21 [130].

Although the number of latches and thus the loading of the clock buffer has doubled, this load does not depend any more on the data activity. In [76] a dummy switch driver additionally ensures constant loading of the VCS generator. The latter not only does provide a reduced off-voltage for the current switches, but it also has a finite output impedance that would lead to data-dependent settling of the switch drivers' supply voltage. In the same work also the number of MSB-cells that switch in each clock cycle is made approximately constant by using a MMS switching scheme [90] for the uppermost unary-coded segment. Thus a very high degree of decorrelation between the input data and the current switching process is achieved.

The constant activity principle can also be extended to the digital data decoders used for the unary segments. In general, each data node in a digital circuit can be complemented with a dummy node that becomes active whenever the principal data node remains inactive [131]. In this way the $\mathrm{d}I/\mathrm{d}T$—noise that may couple via the substrate to the DAC-elements is concentrated at the clock rate. The area overhead in the digital circuitry can become however considerable. Most applications simply try to prevent the digital switching noise from entering the analog DAC-core by using a separate power supply for the digital decoders, by physical separation of the digital and analog circuitry, as well as by proper shielding of sensible nodes.

4.2.6 Return-to-Zero and Track/Attenuate

In a return-to-zero (RZ) converter the DAC-elements are reset in each sample by switching the current to a common node for a fraction D of the sampling period T. This in turn means that the effective output amplitude after baseband filtering is reduced by a factor $1/D$. In most implementations the duty factor D is chosen to be 0.5, resulting in a 6 dB amplitude loss of the reconstructed output signal, as compared to an equivalent NRZ-converter.

The main advantage of an RZ converter is improved dynamic linearity, because all current sources are first connected to a common-mode potential before they are switched to the output nodes. Provided that the voltage at the internal nodes of the current cell, especially across the source node capacitance, is settled during the reset period and remains independent from the output voltage, the charge injected into the output current only generates a linear error. This means that the memory effect, also called intersymbol interference, is, at least in principle, completely eliminated [130].

The same is true for the charge injection coming from the current switches. Because the switches are operated in each sampling period, the resulting error is, to first order, linearly related to the output voltage.[3] Thus no nonlinear distortion is produced by the switches' charge injection.

Figure 4.22 shows the implementation of a Return-to-Zero current cell in a GaAs heterojunction bipolar technology [32]. A similar circuit is described in [24], using a 0.35 μm BiCMOS technology. In both implementations a second switch pair is stacked on the output of each primary current switch. During the reset phase the

[3] This assumes that the charge error generated by turning on the current switch transistor does not depend on the voltage at its drain, which is connected to one of the output nodes.

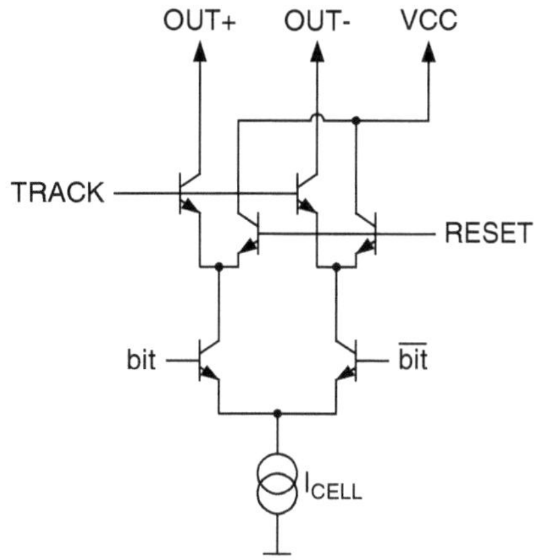

Fig. 4.22 Return-to-zero current cell

current is dumped to a low-impedance power rail, and, after proper reset, steered to the output during the track phase.

The Return-to-Zero switch can also be implemented as a single switch in the output stage, with the advantage that no tight synchronization across the current-cell array is necessary [132]. To ensure a symmetrical settling behavior in both phases, the design described in [41] uses an on-chip dummy load for the converter current during the reset phase, while the output load is discharged with a second (identical) on-chip load. At the beginning of the tracking phase, when all switching errors have settled out, the converter current is steered to the output load, such that the voltage pulse can build up.

In dual-polarity arrays the current cells in track mode work into the virtual ground nodes of an operational amplifier. The common node for the reset of the current cells is therefore best provided by a common-mode buffer that generates a voltage equal to the potential seen at the virtual ground nodes of the transimpedance output stage [133]. In this way the potential at the drain nodes of the current switches remains equal in both operating modes, provided that the systematic offset is sufficiently small.

In oversampled converters the Return-to-Zero output waveform requires a considerable slew rate in the I-V converter of the output stage. An active output stage is therefore problematic, because a signal-dependent slew-rate limitation causes nonlinear distortion [47]. This would completely destroy the benefit of the Return-to-Zero functionality. Current-steering D/A-converters delivering a Return-to-Zero output waveform are therefore virtually only used in conjunction with an inherently linear passive output stage. With active output stages the output waveform must generally remain continuous, like in a NRZ-converter. Solutions to this problem are described in the following sections.

A track/attenuate output stage [92, 104] does not require an additional switch in each current cell. Instead, the output nodes are connected to ground and shorted during the reset period using only three switches, see Fig. 4.23. The beginning of the reset period is aligned with the switching of the current cells. Because the output

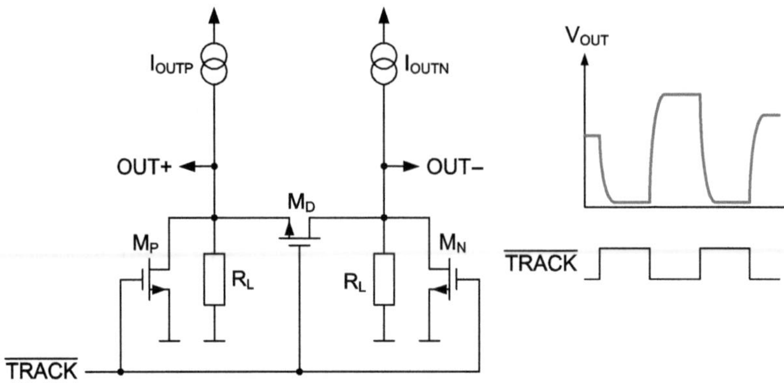

Fig. 4.23 Track/attenuate output stage

voltage is largely attenuated when the current cells are switching, also the dynamic errors are effectively not visible. Only when the current coming from the DAC-array has settled, the output stage is switched back into track mode, and the converter current flows into the load resistors. Although the elimination of the settling period is not perfect because the impedance of the attenuation network is finite, the SFDR is greatly improved. The track/attenuate output stage also eliminates systematic timing errors in the current cells, because the reset period is generated with only three switches in close proximity, operated by a single clock.

A practical limitation of a RZ-architecture is the settling during the reset period. If this settling is not complete, the history of the past samples is not totally canceled and will influence the next output sample, resulting in nonlinear distortion. At very high clock rates the implementation of an effective Return-to-Zero circuitry is therefore not trivial.

Another problem is the jitter sensitivity of a DAC producing a Return-to-Zero waveform. As discussed in Sect. 2.3.3, this poses a very stringent requirement on the spectral purity of the sampling clock.

4.2.7 Double Return-to-Zero

As stated in the preceding section, with an active output stage, it is difficult to reproduce Return-To-Zero pulses without running into slew-rate limitation. The latter makes the settling of the output voltage data dependent and thus introduces nonlinear distortion. To circumvent the problem, *Adams et al.* proposed the so-called Double Return-to-Zero architecture in [47].

The principle is shown in Fig. 4.24. The overall converter consists of two identical interleaved D/A-converters that receive the same input data. Each DAC is alternately active for half a clock cycle and reset for the other half of the sampling period. Therefore an effective half-clock Return-to-Zero is implemented for each

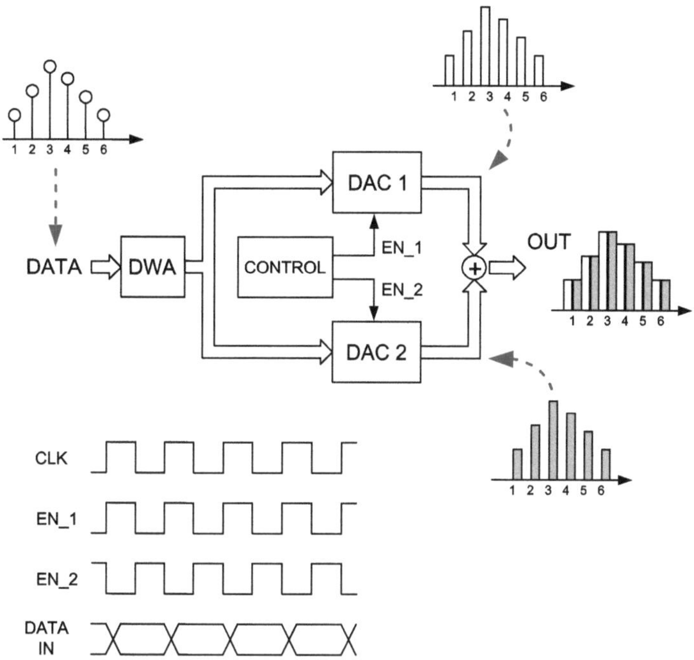

Fig. 4.24 Double return-to-zero architecture

DAC-core. On the other hand, because always one of the two converters is active, the output current is never interrupted. The complete D/A-converter behaves like a NRZ-DAC with the same output amplitude and the same slew-rate requirement for the output stage. In [47], additionally, a segmented DWA algorithm is used to scramble the input data for both converters.

The main drawback of the Double Return-to-Zero architecture is increased power consumption. The switching activity is the same as for the differential quad switching described in Sect. 4.2.4, thus doubled in comparison to a standard RZ-implementation. Also the static power consumption is doubled, because each DAC-core carries the full-scale current. On the other hand, the output waveform is the same as for a NRZ-DAC, such that an active output stage can be employed.

The Double Return-to-Zero architecture has the same limitation with respect to the settling of the current cells in the reset period as the standard half-clock Return-to-Zero. Only when the memory of the circuit is completely canceled during the reset, the nonlinear distortion due to dynamic effects is eliminated.

4.2.8 Full-Clock Interleaved Current Cells

Time-interleaving of converters is extensively used in A/D-converters to maximize the sampling rate [134–137]. A number of identical A/D-converters can be operated

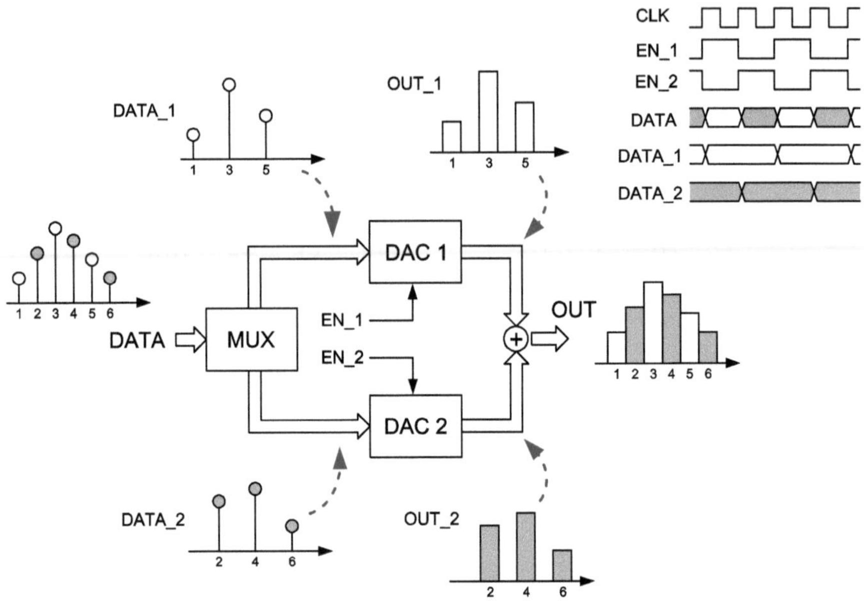

Fig. 4.25 Full-clock interleaved DAC-cores

in parallel and their output appropriately combined to yield a higher effective sampling rate. Also in the D/A direction the effective sampling rate can be increased by time-interleaving two or more converters [138, 139]. This not only relaxes the internal clock speed of the DAC-cores, but, more importantly for linearity, also extends their reset phase, provided that each converter performs a Return-to-Zero cycle in-between its active conversion phases. With two D/A-converter cores, the reset period is already extended to a full clock cycle.

As shown in Fig. 4.25, the overall D/A-converter then consists of two interleaved DAC-cores. Each DAC is alternately active for a full clock period during which it delivers the output current. In the following clock period the DAC-core that was active in the previous sampling period is reset, while the other DAC-core delivers the output current corresponding to the actual input data. Because the two DAC-cores process alternate data samples, a multiplexer is required from a conceptual viewpoint [133]. In a practical realization the multiplexing can be easily accomplished by triggering data latches at the input of the sub-DACs with opposite phases of a divided version of the sampling clock [140, 141].[4] Naturally, this operation can also be distributed among the single DAC-cells [93]. Since current

[4]The design proposed in [141] additionally includes a PLL-controlled, interleaved raised-cosine reference voltage for the resistor-based multiplying sub-DACs, with the goal to additionally suppress the clock-jitter and slew-rate induced error in the output signal during code transitions.

sources ideally do not influence each other when connected together at a sufficiently low-impedance summing node, an interleaved current-based D/A-converter can either be described as two (or more) identical sub-DACs or as a single DAC with as many interleaved individual current cells–both models are completely equivalent from a system perspective.

A conceptually similar implementation is described in [142]. The output currents of the two interleaved DAC-cores are summed by a deglitching circuit. The deglitcher connects the analog output with the active DAC-core half a sampling clock cycle *after* the latter has been updated.Therefore, all data dependent transient errors of the respective sub-DAC occur while its current is dumped to ground and are thus not seen at the analog output node.

The advantage of full-clock interleaving is that the effective switching activity is halved in comparison to the Double Return-to-Zero architecture. This in turn also halves the dynamic power consumption. On the other hand, the reset period is extended to a full clock cycle, and therefore the settling time for the reset is twice as long. This allows the operation at higher clock frequencies with still effective memory cancellation. The NRZ-property of the output waveform is also maintained, such that an active output stage with the same slew-rate limitation can be used.

The concept of interleaved converters can be easily extended to more than two sub-DACs. This allows, e.g., to extend the reset phase for each converter to more than one sampling clock. Any residual regularity related to the Return-to-Zero switching sequence can be eliminated by randomly selecting, in each clock cycle, the active DAC-cores (or DAC-cells) from a "pool" of resetted converters (cells) [143]. Additionally, a complex overall signal transfer function can be constructed by including suitable digital filters, which are run power efficiently on the low clock rate, in front of the individual sub-DACs [144].

The drawback of interleaved converters (A/D and D/A) is the generation of image signal components around fractions of the clock frequency in the presence of a global mismatch between the single converter cores [145, 146]. In noiseshaped converters the image signal components are normally outside the bandwidth of interest, but gain and phase mismatch causes the shaped quantization noise to leak into the baseband and reduces the effective resolution [147]. Therefore, the subconverters in a time-interleaved architecture must be matched to a certain degree so as not to degrade the overall performance.

This topic is further discussed in the hardware realizations described in Sects. 5.2 and 6.2. Both designs employ an interleaved current cell architecture to produce an effective RZ behavior for the single DAC-element. Matching between the two subconverters is achieved by a special floorplan with close proximity of the corresponding DAC-elements. The design described in Sect. 6.2 additionally uses current calibration.

Chapter 5
Noiseshaped D/A-Converters

As described in Sect. 2.3.1, noiseshaping can be used in oversampled data converters to trade off bandwidth with resolution. In a $\Sigma\Delta$-DAC a digital noiseshaper preprocesses the input data *before* the actual conversion to the analog domain. Because the D/A-conversion happens outside of the noise-shaping loop, any nonideality related to this conversion is directly injected into the analog output.[1]

In this chapter we describe two hardware realizations of noiseshaped current-steering D/A-converters that employ dynamic element matching (DEM) to improve the static linearity and the in-band resolution. Both testchip modules are fabricated in a 0.13 μm CMOS technology and use a single 1.5 V supply voltage.

The first design is optimized for an ADSL2+ transmitter in the central office (CO) with a signal bandwidth of 2.2 MHz and an effective resolution of over 14 bits [148]. Strong focus is put on low-power operation to support multichannel transceiver integration [100].

The second module is a general-purpose D/A-converter that can deliver a resolution of 14 bits for ADSL-bandwidths in low-power mode, as well as 12 bits for signal bandwidths close to 30 MHz [133], compliant to the VDSL2-standard. It is targeted at software-programmable multi-mode transceivers that support the various ADSL and very high-speed digital subscriber line (VDSL)-standards.

5.1 A 14-bit Low-Power D/A-Converter

In the central office (CO) the high packing density on the linecard drives the need for multichannel integration of the analog front-end [100, 149]. However, putting multiple transceiver channels on the same silicon die requires to optimize

[1]From a system perspective, this is also true for the feedback DAC in a $\Sigma\Delta$-ADC. In both cases the conversion from digital to analog takes place outside of the noise-shaping loop and thus is likely to determine the performance of the converter.

M. Clara, *High-Performance D/A-Converters*, Springer Series in Advanced Microelectronics 36, DOI 10.1007/978-3-642-31229-8_5,

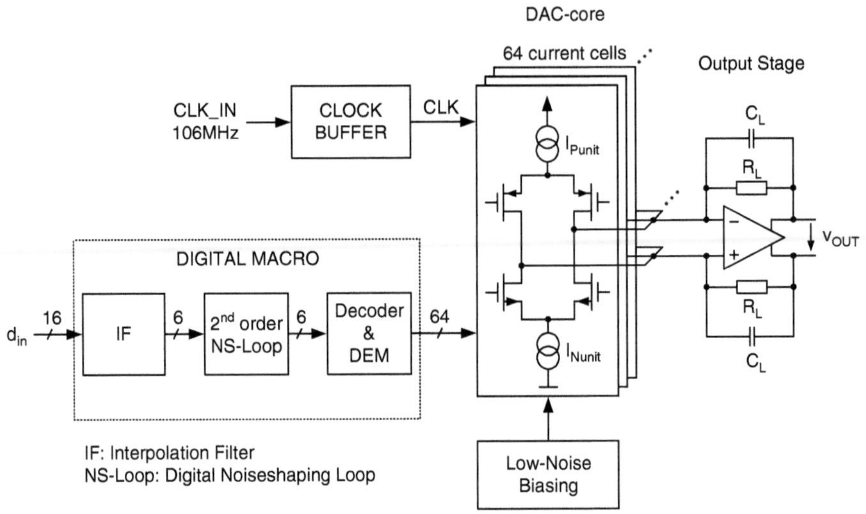

Fig. 5.1 Architecture of low-power ΣΔ D/A-converter

all building blocks with respect to power consumption and silicon area. This is especially the case for high-resolution data converters.

5.1.1 *Converter Architecture*

Figure 5.1 shows the architecture of the D/A-converter optimized for ADSL2+ CO applications. The DAC-core consists of a 64-element unary array with dual-polarity current cells that work into the virtual ground nodes of the transimpedance output buffer. Each unit current cell delivers a nominal differential current of $\pm 12\,\mu$A at its output. The differential full-scale current is thus $\pm 768\,\mu$A. The output current of the DAC-core is converted into a voltage by the feedback resistors R_L. With a resistor value of 911 Ω the full-scale voltage range at the output of the transimpedance stage is set to approximately 1.4 Vpp.

A second-order digital noiseshaper is used to suppress the in-band quantization noise. The sampling clock of the DAC-core is set to 105.98 MHz. For an analog bandwidth of 2.208 MHz[2] this corresponds to an oversampling ratio (OSR) of 24. Using Eq. (2.17) with $L = 2$, the theoretical maximum SNR is calculated to 94 dB, equivalent to an effective in-band resolution of 15.3 bits. This leaves a comfortable 1-bit margin for the actual—digital and analog—implementation.

[2] In DSL-systems clock rates and signal bandwidths are multiples of the 4.3125 kHz symbol rate.

One of the main problems in multibit ΣΔ D/A-converters is the nonlinearity of the conversion characteristic that directly results from the mismatch of the unit elements. To achieve the required static linearity of the current-source array and thus the effective resolution, in this design, a DEM algorithm is implemented.

5.1.2 DEM Selection

To minimize the digital hardware complexity and the power consumption, the simplest possible DEM-algorithm is chosen. Barrel-shift algorithms like data weighted averaging (DWA) and clocked level averaging (CLA) fulfill this requirement, since their implementation basically requires only an element selection logic, an adder, as well as a register to store the index pointer. The only difference between the two algorithms is that in CLA a constant value is used to advance the index pointer in each sampling period, while in DWA, the data sample itself is added to the current pointer.

The DWA algorithm has superior performance compared to CLA, individual level averaging (ILA) or randomization techniques [80]. However, its drawback is a pronounced sensitivity toward asymmetries in the switching transitions of the unit current cells [82]. As described in Sect. 4.1.3, any net difference in the switch-on and switch-off characteristic of the unit current pulse leads to even-order harmonic distortion, although the circuit is fully differential. For large OSRs the even-order distortion is approximately constant across the entire signal bandwidth. As shown in Sect. 4.1.4, the same holds true for the third-order harmonic distortion generated by the charge-sharing effect at the source node of the current switches.

Using Eq. (4.21), we can estimate the maximum rise and fall time mismatch that allows to achieve a linearity of 85 dB, i.e., approximately 14 bits. With a clock frequency of 105.98 MHz the DWA-algorithm requires $|T_{\text{rise}} - T_{\text{fall}}| \leq 2.5\,\text{ps}$. It is highly questionable, whether such a small difference between rise and fall time of the unit current pulse can be reliably implemented in a high-volume product. Consequently, a barrel-shift algorithm with constant pointer increment (CLA) is chosen in this design.

As discussed in Sect. 3.2.1, the element repetition rate $f_{\text{rep}} = \frac{\gcd(N,P)}{N} f_{\text{CLK}}$ of the index pointer produces spurious tones at multiples of these frequencies. Additionally, the signal and its harmonics are translated to $n f_{\text{rep}} \pm k f_{\text{sig}}$. If these tones appear above the signal bandwidth, they do not degrade the in-band resolution and can be eliminated with an analog reconstruction filter. Taking into account the fundamental tone of a sine wave at the maximum signal frequency and its first two harmonics, then with (3.39), the following inequality must be fulfilled:

$$\frac{\gcd(N,P)}{N} f_{\text{CLK}} - 3 \cdot \frac{f_{\text{CLK}}}{2 \cdot \text{OSR}} > \frac{f_{\text{CLK}}}{2 \cdot \text{OSR}} \longrightarrow \gcd(N,P) > 2 \cdot \frac{N}{\text{OSR}}. \tag{5.1}$$

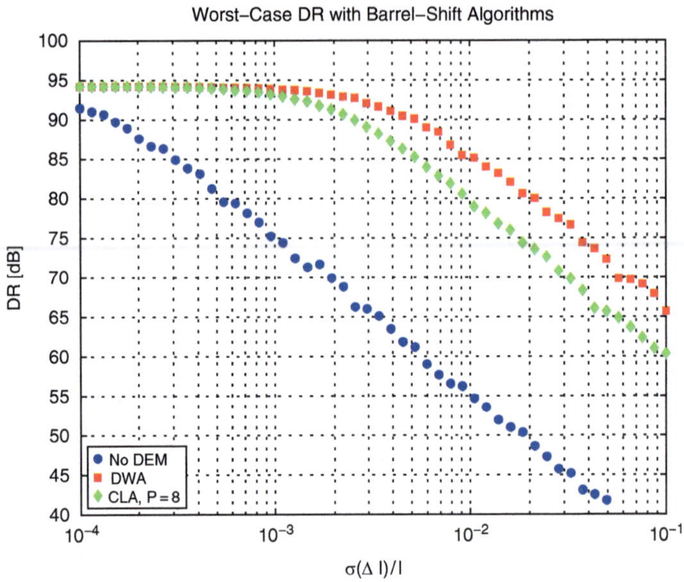

Fig. 5.2 Monte-Carlo simulation of ΣΔ-DAC

With OSR $= 24$ we get $\gcd(N, P) > 5.3$. We choose $P = 8$ and thus expect the first spurious tone around 13.25 MHz. With a maximum signal frequency of 2.2 MHz the fundamental tone mixes to 11 MHz and its third harmonic to 6.6 MHz. These frequencies are conveniently filtered out with the analog reconstruction filter.

From Fig. 5.2 we can estimate the needed matching of the unit current cells that allows to achieve a dynamic range (DR) better than 85 dB. The required accuracy is approximately $\sigma(\Delta I)/I < 0.5\,\%$. Note that with DWA we would only need $\sigma(\Delta I)/I < 1\,\%$, equivalent to an area reduction for the current-source transistors of a factor 4.

5.1.3 Unit Current Cell

Figure 5.3 shows the schematic of the dual-polarity current cell. It consists of two cascoded current sources that generate the unit currents $I_{N\text{unit}}$ and $I_{P\text{unit}}$ of 6 μA each. The current-source transistors $M_{\text{SRC}P}$ and $M_{\text{SRC}N}$ are sized to achieve a matching of $\sigma(\Delta I)/I \approx 0.25\,\%$, leaving a margin of approximately 0.25 % for additional effects, e.g., layout nonidealities and processing gradients.[3]

[3]The matching constants are derived from closely spaced devices with perfect common-centroid layout to exclude long distance effects like processing gradients. It is not possible to maintain such

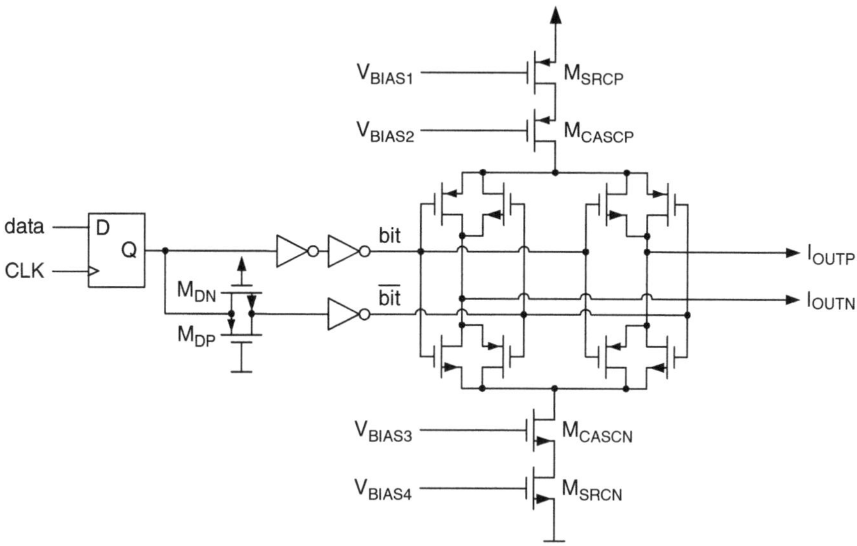

Fig. 5.3 Unit current cell

Current-source and cascode transistors of all current cells are biased by a common low-noise biasing.

Depending on the input data bit, the switching quad steers the unit currents to the virtual ground nodes of the transimpedance buffer. The switches are driven with a pair of delay-equalized inverters. A local resynchronization latch controls the sampling instant of the current cells. The sampling clock must be appropriately buffered and distributed within the array, such that all current cells are updated exactly at the same time.

The current switches are minimum-size transmission gates. Using an equal-sized PMOS and NMOS transistor in parallel leads to first-order cancellation of charge injection and clock feedthrough. The switch drivers are optimized to have a symmetric rise and fall time with a crossing point at half the supply voltage. The transmission gate M_{DP}/M_{DN} equalizes the delay of the inverted control signal ($\overline{bit}$) with respect to the non-inverted signal (*bit*), since the latter is passing through one inverter more. It is important that these two signals have very fast edges, in order to keep the switching transition as short as possible. During the transition the impedance seen at the drain of the cascode transistors is not well defined and the cell current partly discharges this node. The net charge error is signal dependent and results in dynamic nonlinearity that is not suppressed by the DEM-algorithm.

a degree of matching in an extended current-source array, unless a complex matrix-based switching scheme is employed [64].

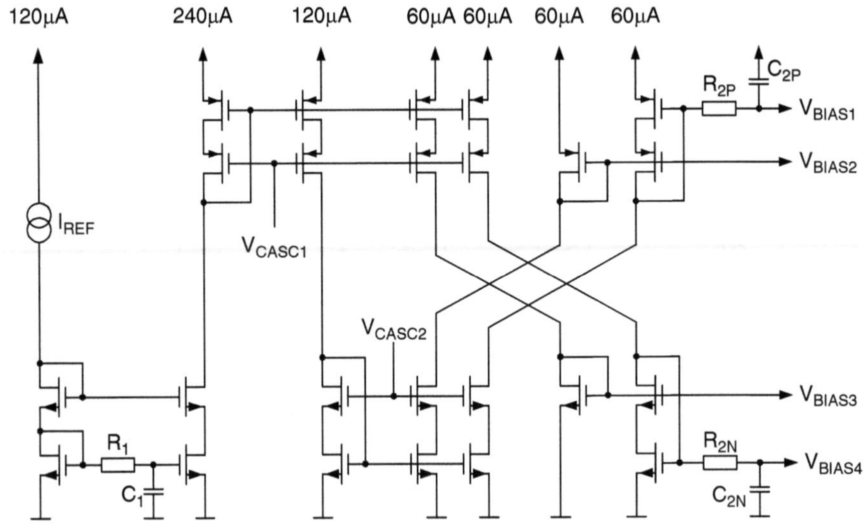

Fig. 5.4 Low-noise biasing

5.1.4 *Low-Noise Biasing*

As discussed in Sect. 2.3.2, the correlated bias noise is multiplied with the input code sequence and appears as noise sidebands of the carrier(s) in the output spectrum.

The circuit of the DAC-biasing is shown in Fig. 5.4. The incoming noise from the reference circuit is filtered with a first RC-low pass in the gate connection of the first current mirror. After mirroring the reference current to the PMOS and NMOS biasing transistors, a second RC-filter suppresses most of the biasing and the remaining reference noise.

The reference current and the current in the biasing transistors are chosen ten times as large as the unit current in the DAC-elements. The factor M_0 in Eq. (2.34) is thus 6.4 at low frequencies. The noise filter capacitors and the mirror transistors in the biasing are sized sufficiently large to suppress the low-frequency noise, so as not to degrade the upstream performance due to the transmitter noise [148].

The current branch after the first noise filter is given a four times larger current (240 μA). Its contribution to the total bias noise is accordingly smaller, because the current noise is attenuated by the same factor.

5.1.5 *Output Stage*

As shown in Fig. 5.1, the differential output currents of the unit cells are summed at the virtual ground nodes of the operational amplifier and converted into a

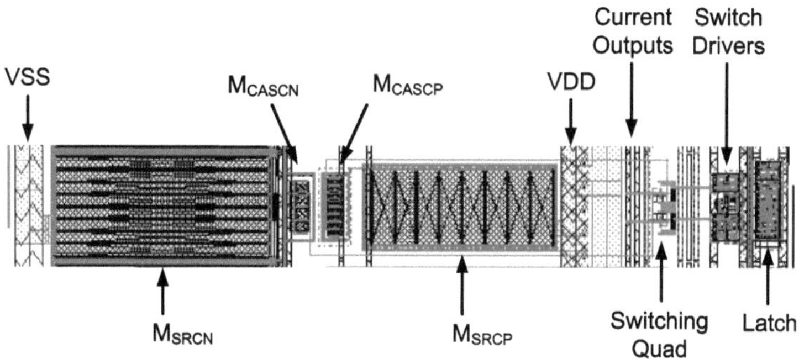

Fig. 5.5 Layout of unit current cell

voltage by the resistor in the feedback. A 12 pF capacitor in parallel with the 911 Ω resistor provides a feedforward path for the sharp current pulses and relaxes the slew-rate requirement for the op-amp. On the other hand, the parallel RC-combination in the feedback of the operational amplifier also introduces an LHP real pole in the transfer function of the converter. In this design the pole is set to approximately 15 MHz, such that it does not introduce a noticeable attenuation within the bandwidth of interest. The first-order roll-off of the output stage is part of the overall reconstruction filter.

The active element is a two-stage Miller-compensated operational amplifier with a simulated DC-gain larger than 43 dB and a unity-GBW of approximately 300 MHz. The slew rate must be quite large in order to prevent slew-rate limitation for larger input code steps, since this would result in code-dependent settling and thus nonlinear distortion [148]. The simulated slew rate in this design is 210 V/ μs.

5.1.6 Layout

Figure 5.5 shows the layout of the unit current cell. The current-source transistors $M_{\mathrm{SRC}P}$ and $M_{\mathrm{SRC}N}$ take up most of the area. The current switches are close to the switch drivers and the resynchronization latch. The cascode transistors are sized to make the drain to bulk capacitances of both polarities equal. Also the connection to the current switches has the same length for both polarities. In this way, the parasitic capacitance at the source node of the switches is also the same. This ensures that the time constant associated with the virtual ground nodes does not dependent on the input code.

Figure 5.6 shows the layout of the DAC-core. The 64 unit current cells are placed in two rows of 32 cells with one pair of dummy cells at each side to guarantee equal surrounding. The supply rails for the NMOS and PMOS current sources are routed as a tree-structure to equalize the voltage drop in the source connection.

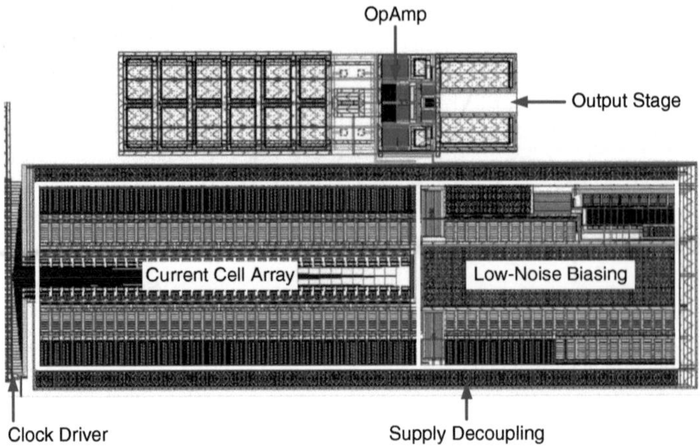

Fig. 5.6 DAC-core layout

Together with the dummy cells, the current sources that finally determine the converter resolution contribute only about 25 % of the total area.[4]

The switching logic, comprising the current switches, the switch drivers, and the resynchronization latches, makes up for 10 % of the converter area. The clock buffer is placed in the lower left corner of the converter module. From there the clock is distributed via two symmetrical clock trees to the current cells.

The low-noise biasing is at the right side, opposite to the digital data interface. It consumes 25 % of the converter area, mostly due to the large noise filter capacitors.

The transimpedance output buffer is placed above the current-cell matrix, mainly due to overall floorplan restrictions. Its area is dominated by the feedback capacitors and is about 20 % of the overall converter area.

The remaining 20 % of the module area are used for the wiring and the supply blocking capacitors, with 5 % accounting for the 64 data lines that have to be distributed between the two rows of current cells.

Table 5.1 summarizes the relative silicon area of the various converter components. The interpolation filter, digital noiseshaper, and barrel shifter are not included, since they are integrated with a larger digital macro common to all channels.

The converter is implemented in a 1P6M 130 nm CMOS process. The feedback capacitors in the output stage are realized as vertical metal-insulator-metal (MIM) capacitors located at the top of the metal stack. The silicon area of the converter core, i.e., without the digital noiseshaper, is 0.35 mm^2. The binary to thermometer decoder and the barrel-shift logic, not shown in Fig. 5.6, together occupy only 0.02 mm^2.

[4]As is evident from Fig. 5.5, in terms of silicon area, the cascode transistors are practically negligible in comparison to the current-source transistors.

Table 5.1 Area contribution of the major building blocks

Converter component	Area (mm^2)	Percentage (%)
Current sources	0.09	25
Switching logic	0.035	10
Low-noise biasing	0.09	25
Output stage	0.07	20
Wiring and blocking caps	0.07	20
Total	0.35	100

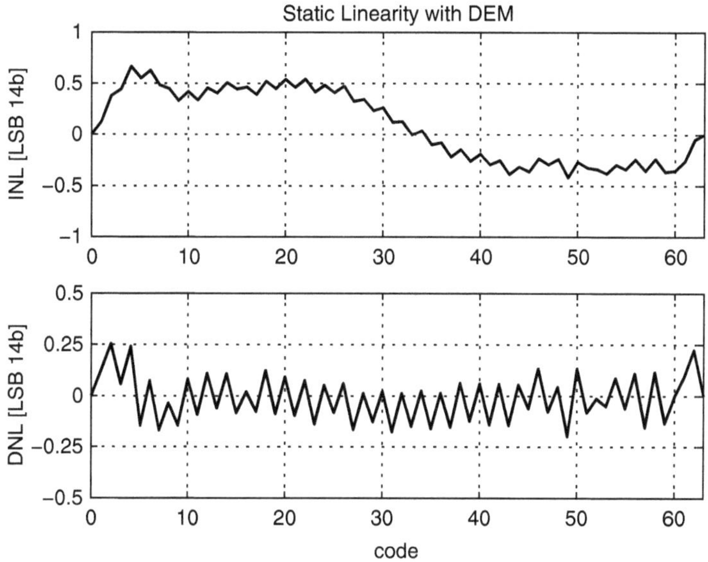

Fig. 5.7 Static linearity with DEM activated

5.1.7 *Experimental Results*

The power dissipation of the converter core, clocked at 106 MHz, is only 9 mW from a 1.5 V supply [148]. The active output stage uses 6 mW, while the static power dissipation of the current matrix together with the biasing is 1.5 mW. The remaining 1.5 mW is dynamic power due to the current-cell switching, the clock buffer, the binary to thermometer decoder, and the barrel-shift logic. The digital noiseshaper, not included in this power figure, is estimated to add another 1.5 mW. This sums up to a total power dissipation of only 10.5 mW for the whole D/A-converter.

Figure 5.7 shows the typical static linearity of the converter with activated DEM. The maximum integral nonlinearity (INL) and DNL at the 14-bit level are 0.67/−0.5 LSB and 0.25/−0.2 LSB, respectively. A full-scale sine wave mapped onto this static characteristic results in a harmonic distortion of better than −90 dBc, indicating the maximum achievable distortion level.

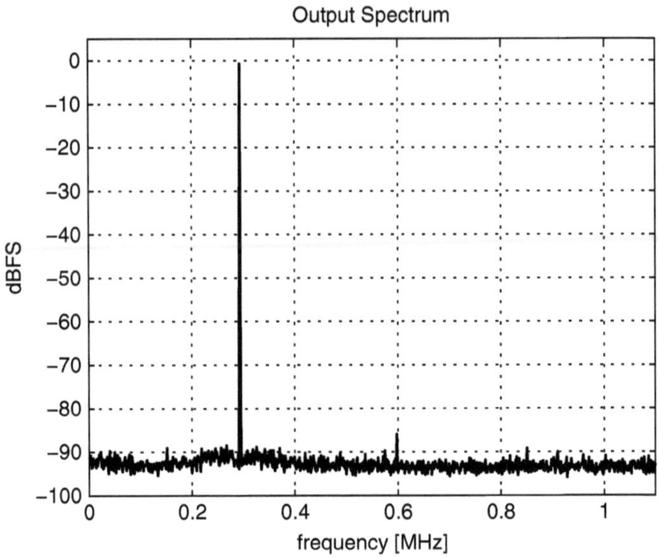

Fig. 5.8 Output spectrum with 300 kHz single tone

Figure 5.8 shows the measured output spectrum for a near full-scale single tone at 300 kHz. The third harmonic is barely visible above the noise floor at approximately −90 dBc, consistent with the static nonlinearity level. However, the nonlinear distortion is dominated by the second harmonic found at −85.3 dBc, most probably limited by a slight imbalance in the output stage, or even by the measurement equipment.

For a single-tone signal at 700 kHz (see Fig. 5.9), the second harmonic rises to −80.7 dBc, while the third harmonic is now at −84.1 dBc. The linearity deterioration at higher signal frequencies seems due to the increased switching activity.[5]

The MTPR as a function of the signal frequency is shown in Fig. 5.10. The Crest Factor of the synthesized DMT-signal is CF $= 5.6$. The mean value of the MTPR, averaged over all frequency bins, is close to 76 dB.

The intermodulation products that are outside the (downstream) transmit signal, but fall into the (upstream) receive band, are measured by the missing band power ratio (MBPR). The MBPR is a measure for the effect of the active near-end transmitter on the receiver. The intermodulation products generated in the receive band, weighted by the echo transfer function, are perceived as noise at the receiver input. An example is shown in Fig. 5.11. The average MBPR for the upstream band (=receive direction) is around 76 dB. The dynamic range (DR) of the converter is slightly over 90 dB in a bandwidth of 1.1 MHz and reduces to 87 dB for the ADSL2-bandwidth of 2.2 MHz.

[5] A rise-fall time mismatch of 200 ps would be sufficient to generate the measured HD_2.

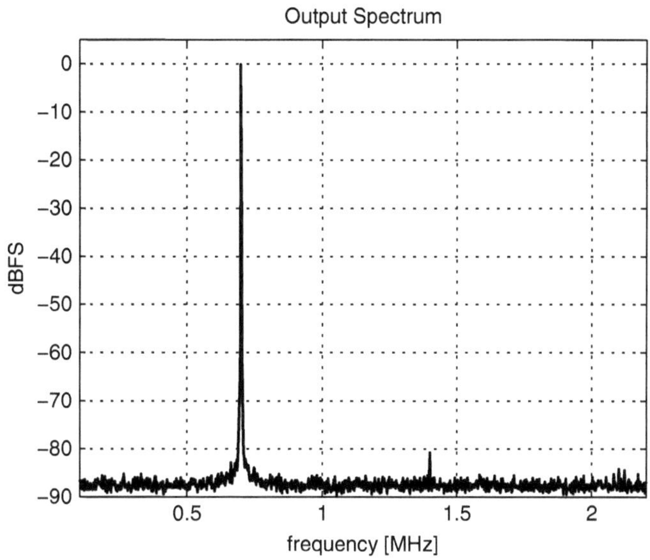

Fig. 5.9 Output spectrum with 700 kHz single tone

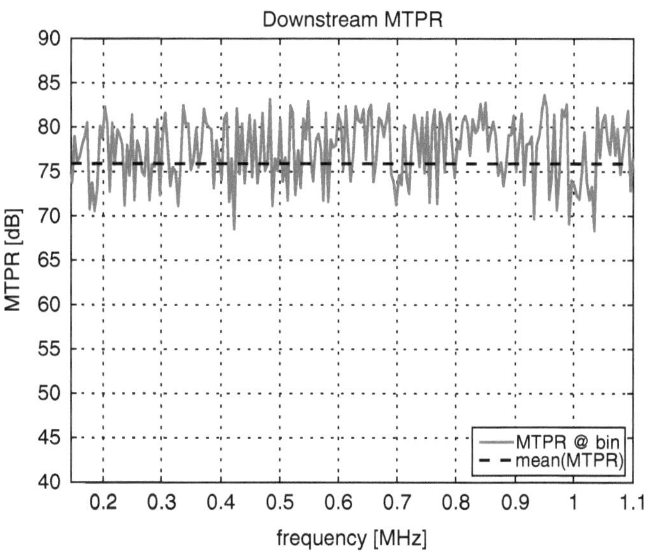

Fig. 5.10 MTPR

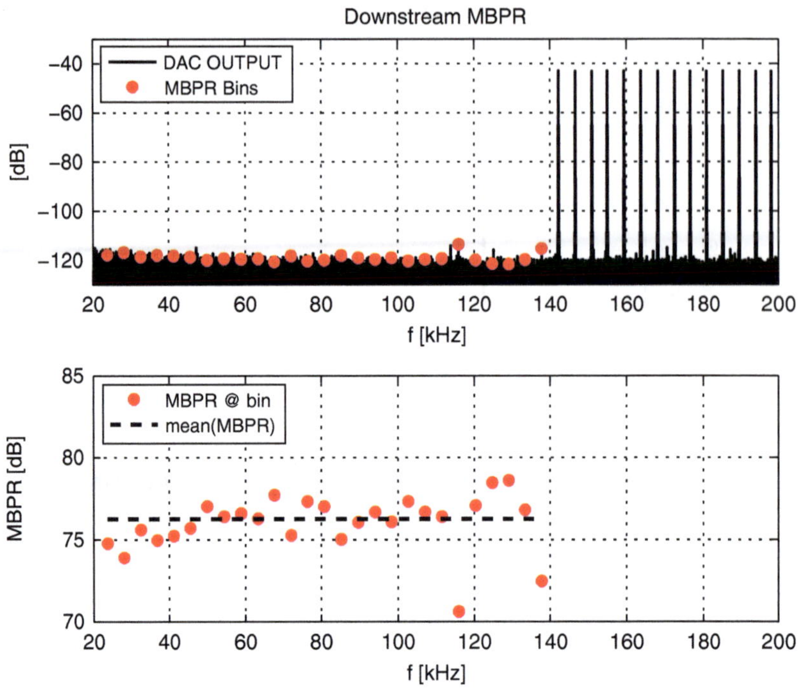

Fig. 5.11 MBPR

5.2 A 12-Bit/14-Bit Multistandard DAC

VDSL is a family of DSL transmission standards that extends the spectrum usage up to 30 MHz. With more bandwidth available, the data throughput increases accordingly. For example, the VDSL2-30a profile uses the full DSL-spectrum from 25 kHz up to almost 30 MHz and allows a symmetrical full-duplex data rate of 100 Mbit/s. This data rate is however limited to a maximum loop reach of approximately 350 m (1.1 kft), because of the considerable attenuation of the twisted-pair line at higher frequencies [29]. A high-quality VDSL-transmitter requires a linearity and resolution in the range of 11.5–12 bits for a signal bandwidth up to 30 MHz. On the other hand, ADSL and ADSL2+ occupy a bandwidth of only 1.1 and 2.2 MHz, respectively, but need a linearity/resolution of 13.5–14 bits in the transmitter.

In VDSL a multitude of different frequency band plans, even with country-specific flavors, is in use. A multi-standard transceiver is able support all different transmission standards with the same hardware. The analog and digital circuits can be configured by software programming to suit the desired operating mode. In general, this means that the analog and digital circuits, as well as the data converters, must include a considerable amount of flexibility that allows them to be tailored to the chosen transmission standard.

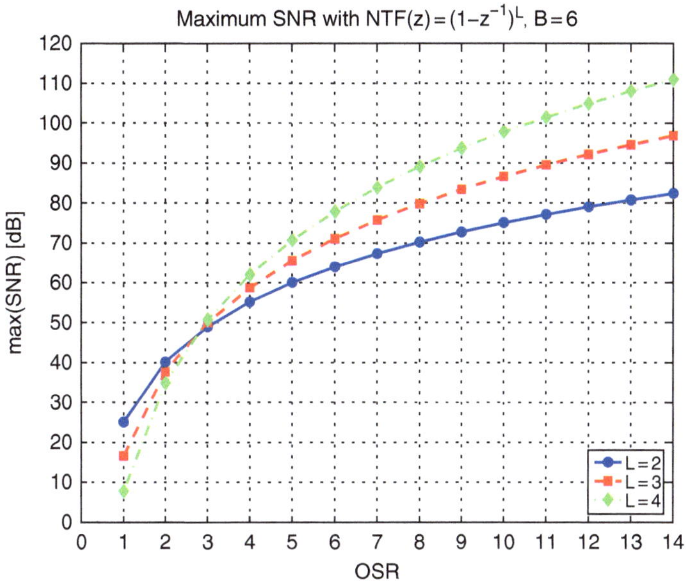

Fig. 5.12 Maximum SNR of a 6-bit ΣΔ-DAC with $\mathrm{NTF}(z) = (1 - z^{-1})^L$

5.2.1 *Low-OSR Noiseshaper*

Equation (2.17) describes the maximum signal-to-noise ratio of an ideal L-th order noiseshaper, whose NTF has L poles at DC, i.e., $\mathrm{NTF}(z) = (1 - z^{-1})^L$. Figure 5.12 shows the achievable SNR as a function of the OSR with a physical resolution of 6 bits for different loop-order L.

As described in the previous section, for ADSL, a second-order noiseshaper and a sampling rate of around 100 MHz are sufficient to achieve a resolution of 14 bits. For VDSL, although we only need a resolution of 12 bits, a minimum oversampling of 12.5 is necessary to achieve an SNR of 80 dB with second-order noiseshaping.[6] Since the analog bandwidth is close to 30 MHz, this results in a minimum sampling frequency of 750 MHz. Although converter sampling speeds well beyond 1 GHz can be handled quite comfortably in the used 0.13 μm CMOS process [150, 151], the dynamic power consumption for 64 unit current cells would be considerably high. Another problem with high sampling rates is that the dynamic errors are in general directly proportional to the clock frequency (see Sects. 4.1.3 and 4.1.4) and it is therefore more difficult to achieve a certain linearity. For a power-optimized high-resolution D/A-converter, it is highly desirable to choose the lowest possible sampling frequency.

[6]Again, we desire to have a design margin of about 1 bit.

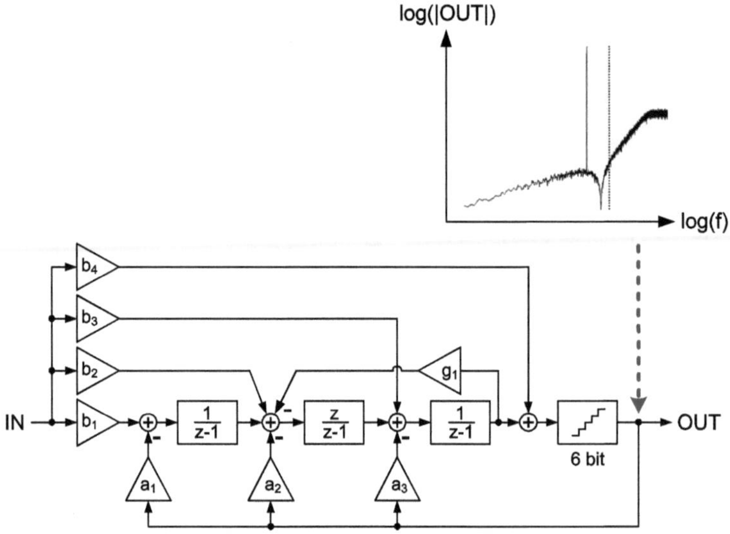

Fig. 5.13 Low-OSR third-order noiseshaper

One possibility to lower the sampling frequency of a $\Sigma\Delta$ converter is to increase its physical resolution. By going from 6 to 8 bits, the minimum OSR with a second-order noiseshaper becomes 7.25, equivalent to a clock frequency of 435 MHz. Because the number of unit current cells is increased by a factor of 4, while the original clock frequency of 750 MHz (for a 6-bit converter) is not even halved, this approach seems rather disadvantageous.

More promising than just increasing the number of output levels is to change the order L of the noiseshaping. With a third-order modulator of the form $\text{NTF}(z) = (1 - z^{-1})^3$ we require an OSR of 8, or, equivalently, a sampling clock of 480 MHz, while with a fourth-order $\text{NTF}(z) = (1 - z^{-1})^4$, we would even come down to $\text{OSR} = 6.35$, or $f_{\text{CLK}} = 380\,\text{MHz}$.

A more aggressive noiseshaping also requires a higher-order reconstruction filter to suppress the out-of-band quantization noise toward the line. As a practical rule of thumb, the filter following the converter should at least have the order of the NTF. In the transmit direction, we therefore wish to limit the noiseshaper order to $L = 3$. This also minimizes the power consumption of the reconstruction filter, since a third-order analog filter can in principle be realized with only one operational amplifier [152, 153].

The block diagram of the chosen third-order noiseshaper is shown in Fig. 5.13. It is a so-called cascade of resonators with distributed feed-back (CRFB) structure [44]. Recall that an NTF with $L > 2$ cannot be realized by simply placing L zeros at DC ($z = 0$), because the resulting modulator would be unstable. Instead, the resonator formed with coefficient g_1 creates a complex in-band zero that allows to maximize the SNR, especially at a low OSR. At the same time, poles are introduced into the NTF to improve the stability of the modulator [154].

Table 5.2 Third-order modulator coefficients

Coefficient	Value
a_1	1.4438
a_2	1.8626
a_3	1.0158
b_1	1.4438
b_2	1.8626
b_3	1.0158
b_4	1.0000
g_1	0.1623

By optimizing the position of the in-band zero and the poles of the NTF the SNR of the modulator can be maximized, while at the same time stability is maintained. This is done with dedicated MATLAB™ functions[7] from the Delta-Sigma Toolbox [155].

For OSR = 6 and a 6-bit internal quantizer the resulting stable NTF with optimized SNR is given by

$$\mathrm{NTF}(z) = \frac{(z-1)\left(z^2 - 1.838z + 1\right)}{(z+0.04068)\left(z^2 - 0.00009464z + 0.3875\right)}. \tag{5.2}$$

The coefficients for the CRFB structure are reported in Table 5.2.

The system clock is chosen to be 350 MHz, resulting in an analog bandwidth of 29.16 MHz for an OSR of 6. This is still perfectly suitable for VDSL2-applications extending up to 30 MHz, since the tones close to the upper frequency band edge are of less importance for the system throughput. Because of the already large cable attenuation at 30 MHz, the transmitter noise floor can be higher in this frequency range.

Figure 5.14 shows the in-band resolution of the digital noiseshaper as a function of the signal amplitude. The maximum SQNR is slightly above 80 dB for a −0.5 dBFS single-tone signal. The modulator can thus be driven very close to the full-scale value before overload occurs. The dynamic range of the digital ΣΔ-modulator is close to 82 dB, resulting in a design margin of more than 1 bit.

5.2.2 *Interleaved Data Weighted Averaging*

CLA in conjunction with low OSR requires a large pointer increment P in order to push the modulation products of the signal and its harmonics out of the frequency band of interest. Using Eq. (5.1) we get the following restriction:

[7]The functions *synthesizeNTF.m*, *simulateSNR.m*, *clans.m*, and *realizeNTF.m* are used to find a stable NTF with optimized in-band SNR.

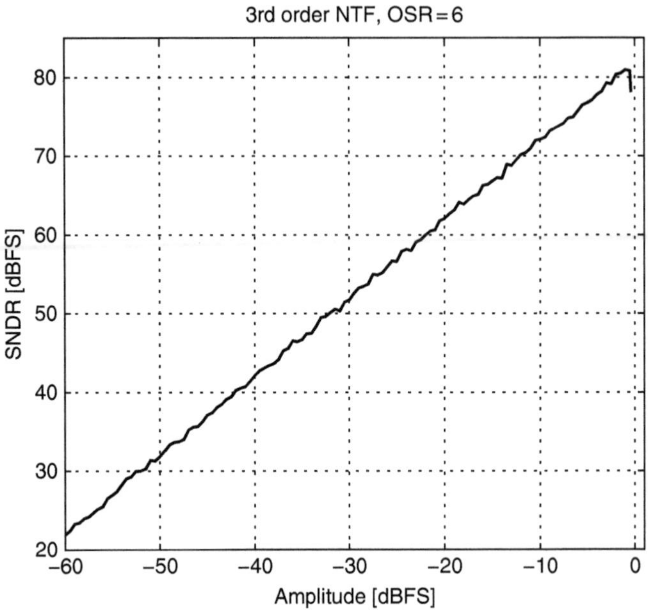

Fig. 5.14 SQNR at output of digital noiseshaper

$$\gcd(N, P) > 2 \cdot \frac{N}{\mathrm{OSR}} = 2 \cdot \frac{64}{6} = 21.33. \tag{5.3}$$

Inequality (5.3) can only be fulfilled with $P = 32$. Since such a large pointer increment does not provide effective averaging of the static linearity errors,[8] a DWA algorithm must be implemented.

DWA is the most effective barrel-shift algorithm in terms of resolution improvement for any given unit current matching, because it has the highest average element repetition frequency. Due to the increased number of switching transitions, which also strongly correlates with the input data, any error associated with the unit element transition is converted very efficiently in harmonic distortion. This fundamental limitation is described in Sect. 4.1.

Return-to-Zero (RZ) is the most effective technique to eliminate the data dependency of the transition errors and allows the implementation of DWA in a current-steering DAC [82]. On the other hand, for an active output stage, we prefer an NRZ output current. The Double Return-to-Zero architecture [47] resets each

[8]CLA is most effective with $P = 1$, because then any static input code c uses all elements of the array in one revolution of the index pointer. With $N = 64$, $P = 32$, any static input code $c \leq P$ uses only $2c$ elements.

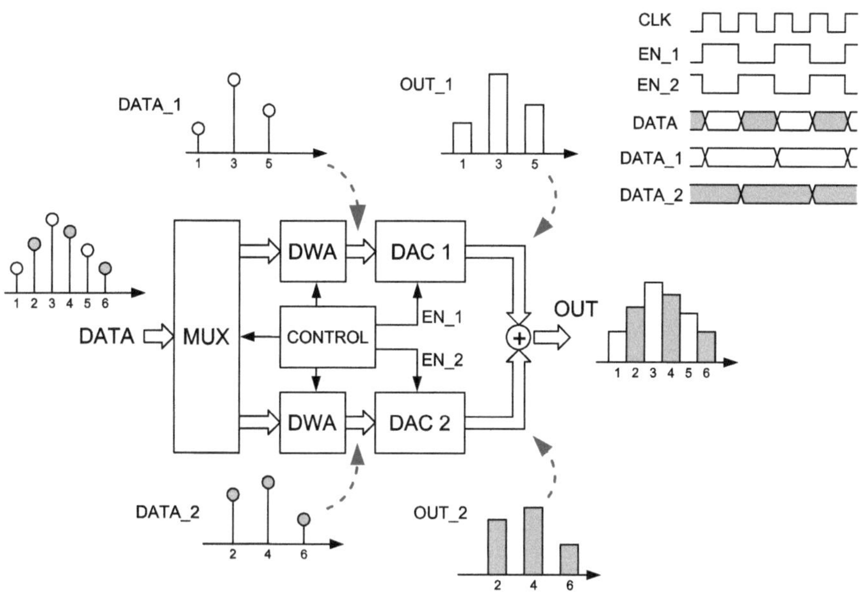

Fig. 5.15 Interleaved D/A-converter with DWA

unit current cell for half a clock cycle and delivers a continuous output current, but it is not very power-efficient at high clock frequencies (see also Sect. 4.2.7).

Figure 5.15 shows the implemented interleaved D/A-converter with DWA [133]. As described in Sect. 4.2.8, interleaving provides a full-clock reset period for each unit element at half the switching activity, as compared to a Double-RTZ structure. This allows to implement an effective Return-to-Zero at high clock frequencies with relatively low power consumption, while delivering an NRZ output current. To increase the static linearity of the two interleaved DAC-cores, the DEM-algorithm operates independently on the two input data streams coming from the multiplexer [156].

Behavioral Monte Carlo simulations (see Fig. 5.16) show that a matching of $\sigma(\Delta I)/I \leq 0.5\,\%$ is required to achieve a DR of 75 dB with an interleaved converter using the third-order noiseshaper described above with OSR $= 6$. With the same unit element accuracy, a second-order noiseshaper at OSR $= 24$ allows to achieve a DR of 86 dB, adequate for ADSL-systems.

Even with ideal subconverters the performance of an interleaved converter array is limited by systematic errors. In a Nyquist converter, a gain or phase mismatch between the two interleaved converter cores gives rise to signal components at $\frac{1}{2} f_{\mathrm{CLK}} \pm f_{\mathrm{SIG}}$, while a DC-offset produces a tone at $\frac{1}{2} f_{\mathrm{CLK}}$. In an oversampled converter these signal components are normally outside the bandwidth of interest and therefore pose no problem. However, in a $\Sigma\Delta$ converter array, the gain and phase mismatch additionally causes the shaped quantization noise to leak into the baseband [147].

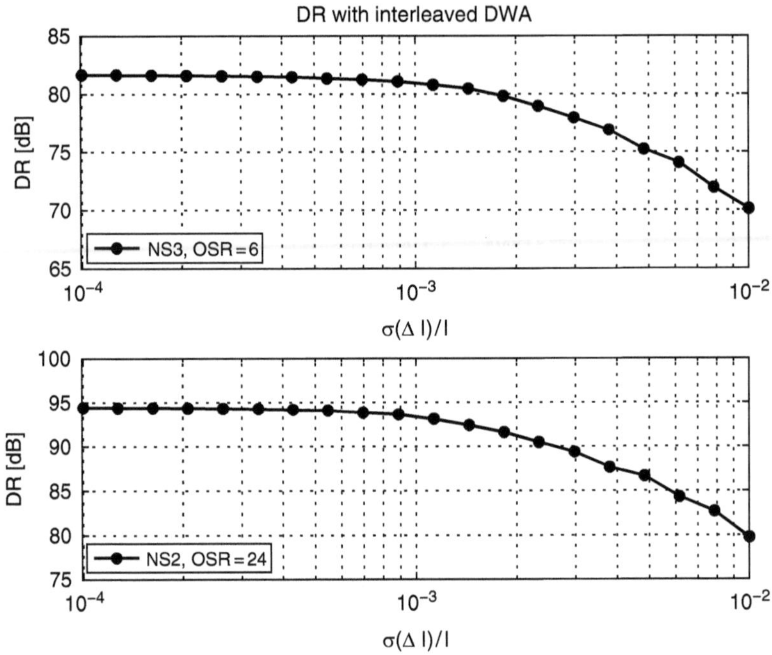

Fig. 5.16 Interleaved-DWA performance limited by random mismatch

Figure 5.17 shows the simulated SNR-loss of an otherwise ideal interleaved D/A-converter as a function of the relative full-scale range (FSR) error and the systematic phase error, e.g., caused by clock skew. A maximum SNR-loss of 1 dB requires an FSR mismatch of better than 0.275 %, or an aggregate phase error of less than 7.5 ps.

5.2.3 Converter Architecture

The converter architecture is shown in Fig. 5.18. It consists of two 6-bit dual-polarity DAC-cores with 64 unit current cells each. The output currents of the unit cells are summed at the virtual ground nodes of the transimpedance stage and converted into the differential output voltage.

During the Return-to-Zero period, the current of all unit cells of the respective DAC-core is shorted to the common node. This node is kept at a potential equal to the voltage seen at the virtual ground nodes of the output stage by a single-ended buffer. A large blocking capacitor C_{BATT} absorbs high-frequency spikes that occur at the common node during the switching transitions at the beginning of each sampling period. Since the PMOS and NMOS current sources are nominally equal, only the difference of the total P- and N-side current must be delivered by the common-mode buffer. Because the switching activity at the common node is, to first order, not data dependent, the speed requirement for the common-mode buffer is very relaxed, such that it can be optimized for very low power consumption [133].

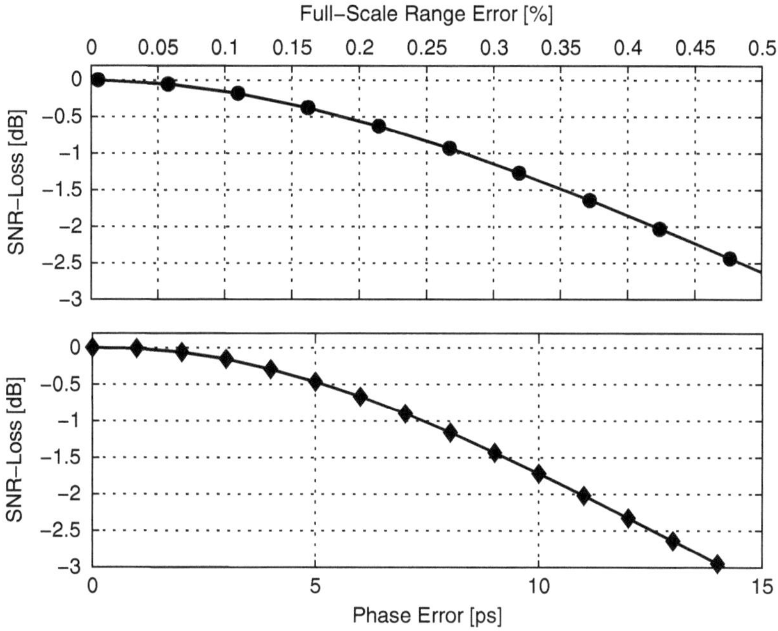

Fig. 5.17 SNR-loss with third-order noiseshaper at OSR = 6

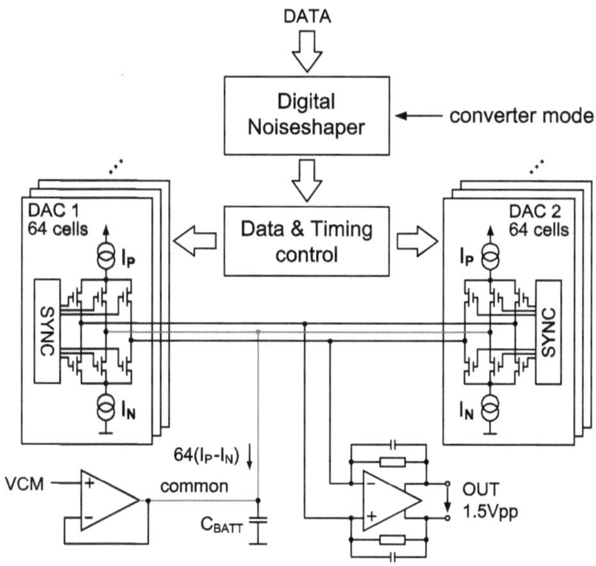

Fig. 5.18 Interleaved D/A-converter architecture

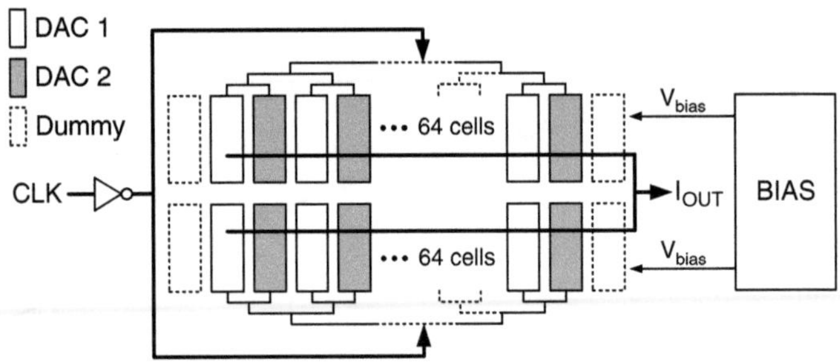

Fig. 5.19 Interleaved D/A-converter floorplan

To maximize the matching between the converter cores, an intermeshed unit cell placement is adopted. Figure 5.19 shows the floorplan of the interleaved DAC-array. The 128 unit current cells of the two DACs are arranged in two rows and placed alternately. A single low-noise bias circuit generates the gate voltages for the NMOS and PMOS current sources. A row of dummy cells is placed at each side of the current-cell array to avoid layout discontinuities at the edges.

A central clock buffer drives the sampling clock over two symmetrical clock trees to the current-cell array. Every subbranch of the clock tree connects to an equal number of unit elements from each DAC-core. The length of the clock line from the clock buffer to every unit element is thus the same. This minimizes the global phase error by averaging the local timing errors over the whole array.

The digital noiseshaper can be reconfigured by software programming. For high-bandwidth applications, like VDSL, the third-order noiseshaper of Fig. 5.13 is selected. With a clock frequency of 350 MHz a signal bandwidth close to 30 MHz is possible. For lower-bandwidth applications with higher OSR the noiseshaper can be reconfigured to a lower order to optimize the out-of-band behavior and reduce the digital power consumption. For ADSL, e.g., a second-order noiseshaper with a clock frequency of 106 MHz (OSR $= 24$) is sufficient.

5.2.4 *Current-Cell Design*

The unit current-cell schematic is shown in Fig. 5.20. Transistors M_{BP}, M_{CP} and M_{BN}, M_{CN} form the complementary cascoded current sources delivering $6\,\mu$A each. The currents are switched to the outputs I_{OUTP}, I_{OUTN} during the active sampling period, or to the common node during the reset period. The operating mode is selected with the global signal ENABLE. With ENABLE=1 the data input DATA selects the polarity of the output current. With ENABLE=0 the unit cell is disconnected from the output nodes and the current is steered to the common node.

DATA and ENABLE are locally re-latched with the rising edge of the sampling clock CLK. Appropriately scaled CMOS-inverters buffer the resynchronized switching signals and drive the gates of the current switches. To avoid a discharge of the cascode transistors' drain nodes, the crossing points of the switching

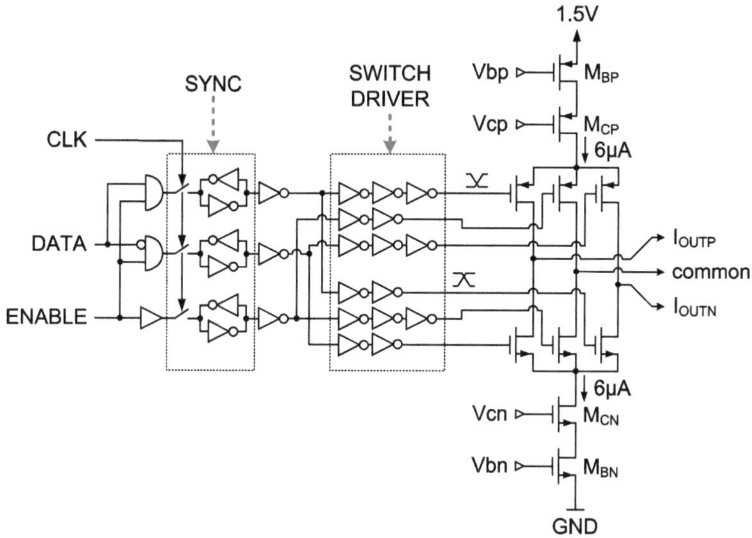

Fig. 5.20 Unit current cell for interleaved DAC

signals are shifted toward the supply rail, such that the current switch transistors are never simultaneously off [117]. This is done by sizing the switch driver inverters appropriately.

As shown above, a relative unit cell accuracy of 0.5 % is sufficient. To have a certain design margin, the current-source transistors are sized and biased for a matching of $\sigma(\Delta I)/I \leq 0.4\,\%$ with respect to the differential output current.

5.2.5 *Low-Noise Biasing*

The biasing circuit for the interleaved converter is shown in Fig. 5.21. The noise of the reference current is first filtered with the low-pass $R1$–$C1$. The reference current is then mirrored to the NMOS and PMOS biasing transistors. The mirror factor in Eq. (2.34) is maintained at $M_0 = 3.2$ to keep the excess noise of the biasing low. The biasing branch for the cascode transistors is given a current of 120 μA, equivalent to 20 unit currents. A low-impedance biasing driving the gate of the cascode transistor helps to better suppress the switching transients coupled via the gate-drain capacitances to the biasing lines V_{CP}, V_{CN}.

Figure 5.22 shows the measured output noise voltage density of the complete D/A-converter for a midscale and full-scale static input signal. By applying Eq. (A.20), the maximum SNR due to circuit noise for a 30 MHz bandwidth is calculated to 77.8 dB. The corresponding dynamic range (DR) is limited by the noise at midscale and amounts to 78.3 dB for this bandwidth.

For an ADSL2-bandwidth of 2.2 MHz, the calculated DR limited by circuit noise is 89 dB, while the maximum SNR for a full-scale sine wave is 86 dB.

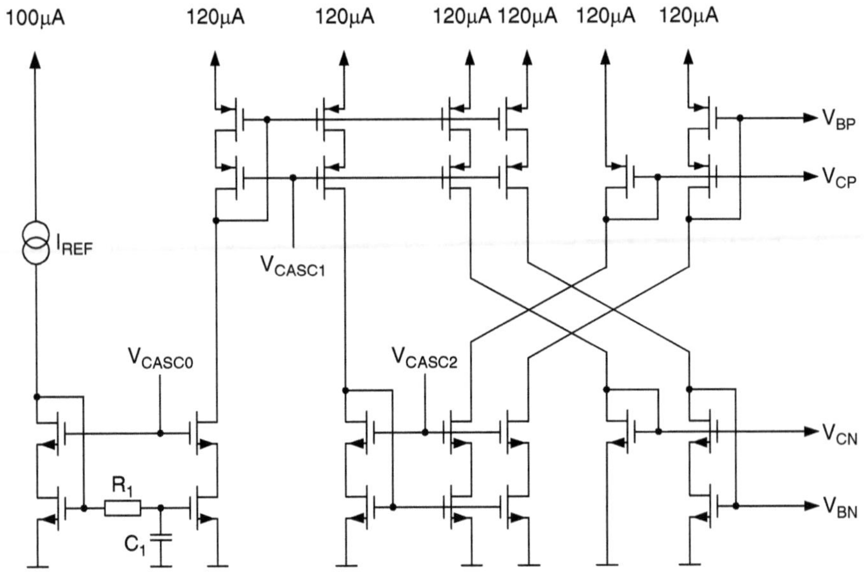

Fig. 5.21 Low-noise biasing

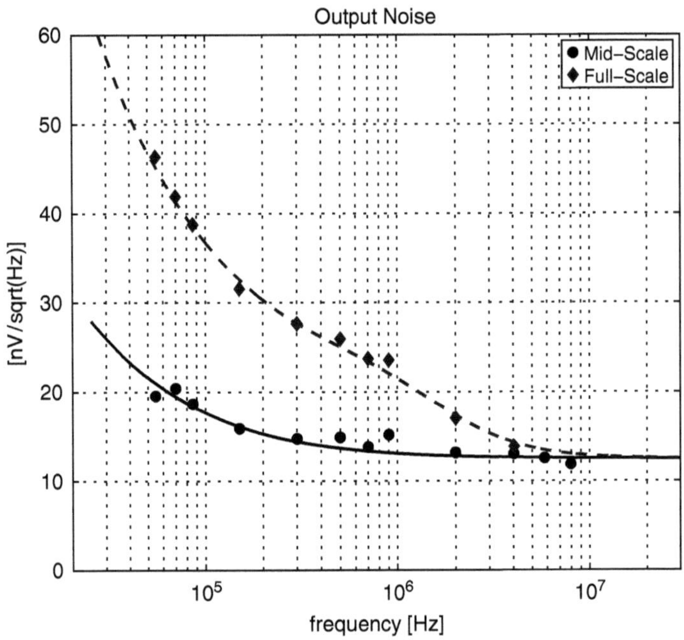

Fig. 5.22 Measured output noise

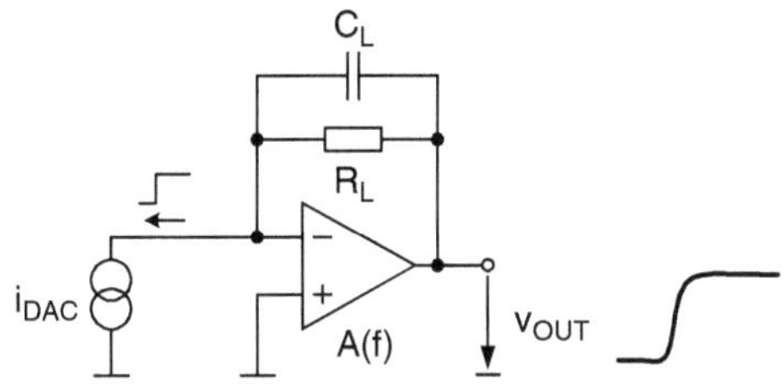

Fig. 5.23 Output stage model

5.2.6 Output Stage Design

The transimpedance output stage converts the current coming from the interleaved current-cell arrays into an output voltage. A single-ended circuit model is shown in Fig. 5.23.

Using the single-pole op-amp model of Eq. (4.8) and assuming a sufficiently large DC-gain $A_0 >> 1$, the frequency-dependent current to voltage conversion can be approximated by

$$T(j\omega) = \frac{V_{\mathrm{OUT}}(j\omega)}{I_{\mathrm{DAC}}(j\omega)} \approx \frac{R_L}{\left(1 + \frac{j\omega}{2\pi \mathrm{GBW}}\right)(1 + j\omega R_L C_L)}. \tag{5.4}$$

If we assume that the GBW of the amplifier is sufficiently large, the feedback capacitor C_L can be used to trim the closed-loop circuit response. By increasing C_L, the LHP-pole on the negative frequency axis generated by the parallel R_L–C_L combination decreases in frequency and ultimately causes an unacceptable attenuation at the upper edge of the signal bandwidth. On the other hand, a larger C_L helps to lower the slew rate of the output voltage for a given input current step and thus relaxes the slew-rate requirement for the amplifier.

The step response of the circuit in Fig. 5.23 to the LSB-current step can be calculated by performing the inverse Laplace transform of $T(s)$ multiplied by I_{LSB}/s:

$$a_{\mathrm{LSB}}(t) = \mathscr{L}^{-1}\left(\frac{I_{\mathrm{LSB}}}{s} T(s)\right) = R_L I_{\mathrm{LSB}} \left(1 + \frac{K_\tau \mathrm{e}^{-\frac{R_L C_L}{2\pi \mathrm{GBW}} \cdot t} - \mathrm{e}^{-2\pi \mathrm{GBW} \cdot t}}{1 - K_\tau}\right). \tag{5.5}$$

Assuming no limitation in the operational amplifier, the slew rate of the output voltage for an LSB-step is given by the first derivative of $a_{\mathrm{LSB}}(t)$ with respect to time t:

$$\mathrm{SR}_{\mathrm{LSB}}(t) = R_L I_{\mathrm{LSB}} \cdot \frac{\frac{2\pi \mathrm{GBW}}{R_L C_L} \mathrm{e}^{-\frac{t}{R_L C_L}} - (2\pi \mathrm{GBW})^2 \mathrm{e}^{-2\pi \mathrm{GBW} \cdot t}}{1 - K_\tau}. \tag{5.6}$$

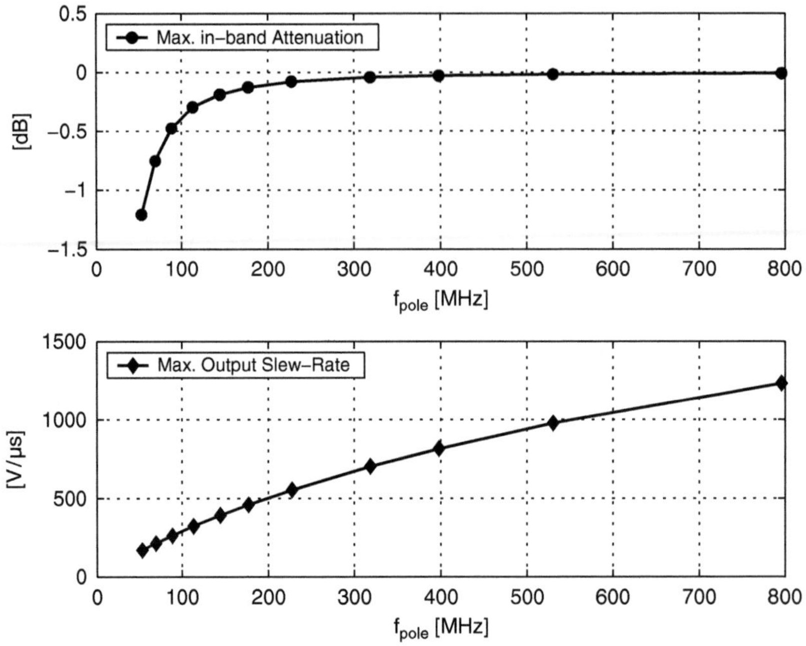

Fig. 5.24 In-band attenuation and output voltage slew rate

As long as the system remains linear, the LSB-step slew rate given by (5.6) will scale with the height of the input code step. The output voltage pulse starting at time $t = 0$ has its largest rate of change at time $t_{\mathrm{SR,max}}$, obtained by equating the derivative of (5.6) to zero and solving for t:

$$t_{\mathrm{SR,max}} = \frac{R_L C_L \cdot \ln(K_\tau)}{K_\tau - 1}. \tag{5.7}$$

Finally, inserting (5.7) into (5.6), the maximum slew rate of the output voltage as a function of the input code step can be determined:

$$\mathrm{SR}_{\max} = (d_k - d_{k-1}) \cdot R_L I_{\mathrm{LSB}} \cdot 2\pi \mathrm{GBW} \cdot \frac{K_\tau^{\frac{K_\tau}{1-K_\tau}} - K_\tau^{\frac{1}{1-K_\tau}}}{1 - K_\tau}. \tag{5.8}$$

The maximum overall slew rate of the output voltage can now be determined from the statistics of the input code sequence. Because it also depends on the noiseshaper characteristics, a behavioral simulation is in general necessary.

Figure 5.24 shows the in-band attenuation and the maximum output slew rate as a function of the feedback-pole frequency. In this example, the maximum frequency of interest is 30 MHz. The output voltage slew rate is calculated with (5.8) for the maximum code step produced by a 30 MHz full-scale single tone, assuming the third-order noiseshaper of Fig. 5.13.

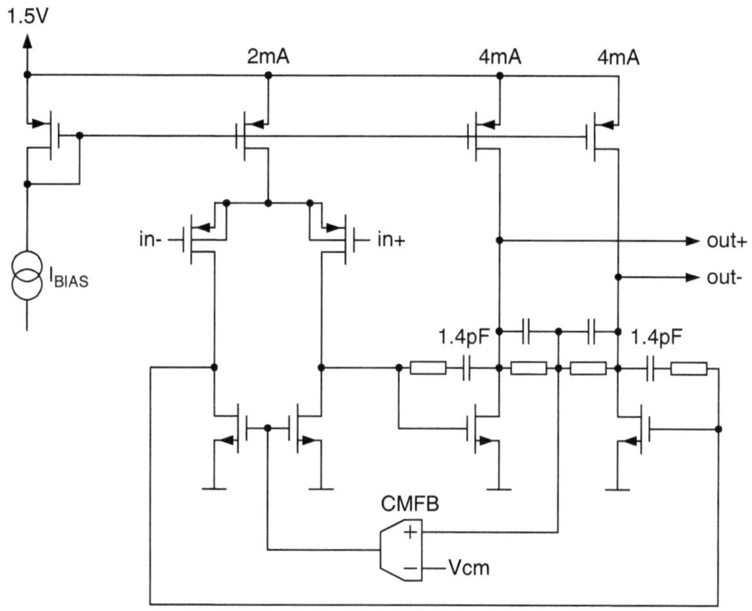

Fig. 5.25 Class-A operational amplifier

In this design we choose the feedback-pole at 100 MHz for a maximum in-band attenuation of less than 0.5 dB at 30 MHz. The resulting worst-case output slew rate is 300 V/μs. Since the amplifier slew rate must exceed this value with sufficient margin, we therefore target an amplifier slew rate of at least 1,000 V/μs.

The amplifier is a standard fully differential two-stage Miller operational amplifier with PMOS input stage and class-A output stage, as shown in Fig. 5.25. The GBW of the amplifier is set close to 1 GHz, roughly a factor three higher than the targeted clock rate. The simulated worst-case slew rate is above 1,400 V/μs, giving almost a factor five margin to the maximum output voltage slew for the largest expected code step.

For lower-bandwidth applications with lower clock frequency and higher OSR the bias current of the operational amplifier is reduced accordingly. Also the feedback capacitor is increased for improved out-of-band filtering.

5.2.7 Layout

The converter is implemented in a 0.13 μm 1P6M standard CMOS process with high-quality vertical MIM capacitors located at the top of the metal stack. Figure 5.26 shows the unit current cell-layout. When comparing with Fig. 5.5, the most remarkable difference is the increased amount of digital logic. This is partly

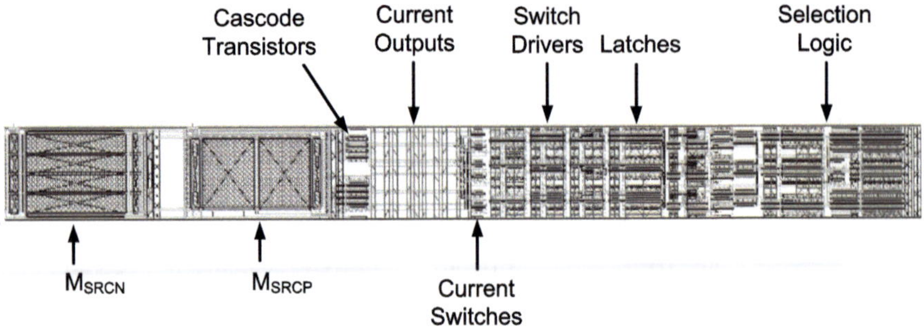

Fig. 5.26 Unit current-cell layout

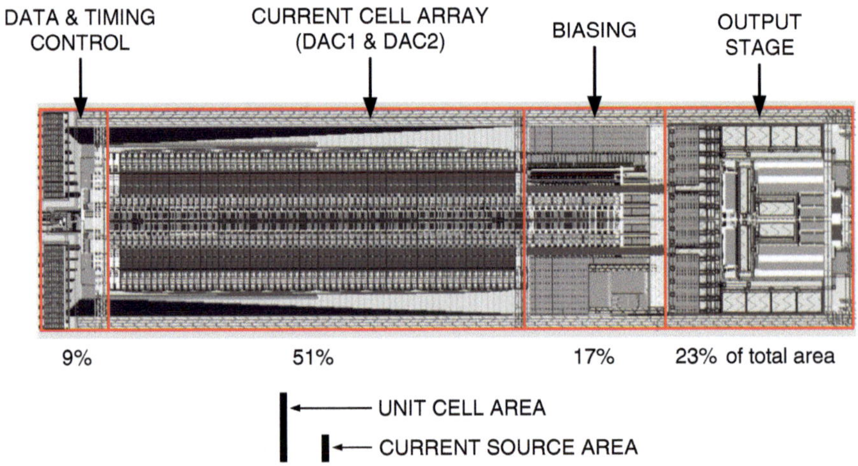

Fig. 5.27 Layout of interleaved D/A-converter

due to the more complex current-cell functionality required by the interleaved architecture. Also, the higher clock frequency inherently requires larger transistor sizes in the digital gates to optimize the switching speed. Note that the current-source transistors now only account for roughly 30 % of the total current-cell area.

The layout of the complete converter module is shown in Fig. 5.27. The current-cell array contains two rows of 64 unit cells each. With additionally two dummy cells at either side of each row, the total number of cells in the array amounts to 136. The active cells belonging to DAC1 and DAC2 are placed alternately to maximize the static and dynamic matching of the subconverters. On the top and bottom of the current-cell array two symmetrical clock trees, driven by a single clock buffer, distribute the sampling clock to the current cells.

Table 5.3 shows the area contribution of the different building blocks. Interestingly, the total current-source area, responsible for the static unit cell matching, occupies less than 10 % of the overall converter area.

Table 5.3 Area contribution of the major building blocks

Building block	Area (mm^2)	Percentage (%)
Current-source transistors	0.042	9.5
Unit current cells	0.14	31
Current-cell array + clock tree + data bus	0.22	51
Biasing	0.075	17
Output stage	0.1	23
Data interface	0.04	9
Overall converter	0.44	100

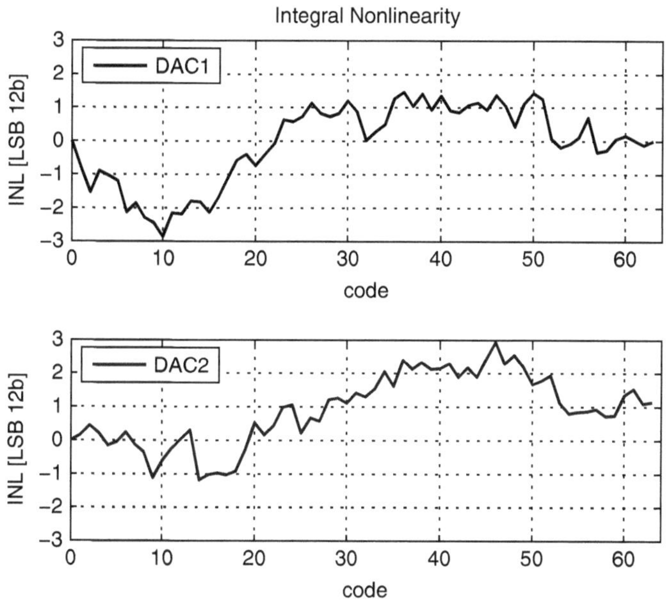

Fig. 5.28 INL of interleaved D/A-converter

5.2.8 *Experimental Results*

The measured INL of a typical device is shown in Fig. 5.28. For the second array, DAC2, the offset is removed, but the gain is not compensated. Therefore, the gain error, or the full-scale range mismatch of the interleaved converters, is directly given by the INL of DAC2 at the full-scale code. In this example it amounts to 1.145 LSB at the 12-bit level, or 0.028 %.

Figure 5.29 shows the measured relative full-scale range error of the interleaved converter arrays for 16 devices. Device 5 has the worst full-scale mismatch of 0.18 %, still well within the limit of 0.275 % for a 1 dB loss in SNR due to this effect.

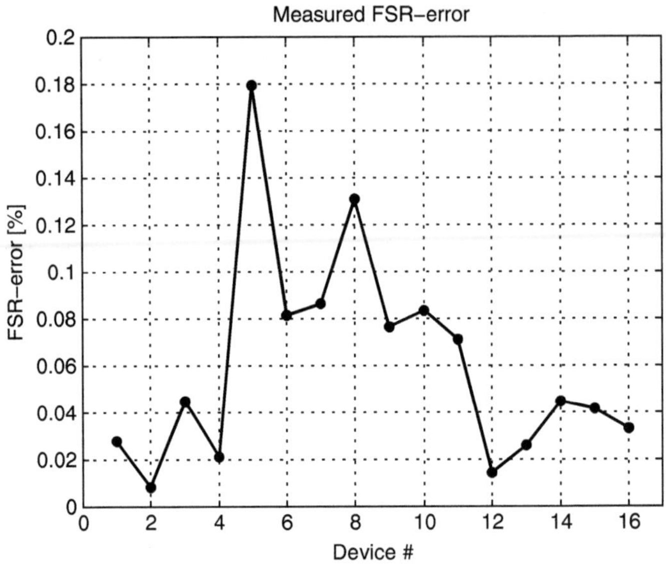

Fig. 5.29 FSR-error for different devices

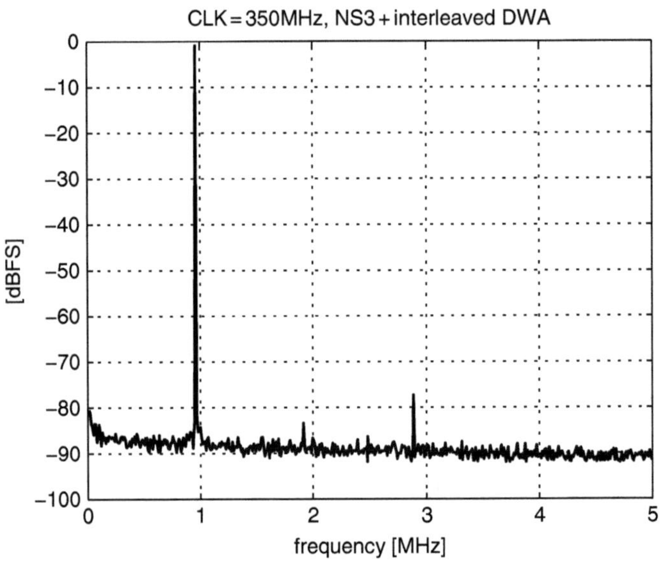

Fig. 5.30 1 MHz single-tone output spectrum

Figure 5.30 shows the output spectrum for a −1 dBFS (dBFS = dB relative to full-scale) 1 MHz sine wave synthesized with a 350 MHz clock. The second harmonic is at −82.4 dBc, while the dominant nonlinearity is the third harmonic at −76.3 dBc.

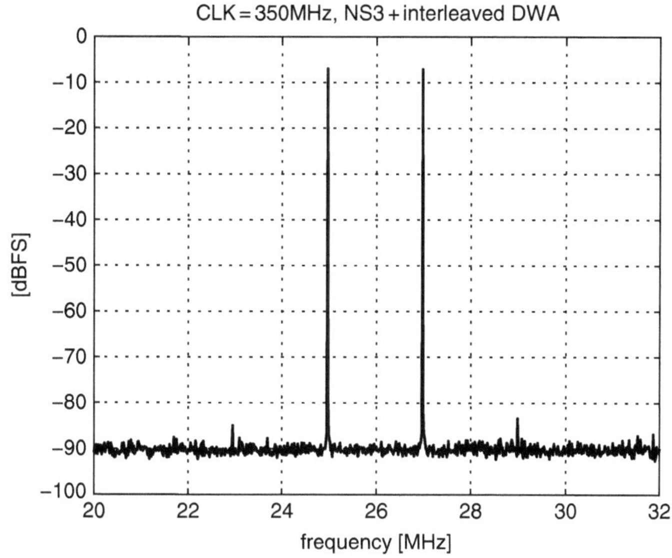

Fig. 5.31 Two-tone output spectrum at upper band edge

A two-tone signal at $f_1 = 25\,\text{MHz}$ and $f_2 = 27\,\text{MHz}$ is shown in Fig. 5.31. Each tone has an amplitude of −7 dBFS. The third-order intermodulation product at $2f_2 - f_1$ is slightly below −76 dBc. This shows that the interleaved-DWA architecture maintains high linearity also for signal frequencies close to 30 MHz. The dominant nonlinearity is the active output stage.

The measured dynamic range of the converter for a clock frequency of 350 MHz and OSR = 6, corresponding to a maximum signal frequency of 29.16 MHz, is shown in Fig. 5.32. The dynamic range (DR) is 73.4 dB, corresponding to 11.9 bits.

Figure 5.33 shows a single-tone output spectrum in a low-bandwidth, low-power mode. The converter is clocked at 106 MHz and the bias current of the active output stage is reduced to 25 %. The 130 kHz, −1 dBFS sine wave is synthesized with a second-order noise shaper. The third harmonic is found at −85 dBc. The DR in this mode of operation is 86 dB for a 2.2 MHz bandwidth.

A corresponding two-tone signal is shown in Fig. 5.34. Both carriers are at −7 dBFS and centered at 1.8 MHz with a spacing of 200 kHz. The odd-order intermodulation distortion is better than −84 dBc.

With a clock frequency of 350 MHz the analog part of the converter draws 45 mW from a 1.5 V supply. The bulk of the analog power consumption is evenly distributed between the output stage (21 mW) and the dynamic switching power of the current cells (22 mW). On the other hand, the static power consumption of the current cells and the biasing, including the single-ended common-node buffer, adds up to only 2 mW. The digital part of the converter, including the DWA-logic and the data multiplexer, consumes 17 mW from a 1.5 V supply. The total power consumption of the converter, clocked at 350 MHz, is thus 62 mW.

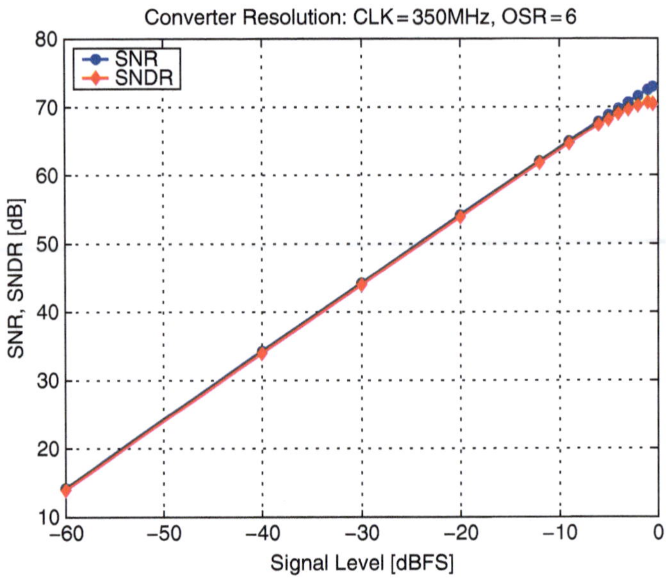

Fig. 5.32 Converter resolution at OSR = 6

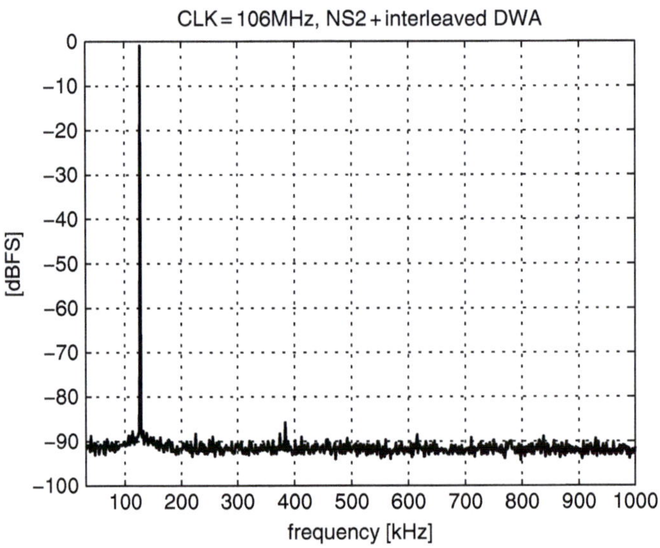

Fig. 5.33 Single-tone output spectrum

When the sampling clock is reduced to 106 MHz, the biasing current of the output stage can be reduced drastically. In this case the analog part of the converter consumes only 15 mW and the digital part 4 mW for a total of 19 mW. Remarkably, this is still a factor 2 higher than the optimized design with comparable performance

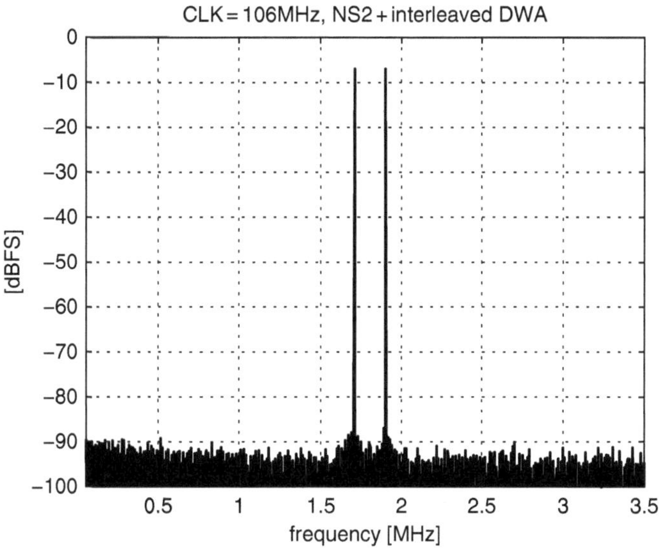

Fig. 5.34 Two-tone output spectrum

described in Sect. 5.1. This is mostly due to the remaining 10 mW of dynamic switching power caused by the interleaved-DWA architecture. A further power reduction in this mode is possible by using only a single subconverter with a simple CLA-algorithm. Although not specifically tested, this would allow to approach the performance efficiency achieved with the design described in Sect. 5.1.

5.3 Literature Comparison of Noiseshaped DACs

In order to compare different designs of the same type, an figure-of-merit (FOM) can be introduced. It is, in principle, a single number which is calculated on the basis of suitably chosen performance parameters. An FOM is usually defined to represent the efficiency of a given device, i.e., the performance relative to the required "investment" in resources. With integrated circuit design the invested resources are usually related, either directly or indirectly, to power dissipation and/or silicon area.

For noiseshaped converters a typical FOM includes the dynamic range DR, the analog bandwidth f_B, and the power drawn from the voltage supply [44, 157]:

$$\mathrm{FOM}_1 = \frac{2^{\frac{\mathrm{DR}-1.76}{3.01}} \cdot f_B}{P} \tag{5.9}$$

FOM_1 has the form Speed · Accuracy^2/Power, meaning that a factor of 2 increase in bandwidth or dynamic range (−3 dB in-band noise power) and halving the power dissipated by the converter are all equivalent improvements. Since the

Table 5.4 FOM_1 comparison

Label	Ref.	Tech. [μm]	Area [mm²]	Power [mW]	DR [dB]	f_B [kHz]	FOM_1 [10^3/pJ]
[Schouw91]	[161]	1.6	7.3	100	115	20	42
[Su93]	[162]	1.2	3	59	94	20	0.57
[Hama96]	[163]	0.6	1.09	22	100	20	6.1
[Adams98]	[47]	0.6	1.25	125	113	20	21
[Falak99]	[114]	0.6	1.7	75	85	5,000	14
[Fuji00]	[164]	0.6	1.77	145	120	20	92
[Gong00]	[88]	0.35	4.7[a]	100[b]	120	20	133.7
[Annov02]	[165]	0.35	0.78	28	98	20	3
[Rueger04]	[166]	0.35	0.9	79.5	110	20	17
[Franc04]	[167]	0.18	0.94	55	80	1104	1.34
[Nguyen08]	[168]	0.18	0.53	1.1	108	24	920
[Giotta04]	[148]	0.13	0.35	9	87	2,200	82
[Clara05,1]	[133]	0.13	0.44	62	73.9	29,166	7.7
[Clara05,2]	[133]	0.13	0.44	19	86	2,200	31
[Colon05]	[169]	0.13	0.22	7.25	97	20	9.2
[Lee06]	[170]	0.13	0.33	4	101	20	42
[Risbo11]	[171]	0.045	0.16	0.875	108	24	1,157

[a] Total chip area reported in [88] is 9.4 mm^2 for two channels, including padframe
[b] Total power dissipation reported in [88] is < 200 mW for two channels

dynamic range, or accuracy, is dimensionless, FOM_1 has the dimension of 1/energy. For analog circuits, whose performance is only limited by device matching, a quantity of the form Speed $\cdot$ Accuracy^2/Power can be shown to be directly proportional to a technology-specific constant [158–160]. Although this property holds for simple enough circuits, it can also be extended, e.g., to low-resolution flash ADCs, where accuracy, speed, and power are all limited by the differential pair in the comparator input stages [160]. It is unclear, however, whether a similar fundamental relationship could be established also for current-steering D/A-converters, since switching speed and power are not directly related to accuracy, especially when calibration or dynamic element matching are employed.

Table 5.4 shows a comparison of recently published designs with respect to FOM_1. Because the number of published current-steering D/A-converters intended for DSL-applications is quite small, also switched-capacitor designs and audio converters are included. A graphical representation of FOM_1 over the inverse silicon area is shown in Fig. 5.35. The x-axis is equivalent to the area efficiency of the converter, while the y-axis is a measure for performance efficiency. Audio converters are marked with "A."

For audio applications a relatively clear trend in performance and/or area efficiency is visible, accompanied by architecture shifts. While early designs have been based on current-steering D/A-conversion, a boost, either in performance or area efficiency, was achieved by switching to SC-architectures, together with a migration to finer line technologies. Recently, current-steering audio DACs again

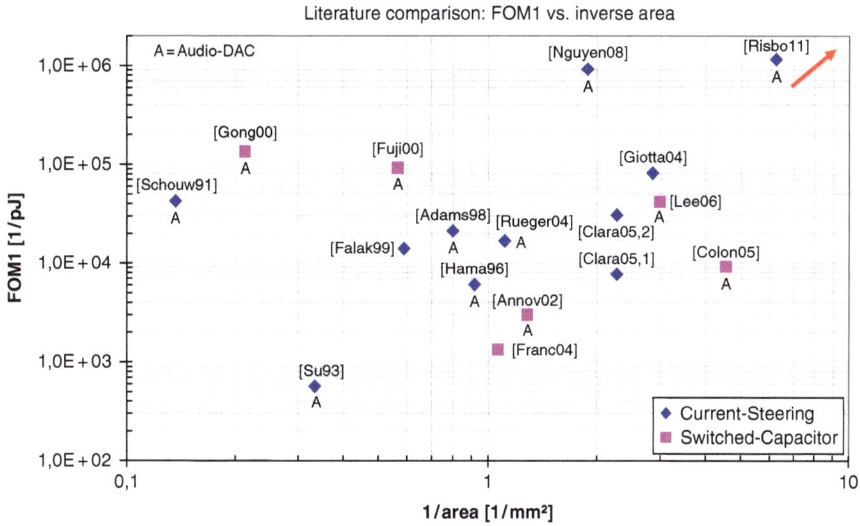

Fig. 5.35 Comparison of ΣΔ-DACs with FOM_1

demonstrate superior performance, due to drastically reduced power dissipation and the employment of very efficient DEM-algorithms. Comparing [168] and [171], the migration to smaller minimum dimensions, while keeping the same performance level, still seems to bear a considerable advantage in terms of area efficiency of the converter. However, such a migration usually entails a smaller output swing due to the reduced supply voltage. This must eventually be recovered by providing more voltage gain in one of the subsequent analog stages, e.g., the headphone driver. Since the full-scale output voltage is not included in FOM_1, the comparison given in Table 5.4 and Fig. 5.35 is not completely fair from a system-level viewpoint, because the need for an eventual amplification to drive the actual load is not reflected.

For wideband applications current-steering topologies seem to be more efficient than switched-capacitor-based designs. The optimized ADSL-converter described in Sect. 5.1 and [148] has a clear power and area advantage compared to the wideband-DAC described in Sect. 5.2 and [133]. Although the power consumption of the latter can be drastically reduced when operated with the same system parameters, the increased dynamic switching power of the interleaved current-cell architecture still causes a 10 mW penalty, resulting in a 2.6 times lower FOM_1. This difference could still be reduced considerably by switching off the interleaved RZ and operating the converter with a CLA-DEM as in the other design. However, even in this case, a 2–3 mW power penalty is expected, mainly because of the overhead in digital logic, which must obviously be designed to support a 3.5 times higher clock rate. These results show that a design, which is optimized for a specific application, always gives the best possible performance and silicon area than a design with multi-standard capability, because the latter necessarily implies having a certain overhead. Again, multi-standard capability is not reflected in FOM_1.

Chapter 6
Advanced Current Calibration

Dynamic current copying is a technique that allows to continuously trim the single DAC-elements in the background. One current source at a time is taken out of the DAC-array and compared with a reference current. Then a negative feedback loop is established, and, by controlling the DAC-element, or a part thereof, in a suitable way, its current is forced to equal the reference current. When the calibration is completed, the trimmed DAC-element is switched back into the DAC-array and continues to take part in the data processing. At the same time the next DAC-element is put into calibration mode. When all current sources of the array are trimmed, the calibration cycle starts anew.

For uninterrupted operation in the foreground, suitable redundancy must be included in the DAC-array. This means that the element under calibration must always be replaced by an identical element that takes over the data processing task. Thus, correct data multiplexing in accordance with the calibration cycle is necessary.

In general, background calibration in a current-steering D/A-converter requires the following elements:

- DAC-element with current-trimming facility
- Reference current
- Current measurement and control circuitry
- Data multiplexing unit
- Redundant DAC-element to replace current source under calibration

In this chapter we describe two hardware realizations of current-steering D/A-converters that employ dynamic current copying for continuous background calibration. Both testchip modules are fabricated in a 0.13 μm CMOS technology and use a single 1.5 V supply voltage.

The first design is a general-purpose 13-bit single-polarity current-steering DAC with passive on-chip termination [42]. The sampling clock can be chosen in the range of 100–200 MHz. A segmented calibration architecture not only trims all the unary elements in the MSB-segment but also the segment boundaries by making the sum of the elements in the lower segments equal to the unit element

M. Clara, *High-Performance D/A-Converters*, Springer Series in Advanced Microelectronics 36, DOI 10.1007/978-3-642-31229-8_6,

in the next higher segment. Furthermore, the characteristic tone generated by the cyclic calibration process is eliminated by introducing a random time base into the calibration control.

The second design is a 13-bit dual-polarity current-steering DAC with active output stage [93]. It employs a sampling clock in the range of 130–300 MHz. A self-calibrated variable reference current is introduced that allows direct trimming of the unary elements in the MSB-segment and in the intermediate segment, while the binary segment is trimmed by boundary matching. A full-clock interleaved D/A-architecture is implemented for improved dynamic performance.

6.1 A Self-calibrated 13-Bit 100–200 MS/s D/A-Converter

6.1.1 Converter Architecture

The converter architecture is shown in Fig. 6.1. It is a 13-bit single-polarity segmented current-steering DAC with resistive on-chip termination. The segmentation of the converter is 6 + 2 + 5 bits, wherein the two upper segments, MSB and ULSB, are unary arrays and the lowest segment (LLSB) is binary coded. The current sources are PMOS-type; thus the terminating resistors are connected to the ground rail.

The choice of segmentation is always determined by conflicting requirements. Although a unary array would offer the best possible dynamic performance, the silicon area required by the digital decoding logic and the number of data lines from this decoder to the current cells grows exponentially with the number of bits assigned to the unary segment. With 6 bits in the MSB-array the area for the decoding logic and the digital wiring is still moderate. To reduce the magnitude of the potentially problematic binary transitions still further, an intermediate unary segment (ULSB) of 2 bits is introduced, adding negligible overhead in digital decoding and bit-line wiring.

The MSB-array consists of 64 unit current sources comprising one redundant element needed for the background calibration. The ULSB (=Upper-LSB) array contains three current cells plus one dummy cell used only during calibration. Because the ULSB-array is trimmed as a whole, a second identical ULSB-array is included. The two ULSB-arrays, ULSB-A and ULSB-B, are calibrated alternately in every second calibration cycle. MSB- and ULSB-arrays are biased with a common biasing, optimized for low-noise performance.

The 5-bit binary-coded LLSB-array (=Lower LSB) has a dynamically adjustable biasing used to trim the sum current in the LLSB-array. An extra dummy cell is included for calibration. Also the LLSB-array together with the LLSB-biasing is implemented twice (LLSB-A and LLSB-B) and calibrated alternately in every second calibration cycle.

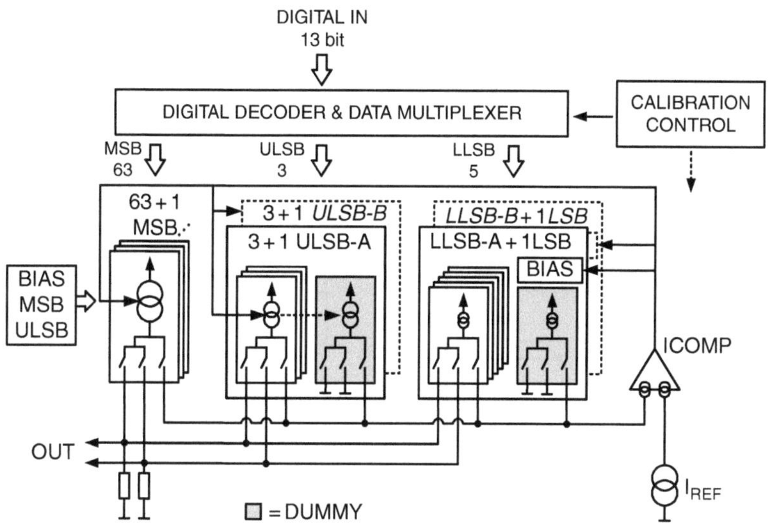

Fig. 6.1 Self-calibrated DAC architecture

During calibration, the current comparator (ICOMP) compares the current from the selected DAC-element, or a combination thereof, with the reference current I_{REF} from the reference cell and forces the current in the DAC-element accordingly. The calibration process and the data multiplexing are controlled with a dedicated digital logic included in the converter macro.

The full-scale current of the converter is 5 mA. With on-chip termination resistors of 100 Ω the differential full-scale output voltage range is 1 V. Since the D/A-converter is calibrated continuously in the background, it must be transparent with respect to the data processing. This means that the DAC-element in calibration is replaced by an equivalent, already calibrated DAC-element.

6.1.2 *Trimmable PMOS Current Cell*

The current cell of the MSB-array is shown in Fig. 6.2. The transistor M_{SRC} with fixed biasing delivers nominally 7/8 of the nominal cell current equal to 78.125 μA. The gate-source voltage of the trim transistor M_{TRIM}, with a nominal current of $1/8 \cdot 78.125\,\mu\mathrm{A} = 9.76\,\mu\mathrm{A}$, is set during calibration to adjust the overall cell current to the correct value. We define the relative trimming range R_{TRIM} of the current cell by the ratio of the nominal drain current flowing in M_{TRIM} and the nominal cell current. The MSB-current can then theoretically be adjusted in the range $\pm R_{\mathrm{TRIM}}$, in this design example $\pm 12.5\,\%$. Note that in practice we try to use only about one half of the maximum possible current range in M_{TRIM}, since we must keep

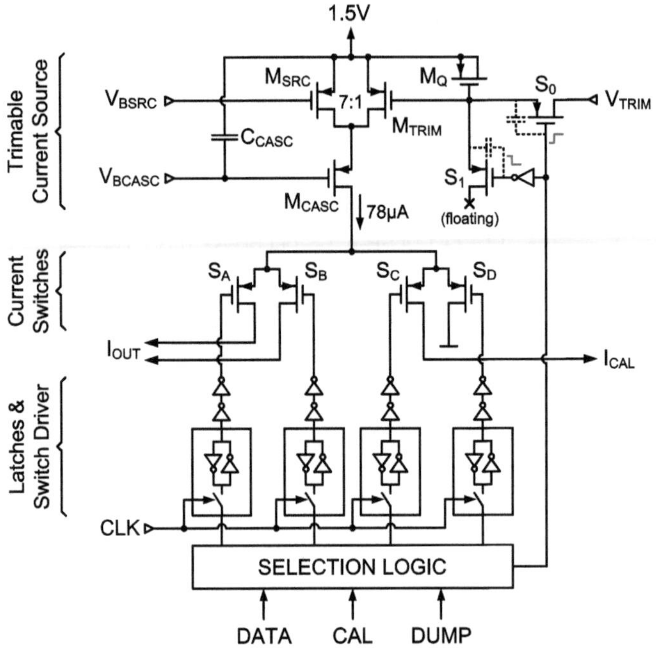

Fig. 6.2 Trimmable PMOS current cell

the current comparator within its linear range for approximately constant gain (and bandwidth) of the calibration feedback loop.

The MOS-capacitor M_Q helps to store the gate-source voltage of M_{TRIM} between two successive calibration slots. M_Q has the same aspect ratio as M_{TRIM} and its area is four times as large. At the end of the calibration, switch S_0 is opened and charge is injected onto the gate node of M_{TRIM}. With an equal-sized transistor S_1 connected as shown and switched in the opposite direction a large part of this error charge can be canceled. This makes the initial current error $\Delta I_{\text{DS}}(0)$ in Eq. (3.41) very small.

The cascode transistor M_{CASC} increases the output impedance of the current cell. A MIM capacitor C_{CASC} of approximately 650 fF is placed above each current cell to decouple the gate node of M_{CASC}. This improves the dynamic performance of the converter by providing a high-frequency short for the switching spikes that are present at the drain node of the cascode transistor and couple to V_{BCASC} via the gate-drain overlap capacitance.

The ULSB-cell is a scaled version of the MSB-cell. The width of all transistors in the analog section, including M_Q and the switches, is made a factor 4 smaller. In this way the current densities remain the same and the matching between MSB-array and ULSB-array is improved. Because the ULSB-cell uses the same resynchronization latches and the same switch driver size as the MSB-cell, dummy current switches are included to equalize the loading for the switch drivers.

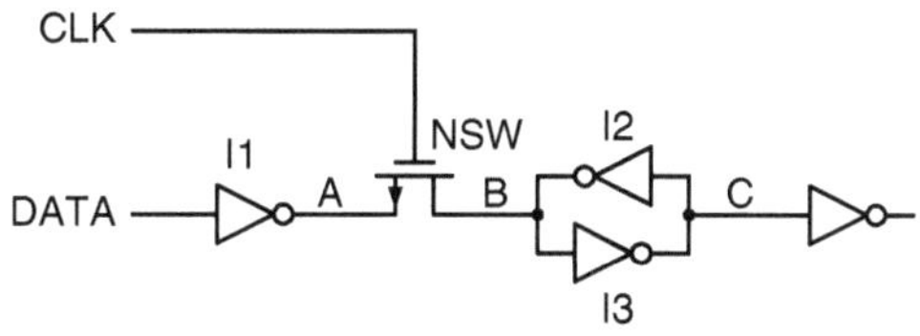

Fig. 6.3 Standard buffer-latch circuit

The LLSB-array has a calibrated biasing; therefore the LLSB-cells do not include transistors M_{TRIM}, S_0, and S_1. Among the LLSB-cells, transistors M_{SRC} and M_{CASC} are binary scaled to deliver the correct output current. Again, the resynchronization latches and the switch drivers are the same as in the MSB-cell. Also in the LLSB-array the switch driver load is equalized with dummy current switches.

The inclusion of the calibration output branch in the current cell as shown in Fig. 6.2 poses a certain problem for the implementation of the resynchronization latch. Standard current cells only have two current switches always switching in opposite directions, in which case a symmetric, differential latch structure can be used [33, 34, 117]. The addition of a third switch to steer, once in a while, the cell current toward the current comparator, makes the use of a differential latch impractical, since each switch must be controlled independently from the others. Single-ended latches are however problematic with respect to symmetrical switching.

The standard buffer-latch circuit is shown in Fig. 6.3. The buffer inverter I_1 drives the latch. When *CLK* goes high the NMOS-switch *NSW* is turned on and I_1 overwrites the latch consisting of I_2 and I_3. Because I_2 is usually a weak inverter, this latch structure is sometimes also called a David-Goliath latch [34]. When *CLK* returns low, I_2 provides positive feedback to regenerate and store the written data. Figure 6.4 displays the result of a SPICE-simulation showing the strongly asymmetric behavior of the standard buffer-latch circuit with respect to the rising and falling data transition.

The dependence of the on-resistance of the NMOS-switch *NSW* on the voltage at nodes *A* and *B* makes the driving point impedance at the input of the latch (node *A*) strongly dependent on the previously stored data. This effect causes a different switch-on and switch-off delay, which is unacceptable for the resynchronization latch in a D/A-converter. Instead, it is necessary to equalize the rising and falling transition without making use of the inverted switch branch.[1]

The improved buffer-latch circuit used in this design is shown in Fig. 6.5. The PMOS-switch *PSW* in parallel to *NSW* helps to keep the on-resistance between nodes *A* and *B* low during the high-transition at these nodes. Because we want to use only a single clock phase to control the latch, the gate of *PSW* is not controlled

[1]In a differential latch structure it is possible to minimize this asymmetry by cross-coupling of the inverters I_2 and grossly oversizing I_1 and *NSW* [34]. This approach cannot be used for the current cell of Fig. 6.2, because the calibration mode is not compatible with differential switching.

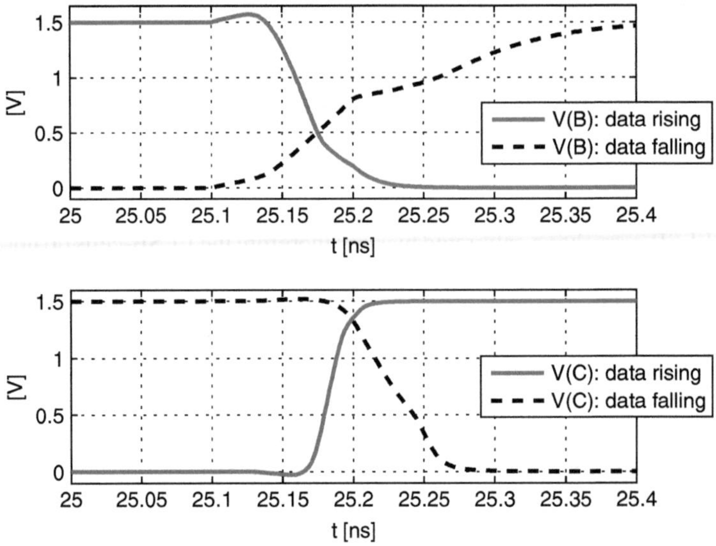

Fig. 6.4 Simulated transitions of standard buffer-latch circuit

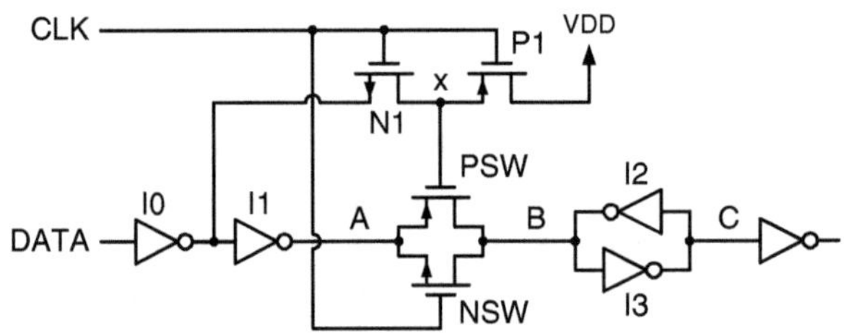

Fig. 6.5 Improved buffer-latch circuit

by an inverted version of the sampling clock but by the input data itself, specifically by inverter I_0, which also drives inverter I_1.

The simulated transition of the improved buffer-latch circuit is shown in Fig. 6.6. The rising and falling data transitions are now, at least in simulation, completely symmetric and cross each other exactly at VDD/2.

6.1.3 Segmented Background Calibration

The MSB-cell in calibration mode is shown in Fig. 6.7. The cell current I_{MSB} is steered to the low-impedance node A of the current comparator. The potential at this node is set approximately equal to the common-mode output voltage, such that the current cell sees the same output voltage during calibration as, on average, during normal operation.

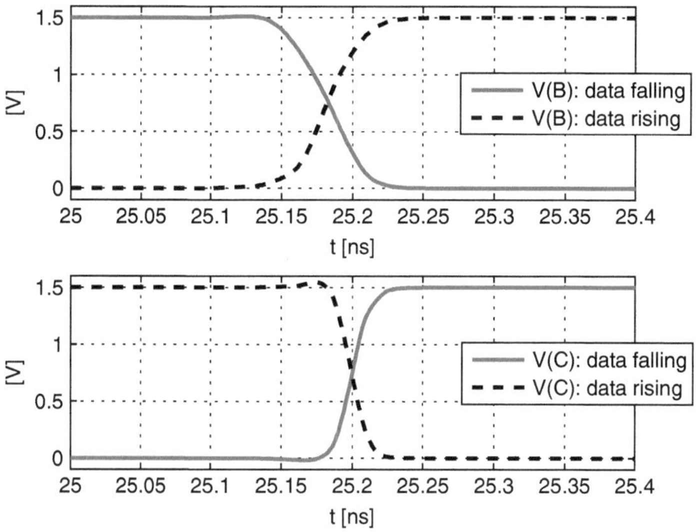

Fig. 6.6 Simulated transitions of improved buffer-latch circuit

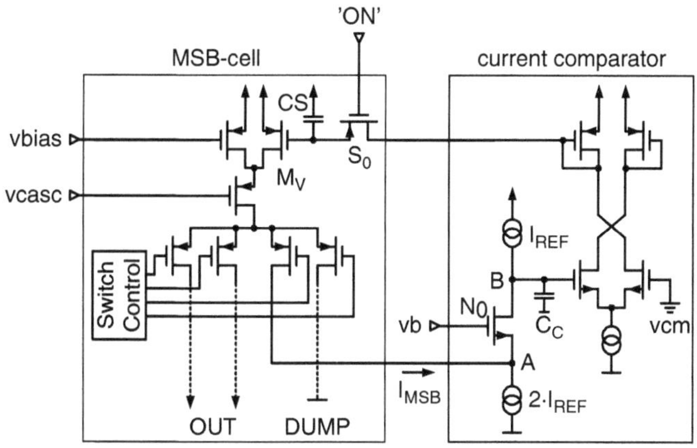

Fig. 6.7 Calibration of the MSB current cell

The difference current $I_{\mathrm{MSB}} - I_{\mathrm{REF}}$ is transferred to the high-impedance node B and develops an error voltage $(I_{\mathrm{MSB}} - I_{\mathrm{REF}}) \cdot R_B$ at the input of the differential amplifier. R_B is the impedance seen at node B. The resulting current imbalance of the differential amplifier is mirrored to the adjustable current source transistor M_V, identical to M_{TRIM} in Fig. 6.2. The DC-gain of the negative feedback loop is given by

$$A_{\mathrm{CAL}} = \frac{1}{2} gm R_B \tag{6.1}$$

gm is the transconductance of the differential pair transistors. The mirror ratio between the PMOS-diode load of the differential amplifier and transistor M_V is 1 in this design. Assuming complete settling of the calibration loop, the initial current error $\Delta I_{\mathrm{MSB},0}$ of the cell is reduced by the DC-gain of the loop:

$$\Delta I_{\mathrm{MSB,cal}} = \frac{\Delta I_{\mathrm{MSB},0}}{1 + A_{\mathrm{CAL}}} = \frac{\Delta I_{\mathrm{MSB},0}}{1 + \dfrac{gmR_B}{2}}. \tag{6.2}$$

We can calculate the minimum required value of A_{CAL} by considering that the residual current error of the MSB-cells must be smaller than 0.5 LSB. With a resolution of M bits in the MSB-segment and an overall converter resolution of B bits we get the following inequality:

$$\max\left(|\Delta I_{\mathrm{MSB,cal}}|\right) = \frac{\frac{1}{2} R_{\mathrm{TRIM}} \cdot 2^{-M} I_{\mathrm{FS}}}{1 + A_{\mathrm{CAL}}} < \frac{1}{2} \cdot \frac{I_{\mathrm{FS}}}{2^B}. \tag{6.3}$$

We have included a factor 1/2 on the right-hand side to account for the fact that we only want to use at most one half of the theoretical trimming range provided by M_{TRIM}. With $R_{\mathrm{TRIM}} = 1/8$, the minimum required DC-gain of the calibration loop to achieve a worst-case calibration error of 0.5 LSB at the 13-bit level, assuming full use of one half of the theoretical trimming range ($\pm 1/2 R_{\mathrm{TRIM}}$), is given by

$$\max\left(|\Delta I_{\mathrm{MSB,cal}}|\right) = \frac{0.0625 \cdot 2^{-6} I_{\mathrm{FS}}}{1 + A_{\mathrm{CAL}}} < \frac{1}{2} \cdot \frac{I_{\mathrm{FS}}}{2^{13}} \rightarrow A_{\mathrm{CAL}} > 14.36 \approx 23\,\mathrm{dB}. \tag{6.4}$$

The negative feedback loop is stabilized by capacitor C_c that introduces a dominant LHP real pole at $-1/(R_B C_c)$. With 6.1 the unity gain-bandwidth of the loop is given by

$$2\pi \mathrm{GBW}_{\mathrm{CAL}} = \frac{gm}{2C_c}. \tag{6.5}$$

Assuming a first-order system, the settling time constant becomes:

$$\tau_{\mathrm{CAL}} = \frac{1}{2\pi \mathrm{GBW}_{\mathrm{CAL}}} = 2 \cdot \frac{C_c}{gm}. \tag{6.6}$$

A nondominant pole in the feedback loop is generated by the on-resistance of switch S_0 and the storage capacitor C_S, the latter also including the gate-source capacitance of M_V. Another nondominant pole occurs at the gate node of the PMOS-diode load in the current comparator.[2] For the phase margin of the calibration loop we target a value of 70° to achieve a settling behavior without

[2]This node is distributed throughout the current-cell array. The parasitic wiring capacitance is thus substantial and exceeds the gate-source capacitance of the PMOS-diode by far.

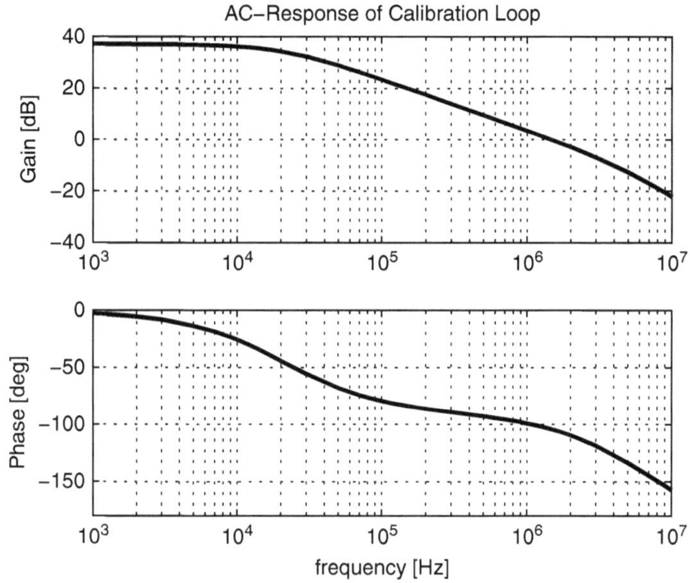

Fig. 6.8 Simulated open-loop AC-response of calibration loop

noticeable overshoot. This requires the first nondominant pole to exceed the unity gain-bandwidth at least by a factor of 3 [48].

Figure 6.8 shows the simulated open-loop AC-behavior of the optimized calibration control loop, including also the parasitic wiring capacitance at the gate of the PMOS-diode load. The simulated DC-gain is 37 dB and the unity-GBW is 1.45 MHz with a phase margin of 76°. If we assume that we need a settling time of $4\ldots5\cdot\tau_{\mathrm{CAL}}$ for the loop, we can expect a minimum calibration time of $T_{\mathrm{CAL,min}} \approx 0.5\,\mu\mathrm{s}$.

The ULSB-array is trimmed by comparing the sum of the output current of four ULSB-cells with the reference cell and adjusting the current in the ULSB-array to ideally achieve the following equality:

$$I_{\mathrm{REF}} = 4I_{\mathrm{ULSB}} \longrightarrow I_{\mathrm{ULSB}} = \frac{1}{4}I_{\mathrm{REF}} = \frac{1}{4}I_{\mathrm{MSB}}. \tag{6.7}$$

The circuit configuration for the ULSB-calibration is shown in Fig. 6.9. Since the normal operation only requires three ULSB-cells, an extra dummy ULSB-cell is included. This dummy cell is only used during calibration, otherwise its current is dumped to the ground rail. The final accuracy of the operative three ULSB-cells with respect to the MSB-cell is limited by the "intra-segment" mismatch of the dummy ULSB-cell and by the calibration error given by Eq. (6.2).

To calibrate the LLSB-array to the size of one ULSB-cell, the total current of the LLSB-array plus one additional current of the size of 1 LSB is summed to three previously calibrated ULSB-currents and compared with I_{REF}; see Fig. 6.10. Again,

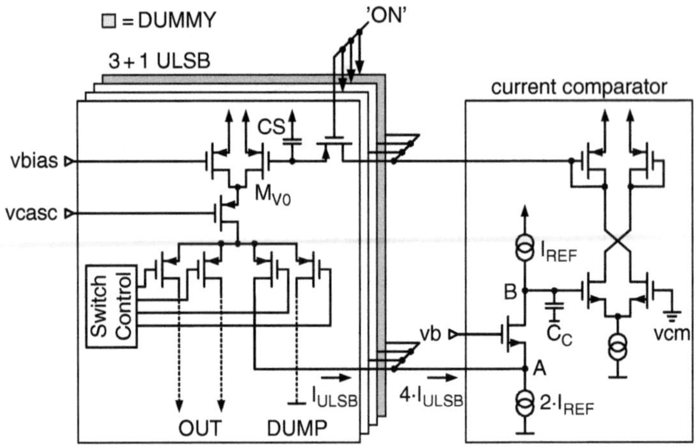

Fig. 6.9 Calibration of the ULSB-array

the additional LLSB-current is only used during calibration and otherwise dumped to the ground rail. Via the bleeding transistor M_{VB}, the LLSB-biasing is adjusted to (ideally) fulfill the following equation:

$$I_{\text{REF}} = 3I_{\text{ULSB}} + \sum I_{\text{LLSB}} + 1I_{\text{LSB}} \longrightarrow \sum I_{\text{LLSB}} = I_{\text{ULSB}} - 1 \cdot I_{\text{LSB}}. \quad (6.8)$$

Together, Eqs. (6.7) and (6.8) describe the ideal segment boundary trimming provided by the described calibration method. The accuracy within the lower segments must be provided by intrinsic matching and thus requires proper sizing of the current source transistors and identical surroundings.

The complete calibration cycle is shown in Fig. 6.11. First, all 64 MSB-cells are trimmed to be equal to the reference current. Then one of the ULSB-arrays is calibrated while the second ULSB-array is assigned to data processing. In the next calibration slot the previously trimmed ULSB-cells are used to adjust one of the LLSB-arrays. At this point the calibration cycle is finished and starts again from the first MSB-cell. The two identical ULSB-arrays and LLSB-arrays are calibrated in turn in every second calibration cycle. The trimmed array is immediately used for data processing, while the current of the other array is dumped to the ground rail during the following MSB-array calibration.

6.1.4 Randomized Calibration Cycle

The periodicity of the background calibration generates tones in the output spectrum. Suppose that each calibration slot has a fixed duration of T_{CAL}. With a total of N cells or sub-arrays in the calibration loop the fundamental refresh period is given by

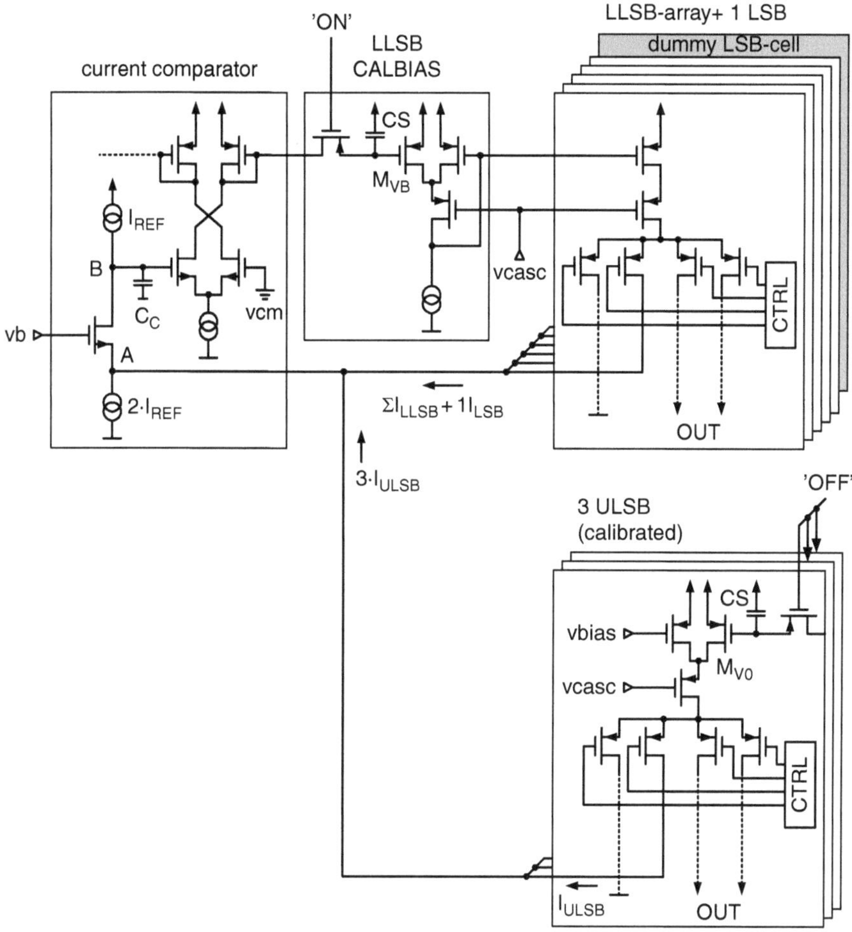

Fig. 6.10 Calibration of the LLSB-array

$$T_{\text{refresh}} = N \cdot T_{\text{CAL}} \longrightarrow f_{\text{refresh}} = \frac{1}{N \cdot T_{\text{CAL}}}. \quad (6.9)$$

The periodic switching of current cells into calibration mode and back, as well as dynamic mismatches associated with the dynamic current copying from cell to cell, introduces an error into the output signal. This error repeats itself in the rhythm of the calibration cycle and thus introduces spectral components at multiples of the fundamental refresh frequency f_{refresh} [104].

In [107] the fundamental refresh frequency is chosen to be $f_{\text{refresh}} = 10.25\,\text{Hz}$ and falls outside the frequency band of interest ranging from 20 Hz to 20 kHz. Still, tones at multiples of f_{refresh} are below $-110\,\text{dBFS}$. Generally, in newer CMOS

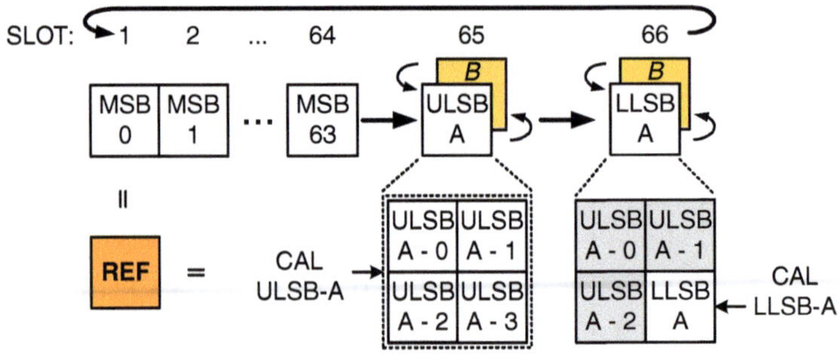

Fig. 6.11 Complete calibration cycle

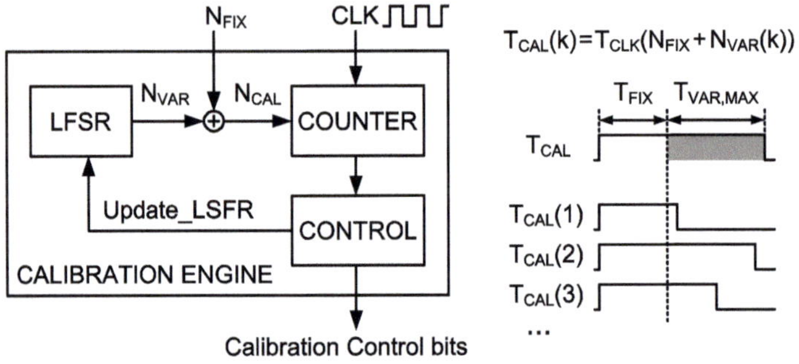

Fig. 6.12 Generation of random calibration cycle

technologies, such a low calibration frequency is not possible due to higher leakage currents.

Floating current sources with backside measurement and the possibility to trim during normal operation do not require a redundant element and make the periodic switching of current cells into calibration mode and back unnecessary [92, 104]. As a result, the calibration tones are greatly attenuated, e.g., in [92] the spurs are below −93 dBFS. As described in Sect. 3.3.5, the implementation of the floating current source becomes increasingly difficult with lower supply voltage. Also, the required folding of the output current at least triples the static power consumption of the converter.

A different approach is taken in this design. Instead of trying to minimize the error source responsible for the generation of the calibration tones, we attempt to destroy the periodicity of the calibration process itself. By doing so, the error energy injected by the calibration should spread out spectrally and eventually merge with the circuit noise floor.

Figure 6.12 shows the calibration control with randomized time base. At the beginning of the calibration slot the down-counter is loaded with a number N_{CAL} that represents the length of the calibration slot in units of clock cycles. N_{CAL} is the sum of a fixed number (N_{FIX}) and a random number (N_{VAR}), the latter coming from a random number generator. When the counter expires the calibration of the cell

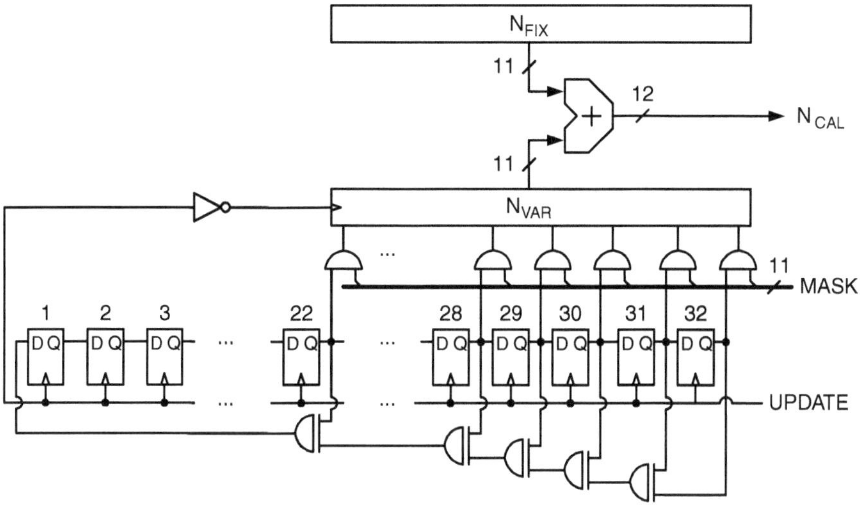

Fig. 6.13 Generation of the random time base

is finished. At the same time the random number generator is triggered to deliver a new random number for the next calibration slot.

Given the period T_{CLK} of the counter clock, usually the sampling clock or a divided version thereof, the duration of a calibration slot with index k is given by

$$T_{\mathrm{CAL}}(k) = (N_{\mathrm{FIX}} + N_{\mathrm{VAR}}(k)) \cdot T_{\mathrm{CLK}}. \tag{6.10}$$

Since the random number generator can assume integer values in the range $0 \cdots N_{\mathrm{VAR,\,max}}$, the minimum and maximum calibration slot length is thus

$$\begin{aligned} T_{\mathrm{CAL,\,min}} &= N_{\mathrm{FIX}} \cdot T_{\mathrm{CLK}} \\ T_{\mathrm{CAL,\,max}} &= (N_{\mathrm{FIX}} + N_{\mathrm{VAR,\,max}}) \cdot T_{\mathrm{CLK}}. \end{aligned} \tag{6.11}$$

A maximum-length linear feedback shift register (LFSR) is used as random number generator in this design. With R shift register stages, the random number sequence repeats after $2^R - 1$ cycles and has approximately a uniform probability distribution [172]. Therefore the average calibration slot length is given by

$$\overline{T_{\mathrm{CAL}}} = \left(N_{\mathrm{FIX}} + \frac{N_{\mathrm{VAR,\,max}}}{2} \right) \cdot T_{\mathrm{CLK}}. \tag{6.12}$$

The implementation of the random number generation with fixed offset is shown in Fig. 6.13 [173]. The LFSR has a length of $R = 32$ and uses a primitive polynomial of the form $P(X) = X^{31} + X^{30} + X^{29} + X^{28} + X^{27} + X^{21} + 1$, known to generate a maximum-length sequence. The fixed calibration slot length is programmable

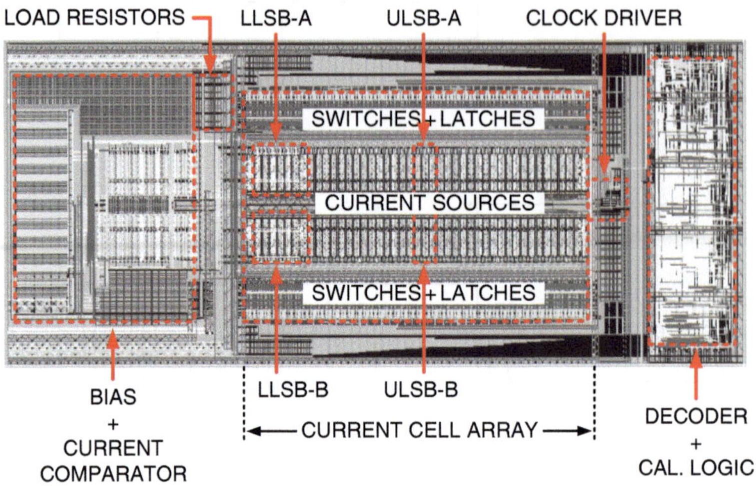

Fig. 6.14 Layout of D/A-converter

with the 11-bit register N_{FIX}. The uppermost 11 bits of the LFSR are tapped with AND-gates and transferred to the 11-bit register N_{VAR}. The "amplitude" $N_{\mathrm{VAR,max}}$ of the variable calibration slot length can be set with the 11-bit word *MASK*. The two register contents are summed together and the resulting 12-bit word N_{CAL} is subsequently loaded into the counter. When the counter expires, indicating the end of the calibration slot, a pulse is generated at the shift-register clock line (*UPDATE*). Subsequently, the LFSR is updated and the next random number is calculated and loaded into the counter.

Note that a maximum-length LFSR with $R = 32$ and an average calibration slot length of $\overline{T_{\mathrm{CAL}}} = 1\,\mu\mathrm{s}$ guarantees that the generated pseudorandom sequence repeats itself only after 1 h:11 min. Thus, for practical purposes the periodicity of the calibration process is eliminated.

6.1.5 Layout

The layout of the D/A-converter is shown in Fig. 6.14. From left to right we find the bias block including the current comparator, the on-chip termination resistors, the current source array, the clock driver, and the digital decoder comprising the calibration control, as well as the data multiplexer. The sampling clock is buffered by a strong clock driver placed exactly in the symmetry axis of the current-cell array. The clock is then distributed symmetrically over two identical clock trees to the current cells in the array. In this way all DAC-elements are nominally updated at the same time. The signal current is converted into an output voltage by the on-chip load resistors placed near the upper right corner of the biasing (in Fig. 6.14).

Table 6.1 Area contribution of the major building blocks

Building block	Area (mm^2)	Percentage (%)
Current-cell array	0.2	40
Biasing	0.12	25
Digital calibration control, decoders	0.07	15
Supply, data and clock routing	0.1	20
Overall converter	0.495	100

Table 6.2 Power drawn from 1.5 V supply

Building block	CLK=100 MHz (mW)	CLK=200 MHz (mW)
Analog	15	19
Digital	3	6
Total	18	25

Table 6.1 shows the area contribution of the most important converter building blocks and elements. The current-cell array accounts for 40 % of the overall converter area, while the current-source biasing makes up for 25 %. The major contributor to the biasing area is the large MOS-capacitor used for filtering the correlated bias noise.

The converter is implemented in a 0.13 μm 1P6M standard CMOS process and operates from a 1.5 V supply. Capacitors C_{CASC} in Fig. 6.2 are implemented as MIM capacitors placed above the active area of the current-source transistors. Each DAC-cell contains a 650 fF capacitor, so the total decoupling for the cascode bias adds up to 52 pF.

6.1.6 Experimental Results

The power consumption of the converter is summarized in Table 6.2. The analog part includes the current-cell array, the biasing and the clock driver, whereas the digital part consists of the data decoders and multiplexers, as well as the calibration control. The total power drawn from a split 1.5 V supply[3] is 18 mW at 100 MS/s and 25 mW at 200 MS/s. These numbers already include the full-scale signal current of 5 mA.

Figure 6.15 shows the measured second harmonic distortion as a function of the average calibration slot length $\overline{T_{\mathrm{CAL}}}$. In order to exclude dynamic effects, a low frequency, near full-scale single-tone signal is used and the average calibration slot length is varied.

[3] In a mixed signal circuit it is good practice to separate the supplies of analog and digital modules on chip level to minimize crosstalk. However, on PCB-level there is usually only a single regulator for each supply voltage.

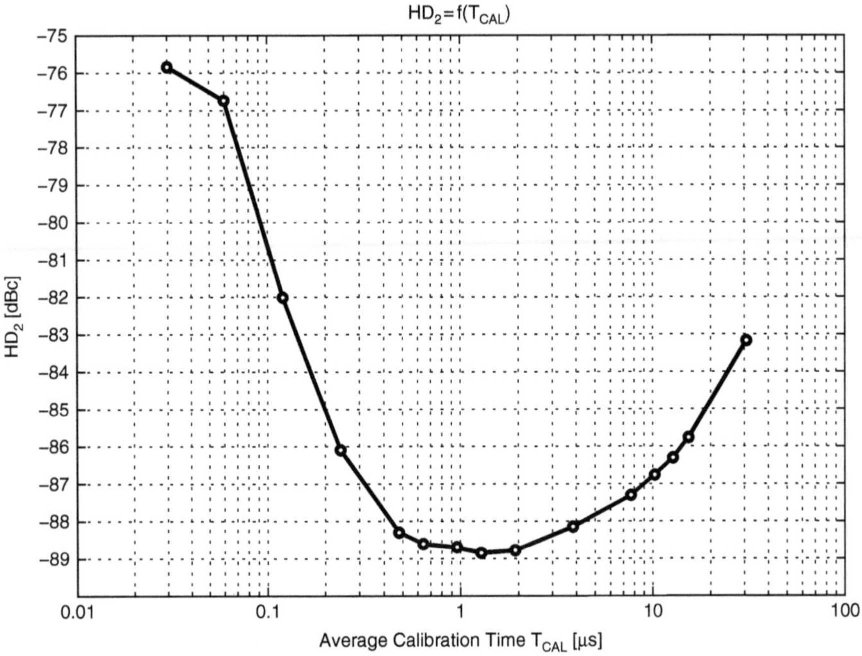

Fig. 6.15 HD_2 as function of the calibration slot length

For $\overline{T_{CAL}} < 0.5\,\mu s$ the calibration loop is not given sufficient settling time and the trimming error is still large. This is consistent with the simulated AC-behavior of the loop shown in Fig. 6.8. The optimum calibration slot length is found in the range of $\overline{T_{CAL}} \approx 1\,\mu s$. For $\overline{T_{CAL}} > 2\,\mu s$ the linearity starts to degrade again, because the discharging of the storage capacitors in the current sources due to leakage currents becomes too large.

The spectral effect of the randomization of the calibration slot length is shown in Fig. 6.16. Without randomization (top), the fundamental calibration tone at $f_{refresh}$ appears slightly below -80 dBFS. When the random time base is activated (bottom), the energy residing in the calibration tone is spectrally spread out and appears as noise.

The maximum spectral magnitude in the vicinity of $f_{refresh}$ as a function of the normalized randomization "amplitude" $N_{VAR,\,max}/N_{FIX}$ is shown in Fig. 6.17. We use $N_{FIX} = 128$, equivalent to $T_{CAL,\,min} = 0.64\,\mu s$ with a system clock of 200 MHz, as seen in Eq. (6.11). This minimum calibration slot length is required to achieve proper settling of the calibration loop; see also Fig. 6.15 above. The optimum amount of randomization is found in the region $N_{VAR,\,max} \approx 1 \cdots 2 \cdot N_{FIX}$. The maximum suppression of the fundamental calibration tone is 21.8 dB as compared to the periodic calibration cycle without randomization.

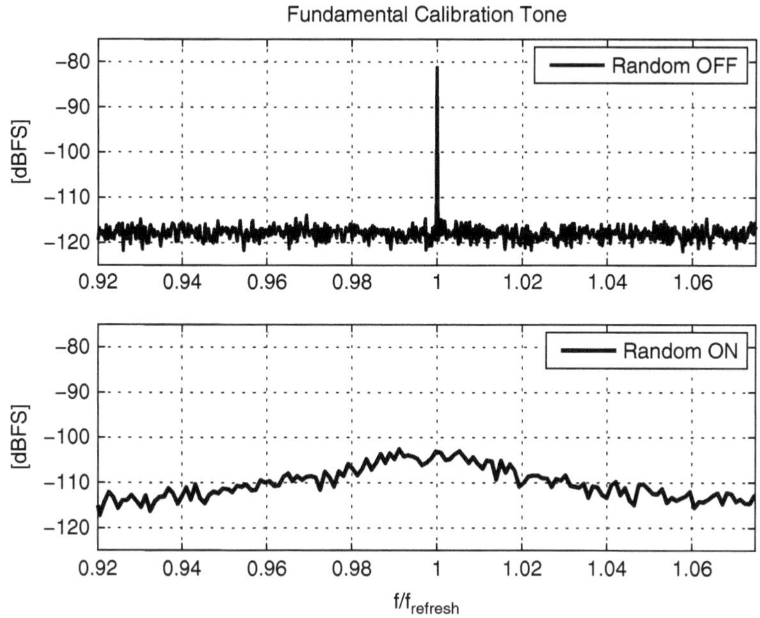

Fig. 6.16 Effect of randomization on calibration spur

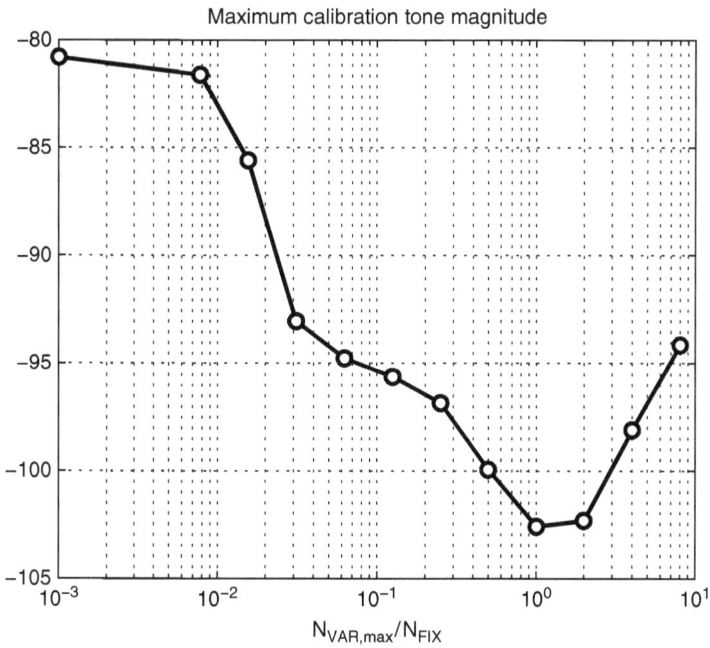

Fig. 6.17 Calibration spur magnitude vs. randomization

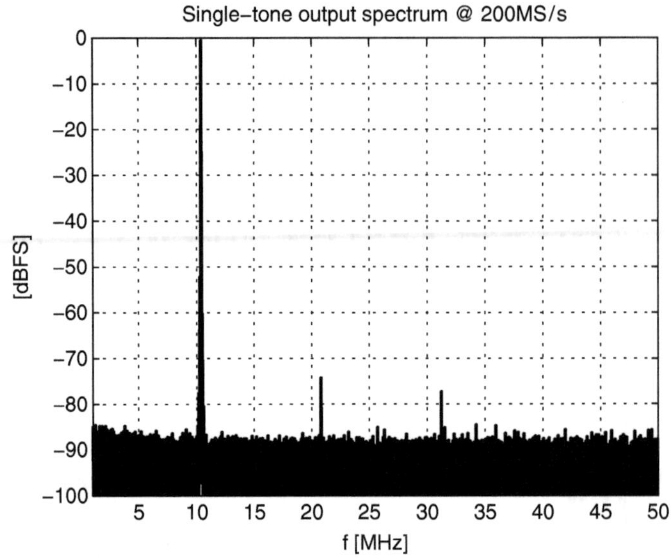

Fig. 6.18 10 MHz single-tone output spectrum

For larger $N_{\mathrm{VAR,max}}$ the average calibration slot length again starts to become excessively long and the degrading current source accuracy causes the noise floor around f_{refresh} to rise.

A 10 MHz single-tone output spectrum is shown in Fig. 6.18. The spurious free dynamic range (SFDR) is limited by the second harmonic distortion found at −73.6 dBc, while the third harmonic distortion is at −76.6 dBc.

The SFDR as a function of the signal frequency is shown in Fig. 6.19 for a sampling rate of 100 and 200 MS/s. At low signal frequencies the SFDR is 83.7 dB for both update rates. Without calibration the low-frequency SFDR is typically in the range of 65–67 dB, limited by harmonic distortion due to current source mismatch. Close to the Nyquist frequency the SFDR is limited by dynamic effects and drops to 67.5 and 54.5 dB for a sampling frequency of 100 and 200 MHz, respectively.

The second and third harmonic distortion product as a function of the synthesized frequency is displayed in Fig. 6.20 for a sampling frequency of 200 MHz. If we assume a rise and fall time mismatch of 25 ps a good correspondence of the measured HD_2 with Eq. (4.18) is achieved for signal frequencies above a few MHz. At 200 MS/s a rise/fall time mismatch of 25 ps corresponds to a net charge error of 0.25 % of the unit current pulse area. Of course, also any other asymmetry in the switch-on and switch-off transition of the unit current cell that produces an equivalent amount of error charge can be the root cause of the perceived even-order harmonic distortion. Likewise, the frequency dependence of HD_3 can be described with reasonable accuracy by Eq. (4.28). The circuit parameters $C_0 = 50\,\mathrm{fF}$ and $\rho = 0.085$ have been obtained from layout extraction and SPICE-simulation of the current cell.

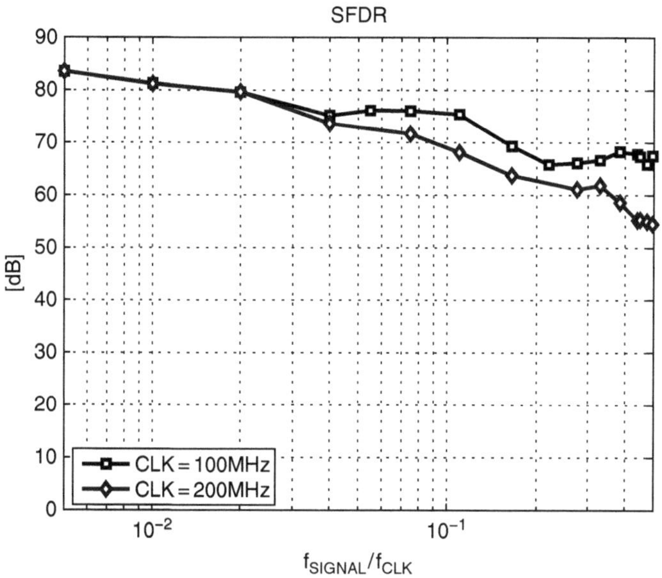

Fig. 6.19 Spurious free dynamic range (SFDR)

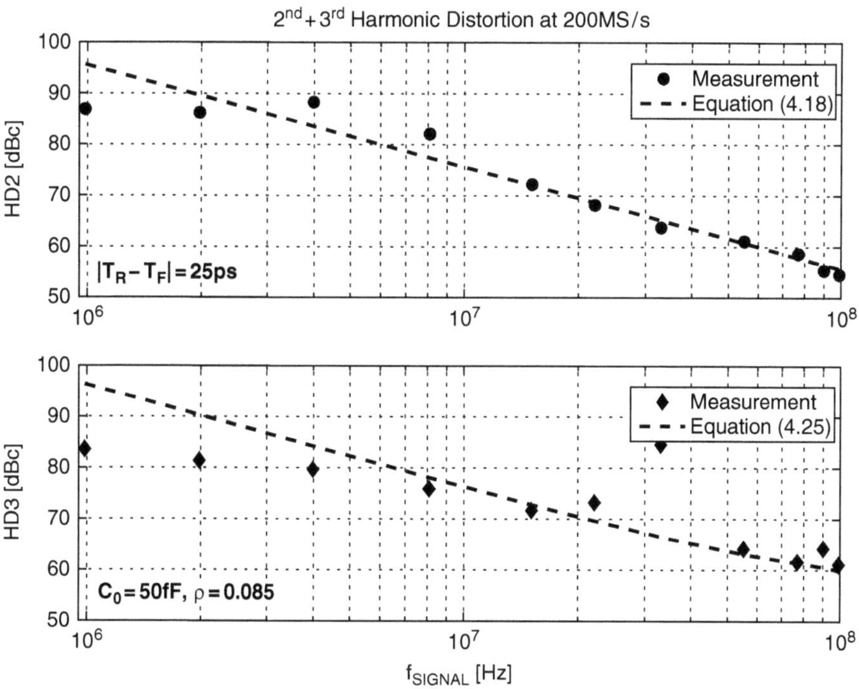

Fig. 6.20 HD_2 and HD_3 for 200 MS/s

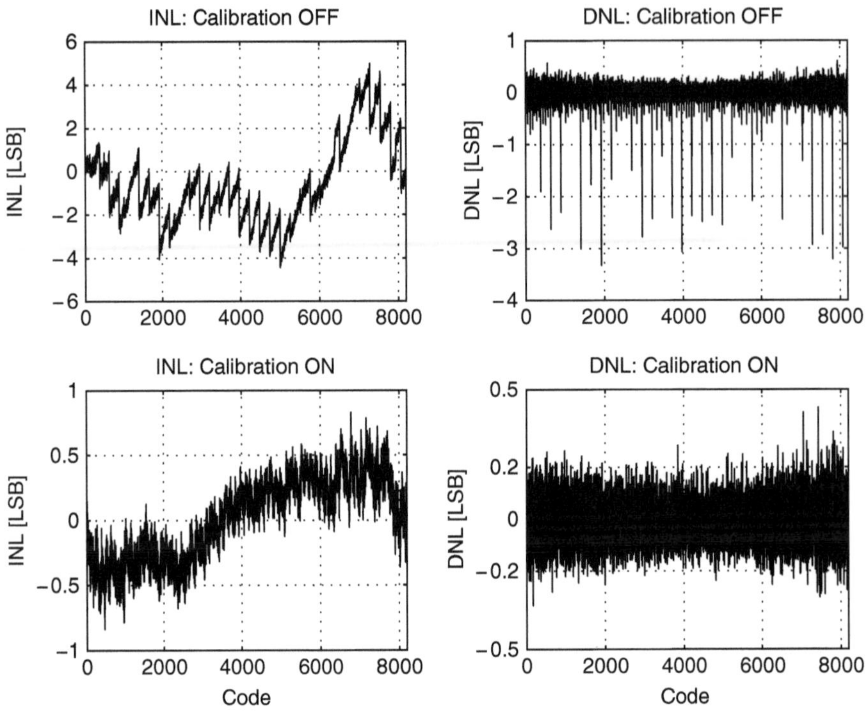

Fig. 6.21 Static linearity with calibration ON/OFF

The measured static nonlinearity of a typical device is shown in Fig. 6.21. The INL and DNL of the uncalibrated converter ranges between $-4.5/5$ and $-3.3/0.6$ LSB, respectively. With the background calibration activated, the maximum INL is ± 0.8 LSB, while the maximum DNL reaches $0.4/-0.3$ LSB.

The MTPR-performance of the converter with a homogeneous multitone-signal sampled at 200 MHz is shown in Fig. 6.22. The signal is located in the frequency band of 500 kHz to 25 MHz and has a crest factor (CF) of 3.5. The average MTPR is slightly below 69.3 dB. A multiband-DMT example is given in Fig. 6.23. The signal has a carrier spacing $\Delta f = 8.625$ kHz and a Crest Factor CF $= 5.6$. The MTPR is measured at the output of a complete analog front-end (AFE), where the D/A-converter described above is integrated together with a third-order active reconstruction filter and an off-chip driver delivering 3 Vpp into a 400 Ω load.[4] The average MTPR/MBPR in this example is 65.2 dB, which makes this AFE suitable for a high-performance VDSL2-modem.

[4]Since the D/A-converter has a full-scale range of 1 Vpp, the reconstruction filter and the off-chip driver must amplify the signal by 3 (10 dB). The limiting nonlinearity for the MTPR in this example is believed to be the off-chip driver.

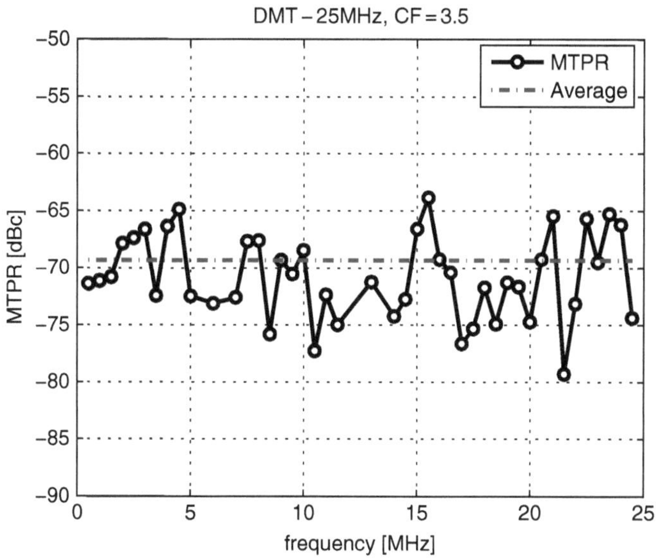

Fig. 6.22 MTPR with homogeneous 25 MHz DMT-signal

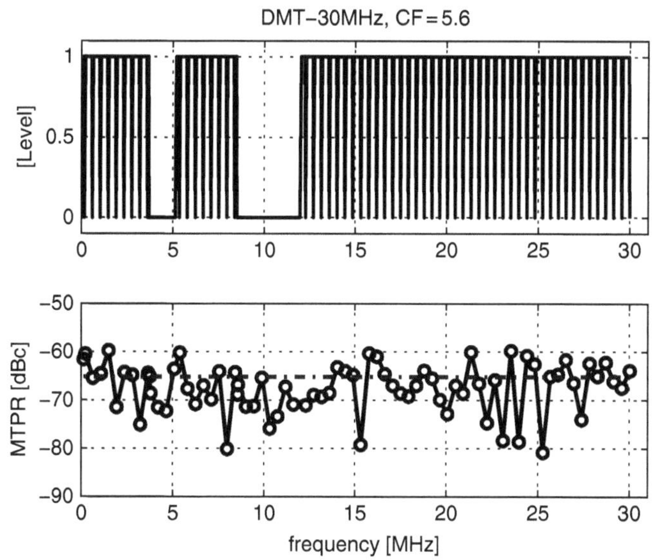

Fig. 6.23 30 MHz DMT-performance of full TX-chain

Noiseshaped Converter Mode

For low-bandwidth high-resolution applications, e.g., ADSL, a noiseshaped unary converter generally has performance advantages, especially in terms of multitone linearity. The main reason is the inherent static and dynamic homogeneity of a

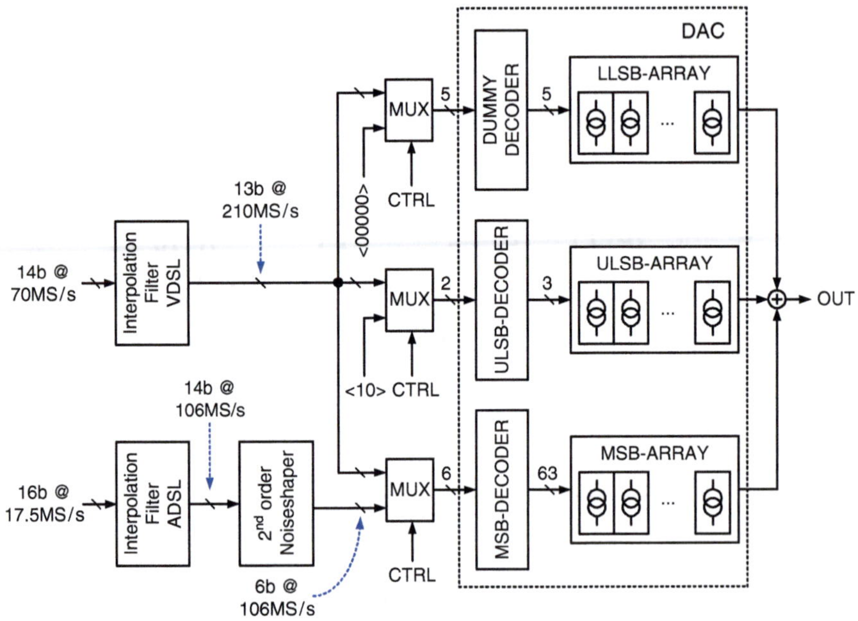

Fig. 6.24 Multimode converter block diagram

unary array as compared to a segmented converter. On the other hand, a multimode transceiver may require to process also signals with much higher bandwidth (e.g., VDSL), albeit in most practical cases with somewhat reduced linearity performance.

In this design, the 6-bit MSB-array of the segmented converter is used as a low-resolution unary DAC, driven by a compatible digital noiseshaper, while the lower segments are switched to a balanced input code to avoid DC-offset. To implement a converter that can process different input data formats, appropriate digital multiplexer circuits are added [174]. A block diagram of the multimode DSL-DAC is shown in Fig. 6.24.

A 1.4 MHz single-tone signal synthesized with the 6-bit MSB-segment, operated at 106 MS/s using a second-order noiseshaper, is shown in Fig. 6.25. With the fundamental at −6 dBFS the third harmonic distortion is found at −86 dBc, indicating that the background calibration scheme achieves a linearity of at least 14 bits.

The measured MTPR for a downstream ADSL-signal with a 1.1 MHz bandwidth and a Crest Factor of CF=5.4 is shown in Fig. 6.26. The average MTPR calculates to 76.7 dB, comparing favorably with the measurement shown in Fig. 5.10.

The MTPR for a downstream ADSL2+ signal with CF=5.2 is shown in Fig. 6.27. In the lower band, the average MTPR is 75.2 dB. Due to the 10 dB smaller carrier amplitude in the higher band, there the average MTPR reduces to 67.6 dB.

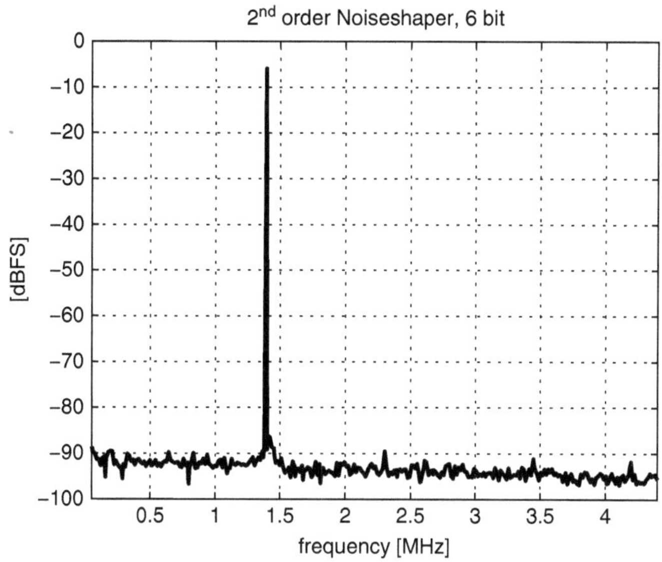

Fig. 6.25 1.4 MHz single-tone with noiseshaped MSB-segment

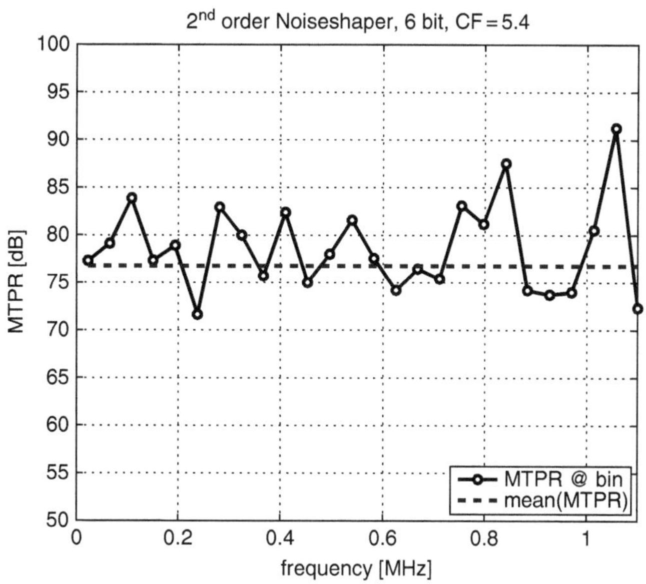

Fig. 6.26 MTPR for ADSL downstream signal

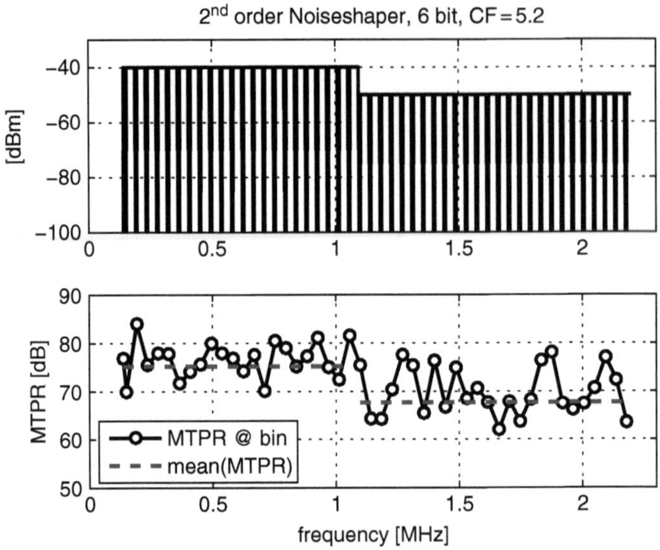

Fig. 6.27 MTPR for ADSL2+ downstream signal

6.2 A 13-Bit 130–300 MS/s DAC with Active Output Stage

As already stated in Sect. 1.4.2, the drawback of a single-polarity current-steering D/A-converter with resistive load is the reduced output voltage range, typically only about 25 % of the supply voltage for a single-ended output. In a fully differential architecture we thus can achieve roughly 50 % of the supply voltage as the full-scale voltage range. Because the load resistors are connected to the rail, the common-mode level at the output of the DAC is typically off-centered at about 12.5 % of the supply voltage.

The input resistors of an active circuit, e.g., a reconstruction filter, connecting to the converter will also load the DAC output. Moreover, since the operational amplifier of the filter operates with a common-mode voltage approximately halfway between the supplies, the connecting resistors will inject DC-current and shift the common-mode voltage at the output of the DAC toward the opposite supply rail. In order to maintain high linearity, we must either ensure that these "tapping" resistors are much larger than the load resistors of the converter or the full-scale range of the DAC must be reduced accordingly. Also, an operational amplifier circuit connected to the D/A-converter output must be able to handle the common-mode level that develops at its virtual ground nodes.[5]

[5] In case the DC-gain of this circuit block varies, the common-mode level also changes with the gain.

With an active transimpedance output stage, on the other hand, the converter can interface naturally with other active circuit blocks, because the output voltage is buffered by the operational amplifier and centered at the correct common-mode level. Moreover, the full-scale range can be maximized to fit the linear range of the other op-amp-based analog circuits in the analog front-end, typically twice as large as for a single-polarity converter operating from the same supply voltage.

In this section we present a dual-polarity current-steering DAC with push-pull output stage that is able to operate with high linearity up to 50 MHz signal bandwidth [93]. The converter can be clocked up to 300 MHz, albeit at this sampling rate only with an on-chip interpolation filter having an upsampling factor of 3, again equivalent to a 50 MHz signal bandwidth. The segmented calibration principle with segment boundary trimming, described in the previous section, is replaced partly by a direct segment calibration architecture. Using a variable and self-calibrating reference cell, we can directly trim the DAC-elements in different segments having different weights.

Again, calibration spurs are attacked by randomizing the calibration time slot, thus eliminating the short-term periodicity of the calibration process. It will be shown that randomization not only eliminates the calibration refresh tones but also suppresses the image tone resulting from the residual dynamic calibration mismatch in the time-interleaved current-cell architecture.

6.2.1 Converter Architecture

The converter architecture is shown in Fig. 6.28. The 13-bit DAC-core consists of trimmed dual-polarity current cells with a segmentation of 6+2+5 bits. This choice is again driven by practical considerations. The uppermost (MSB) and intermediate (ULSB) segment are unary arrays, while the lower segment (LLSB) is a binary array. A full-clock interleaved current-cell architecture is used in the MSB- and ULSB-segment that allows to implement an effective RZ-behavior with NRZ output current waveform.

The current cells are continuously calibrated in the background, necessitating the inclusion of redundant DAC-elements that take over the data processing when needed. The calibration process together with the correct data multiplexing is controlled by a dedicated calibration control logic.

Dynamic current copying is employed to trim the dual-polarity current cells. Accordingly, two independent analog control loops set the P- and N-current in the DAC-cells equal to the current of the dual-polarity reference cell.

The output current of the DAC-core is summed at the virtual ground nodes of the operational amplifier and converted into a differential output voltage by the feedback impedance. The amplifier has a push-pull output stage that can drive 1.5 Vpp into a 1 kΩ on-chip load. Since the nominal feedback resistance is 400 Ω, the differential full-scale output current is set to ± 1.875 mA.

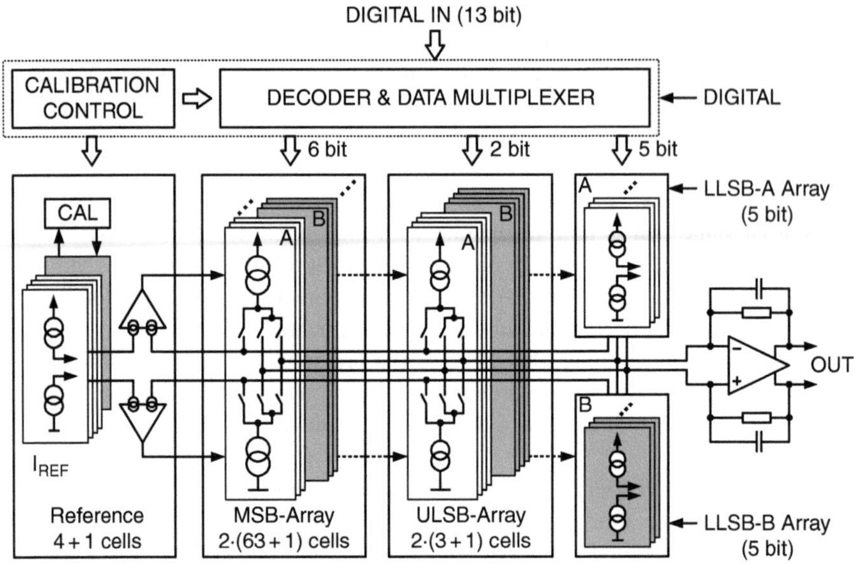

Fig. 6.28 Self-calibrated DAC architecture

6.2.2 The Current Cells

The calibrated dual-polarity MSB current cell is shown in Fig. 6.29. Trimming of the cascoded current sources is done with the calibration transistors P_{TRIM} and N_{TRIM} with a ratio of 1:15, resulting in a theoretical trimming range of $\pm 6.25\,\%$. Again, we restrict the "usable" trimming range to approximately $\pm 3\,\%$.

MOS-capacitors M_{QP} and M_{QN} act as additional storage elements for the gate-source voltage established by the calibration loop and also reduce the effect of error charge injected when the MOS-switches $S_{0,P}$ and $S_{0,N}$ are turned off at the end of the calibration slot. A first-order cancellation of the charge error is provided by the equal-sized dummy switches $S_{1,P}$ and $S_{1,N}$ that are switched in the opposite direction.

In normal operation the current of the cell is switched either to the output nodes by the control signals A and B, or to the common-node VCM by signals C and D. For trimming, the N-side and P-side currents are steered to the calibration outputs $I_{\text{CAL}P}$ and $I_{\text{CAL}N}$, while the control loop is closed by switches $S_{0,P}$ and $S_{0,N}$. Bypass capacitors C_{CASC} are connected from the gate node of the cascode transistors to the supply to absorb voltage spikes generated by the current switching process and to prevent a data correlated disturbance of the cascode bias line.

In the unit current cell of the ULSB-array the widths of all transistors are scaled down by a factor 4 with respect to the MSB-cell. Only the current switches and the

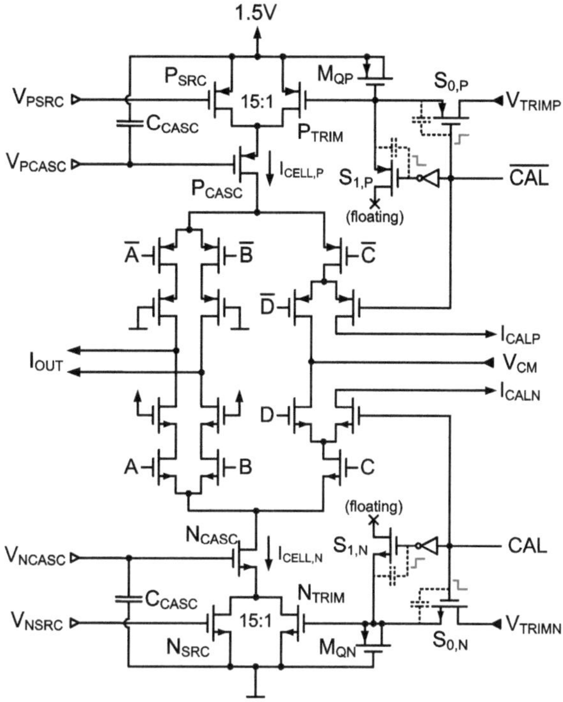

Fig. 6.29 Calibrated dual-polarity current cell

transistors that are part of the calibration loop are not changed in size.[6] The (linear) trimming range for the ULSB-cell is therefore increased to $\pm 12.5\,\%$. Although such a large trimming range is not necessary, the advantage of this sizing is that the dynamics of the calibration loop remain unchanged.

To eliminate dynamic errors, each DAC-element in the unary MSB- and ULSB-array consists of two independent current cells (A and B) that are operated in full-clock interleaved mode; see Fig. 6.30. The cells alternately deliver the output current for a full clock cycle and are reset for the following clock cycle. During reset the current is dumped to the common-node *VCM* that replicates the voltage of the current-summing nodes (= virtual grounds) of the output stage. Each of the cells therefore exhibits a full-clock RZ behavior, while the effective output current of the DAC-elements is never interrupted, i.e., leading to an effective NRZ behavior. The improved buffer-latch circuit with symmetrical rise/fall-time of Fig. 6.5 is used for resynchronization of the input data. The switch drivers are sized to provide an optimum crossing point for the switch control signals.

[6]This applies to the calibration transistors P_{TRIM} and N_{TRIM}, the storage capacitors M_{QP} and M_{QN}, and the calibration switches and dummy switches $S_{0,P}$, $S_{0,N}$, $S_{1,P}$ and $S_{1,N}$.

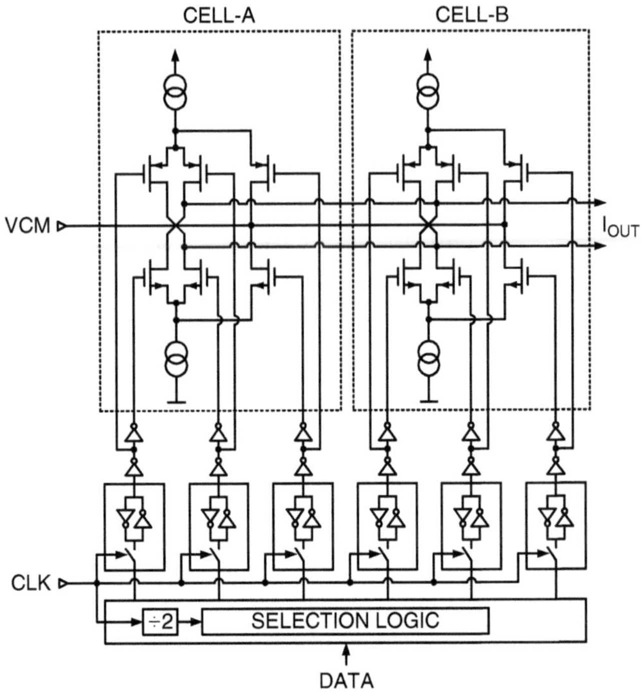

Fig. 6.30 Interleaved current cell (calibration branch not shown)

The 5 + 1 LLSB-cells are binary scaled and not interleaved. Since the trimming of the LLSB-array is performed in the corresponding current mirror, no calibration transistor is necessary in the binary-scaled current cells. In the LLSB-cells (and also in the ULSB-cells) dummy switches are added to the switch control lines to provide identical capacitive loading. This allows to use a single optimized design for the buffer-latch and switch-driver circuits.

6.2.3 *Direct Segment Calibration*

In the converter of Sect. 6.1 only the DAC-elements of the MSB-array are trimmed directly by comparing their current with a reference current of the same nominal value. The calibration control then adjusts the gate-source voltage of the calibration transistor in the MSB-cell to make the two currents equal. The lower segments are appropriately grouped together, and the sum of their currents is trimmed to be equal to the reference current. In this way the segment boundaries are successively matched to each other, limited only by the mismatch of the extra cells that are needed

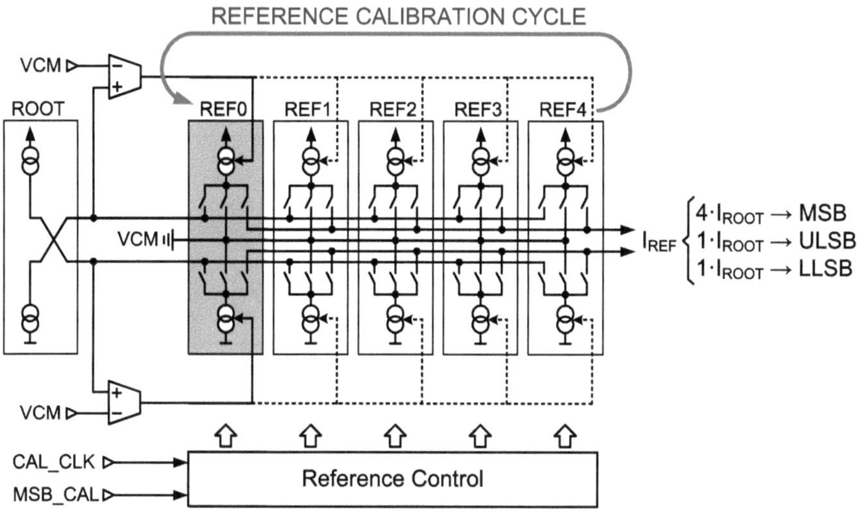

Fig. 6.31 Self-calibrating segmented reference cell

to achieve the nominal reference current value and by the steady-state error of the fully settled control loop (assuming $T_{CAL} >> \tau_{CAL}$).

In this design the variable self-calibrated reference cell shown in Fig. 6.31 is used. It consists of five (dual-polarity) unit reference currents that are continuously calibrated in the background using the dynamic current copying principle. To this aim, the so-called root cell provides two currents with opposite polarity and equal size. The root cell sets the current in the single unit reference elements equal to the nominal current of an ULSB-cell. In each calibration slot one of the five reference cells is compared with the root cell and adjusted dynamically. The four remaining (calibrated) reference cells can be used to trim the current sources in the main DAC-array.

All four available reference cells are combined to trim the MSB-elements; see Fig. 6.32. For the ULSB-cells only one reference cell is switched to the calibration current output I_{REF}, while the current of the unused reference cells is dumped to the common-node VCM. Given the reduced weight of the LLSB-array, a segment boundary match is performed by trimming the bias current mirror of the LLSB-array. The sum of the LLSB-array (plus one additional 1-LSB dummy cell) is made equal to one reference element. This matches the LLSB-segment directly to the single ULSB-cell without the need to include previously trimmed ULSB-elements in the comparison. The LLSB-array together with the biasing during calibration is shown in Fig. 6.33.

A small digital state machine controls the calibration of the reference cells and the amount of reference current switched to the calibration output I_{REF}. It requires as input the calibration clock *CAL_CLK*, signaling the beginning of a new calibration

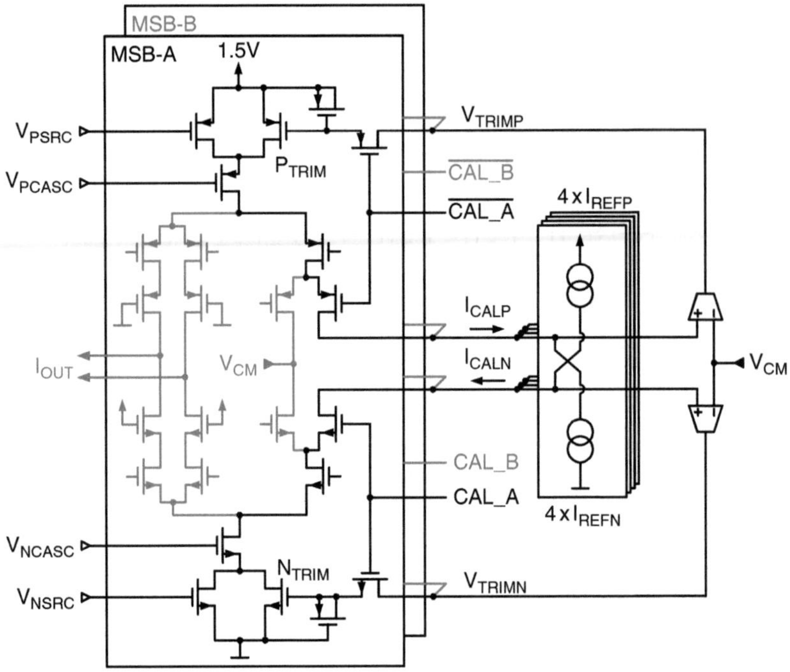

Fig. 6.32 MSB-cell during calibration

slot, and the signal *MSB_CAL* that indicates whether a MSB-cell must be calibrated in the main DAC-array.

Figure 6.34 shows the complete calibration cycle. In the first 64 slots the MSB-cells are trimmed by comparing their current with the sum of 4 reference cells. In the next four calibration slots the ULSB-cells are made equal to one reference cell, while the remaining three available reference cells are idle. These unused (idle) reference currents are dumped to the common-node VCM. In the last slot the LLSB-array is boundary matched to the ULSB-segment, again by comparison with a single reference cell. The unit elements in the MSB- and ULSB-segment each consist of two interleaved current cells. These two cells are calibrated alternately in every second calibration cycle. Since also the LLSB-segment consists of two identical arrays that are calibrated in turn, the complete refresh cycle is 138 slots long.

Randomization of the calibration slot length is applied to suppress the effect of regular patterns generated by the calibration process. This also applies to the somewhat more subtle effect of the reference repetition cycle, i.e., the period after which a given cell in the DAC-array is calibrated again with the same set of reference cells, without randomization also a possible source of tone generation. The complete refresh cycle of the DAC-array consists of 138 slots, while the reference calibration cycle comprises five slots. Since these two numbers are relatively prime,

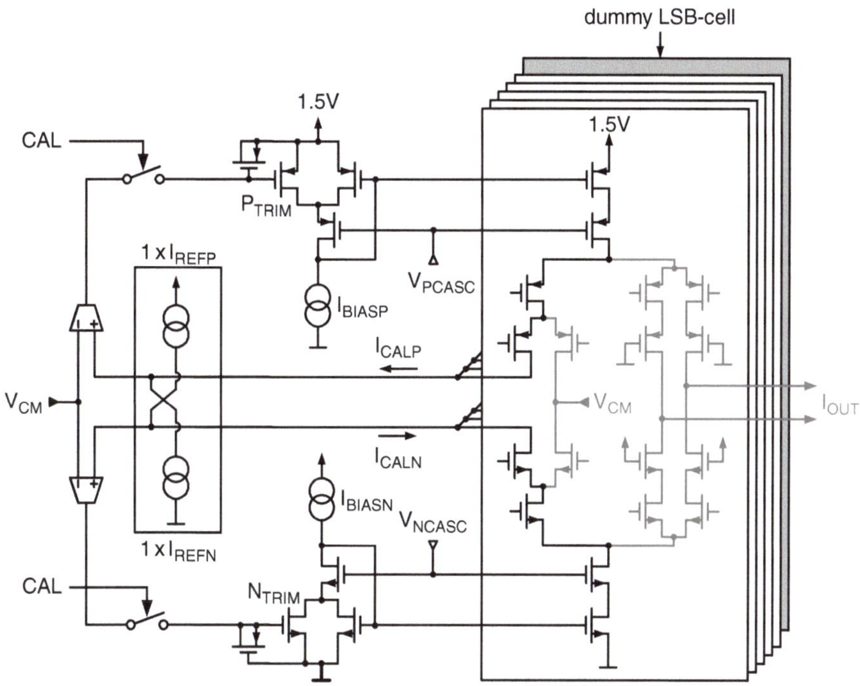

Fig. 6.33 LLSB-array with calibrated bias mirror

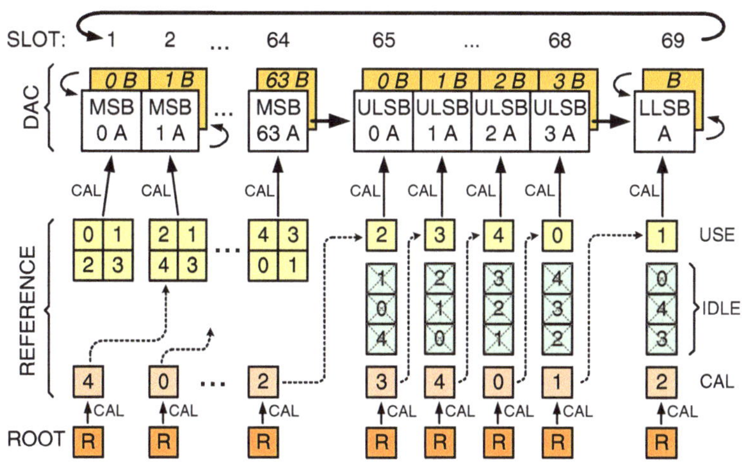

Fig. 6.34 Complete calibration cycle

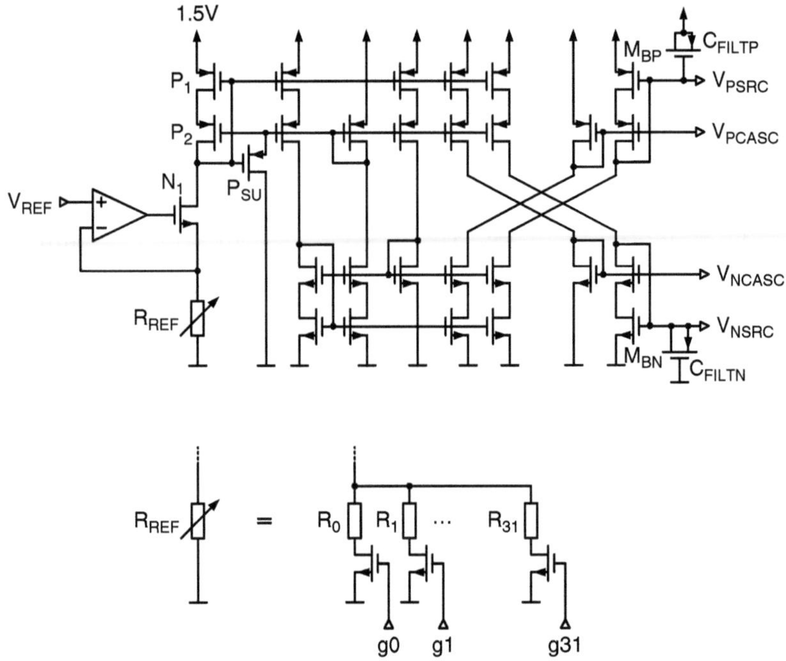

Fig. 6.35 Programmable converter biasing

the reference "repeats itself" only after $138 \times 5 = 690$ slots. Assuming a minimum calibration slot length of $0.5\,\mu$s, without randomization the reference repetition pattern would appear as a spur below 3 kHz, in most applications far outside of the frequency range of interest.[7]

6.2.4 Programmable Biasing

The DAC-biasing is shown in Fig. 6.35. The operational amplifier controls the gate of transistor N_1 to keep its source node at the voltage V_{REF}. The drain current of N_1, given by $V_{\mathrm{REF}}/R_{\mathrm{REF}}$, is then mirrored to the bias transistors M_{BP} and M_{BN}. The gate nodes of the current source transistors ($V_{P\mathrm{SRC}}$ and $V_{N\mathrm{SRC}}$) are filtered with MOS-capacitors $C_{\mathrm{FILT}P}$ and $C_{\mathrm{FILT}N}$. Additional currents are generated to bias the cascode nodes ($V_{P\mathrm{CASC}}$ and $V_{N\mathrm{CASC}}$).

The PMOS-transistor P_{SU} is included to reliably start-up the bias circuit at power-on. If the gate of the cascode transistor P_2 remains clamped to the supply

[7]Obviously from the viewpoint of most modern communication systems. Voice and audio codes of course operate in this frequency range.

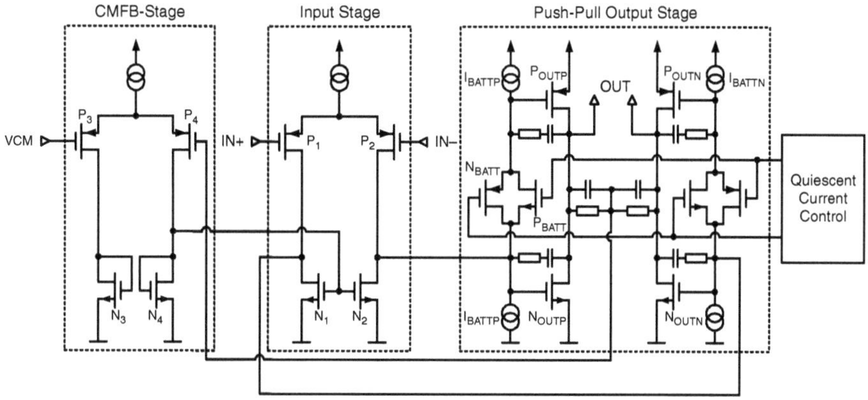

Fig. 6.36 Operational amplifier of transimpedance stage

rail, e.g., due to leakage currents, then no current can flow into the PMOS-diode P_1. In this case the drain current of N_1 starts discharging the gate node of P_{SU} toward ground. Eventually, P_{SU} opens and pulls the gate of P_2 downwards, allowing I_{REF} to flow into the PMOS-diode. Once current equilibrium in the mirror system is established, the voltage at the gate of P_2 will be lower than at its drain, reliably turning off P_{SU}. The start-up transistor P_{SU} is almost minimum size and does not load the current mirror nodes.

The programmable resistor R_{REF} allows to trim the full-scale value of the converter [175]. In this design R_{REF} consists of 32 different resistors that can be selected digitally. This resistor array implements a 20 dB range of the full-scale current in 32 linear steps.

6.2.5 *Push-Pull Operational Amplifier*

The operational amplifier employed in the transimpedance output stage is shown in Fig. 6.36. It is a fully differential two-stage design with PMOS input stage P_1, P_2, active NMOS-load N_1, N_2, and push-pull output stage $N_{OUTP/N}$, $P_{OUTP/N}$. The common-mode output voltage is sensed with a compensated resistive divider and fed back to the common-mode differential amplifier P_3, P_4. The NMOS-diode N_4 controls the gates of the input stage load transistors to set the output common-mode voltage equal to *VCM*. Miller-compensation with zero-nulling is applied as frequency compensation around the output stage transistors.

The class-AB type biasing of the output stage is performed with the floating battery N_{BATT}, P_{BATT}, which are biased with fixed current sources $I_{BATTP/N}$. For small signals the floating battery remains transparent and the gate of $P_{OUTP/N}$ follows the gate of $N_{OUTP/N}$, thus doubling the effective transconductance of the output stage for the same current. This almost doubles the power efficiency of the second stage. Only under heavy drive conditions, while one half of the output stage is driven increasingly hard, the output transistor sitting at the opposite supply rail is pushed toward a constant bias point and delivers a reduced drain current [176]. This region of operation is however never reached in this design, because the ratio of maximum load current to quiescent current is set to approximately 1:6.

The standard translinear-loop biasing of the floating battery with a MOS-diode stack [176, 177] has the drawback that the sensitivity of the quiescent current toward process, supply voltage, and temperature variations[8] is rather large. This problem becomes especially cumbersome in newer CMOS technologies when high-bandwidth requirements must be fulfilled since in this case the output stage and floating battery transistors have quite small channel lengths and cannot be matched very well with the stacked diodes of the translinear biasing branch, mostly due to the considerably different drain-source voltages.

Figure 6.37 shows the implemented quiescent current control (QCC) that tries to improve the poor PVT-sensitivity of the classical push-pull type output stage with translinear biasing. A negative feedback loop forces the drain-source voltage of the replica transistor N_{REP}, biased with the reference current I_{REP}, to be equal to the output common-mode voltage. To that aim, the voltage at the drain of N_{REP} (node R) is compared with the reference voltage VCM, and the bias current in the PMOS- and NMOS-diode stack, P_{D1}, P_{D2} and N_{D1}, N_{D2}, is controlled accordingly. Transistor P_3 helps to further increase the output impedance of the replica current source I_{REP}. The negative feedback loop is symmetrically frequency compensated with capacitors C_C. Since N_{REP} is a scaled version of N_{OUT} and has the same drain-source voltage VCM set by the control loop, the quiescent current in the output stage is nominally given by the replica current I_{REP}, multiplied with the ratio of the width of N_{OUT} and N_{REP}:

$$I_Q = I_{\mathrm{REP}} \cdot \frac{W_{N\mathrm{OUT}}}{W_{N\mathrm{REP}}}. \tag{6.13}$$

6.2.6 *Output Stage Optimization*

The operational amplifier described in the previous section must be optimized for large GBW and large slew rate. From simulation, the following parameters can be retrieved:

- DC-gain $A_0 = 46\,\mathrm{dB}$
- Unity gain-bandwidth GBW $= 2\,\mathrm{GHz}$
- Slew rate SR $= 2400\,\mathrm{V}/\mu\mathrm{s}$

[8] Also called PVT-sensitivity.

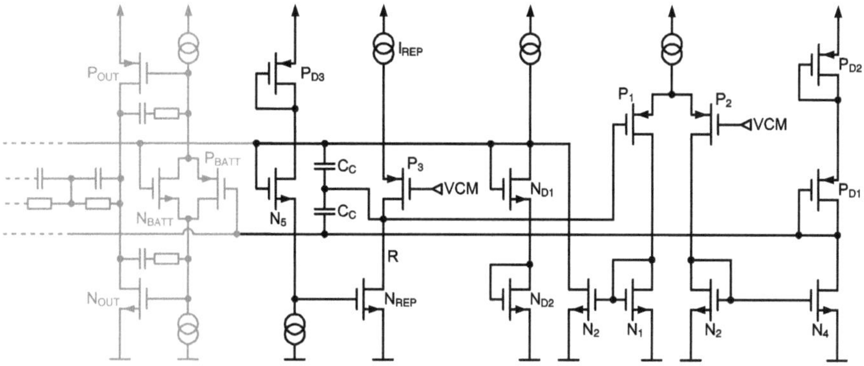

Fig. 6.37 Output stage quiescent current control

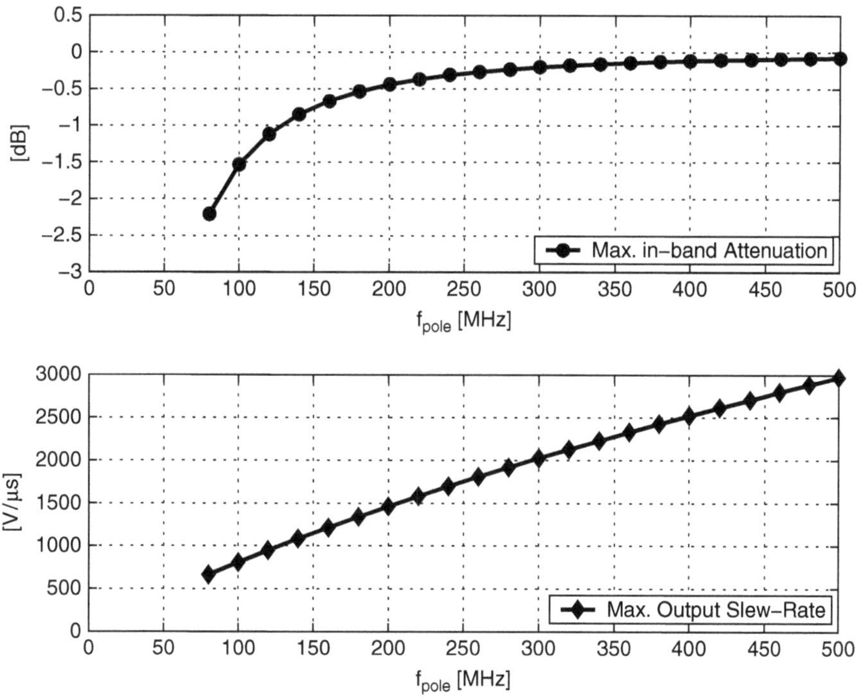

Fig. 6.38 In-band attenuation and output voltage slew rate

For Nyquist operation the maximum input code step is equal to the full-scale range. Figure 6.38 shows the calculated in-band attenuation occurring at the Nyquist frequency (65 MHz) and the worst-case output voltage slew rate as a function of the LHP-pole frequency implemented by the feedback impedance (f_{pole}).

The operational amplifier should not limit the slewing of the output voltage pulse, since this introduces distortion due to the code-step-dependent settling. To allow

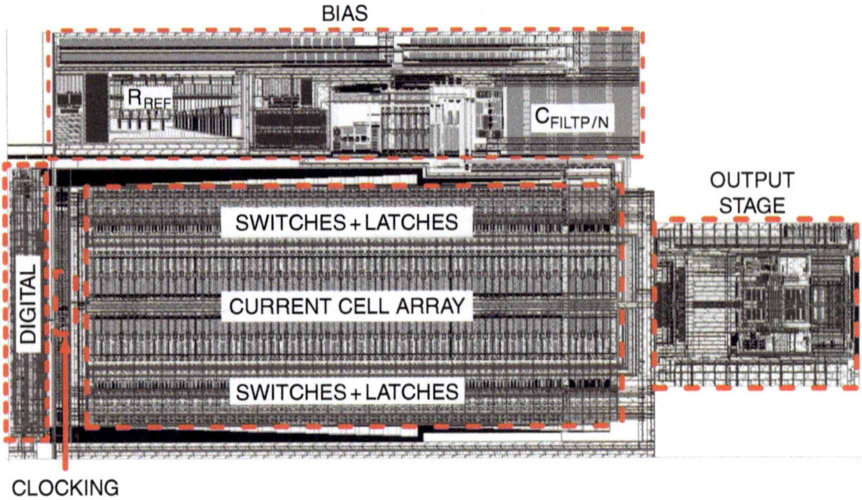

Fig. 6.39 Converter layout

for enough design margin, we choose $f_{\text{pole}} = 100\,\text{MHz}$, resulting in a maximum output voltage slew rate according to the linear model (Eq. (5.8)) of $800\,\text{V}/\mu\text{s}$. As a compromise, the maximum excess in-band attenuation amounts to 1 dB at 50 MHz and 1.5 dB at 65 MHz. Note, that in a NRZ-converter the sinc function introduces almost 4 dB of additional attenuation at the Nyquist frequency [35].

6.2.7 Layout

Figure 6.39 shows the layout of the converter core. The current matrix is arranged in two rows with the current sources in the center. The resynchronization latches and current switches are placed to both sides of the current source array. The sampling clock is buffered by a central clock driver and distributed symmetrically to the latches in the current cells.

The prototype converter is fabricated in a 1P6M 0.13 μm CMOS process using thin gate-oxide (= core) devices only. MIM capacitors are available and can be placed above active area on the two topmost metal layers. This option is used for the cascode biasing bypass capacitors C_{CASC} (see Fig. 6.29), placed on top of the current cells, in the output stage for the Miller-compensation of the operational amplifier, as well as for the transimpedance feedback capacitors. The complete, not area optimized,[9] converter module measures $0.9\,\text{mm}^2$.

[9] For example, $0.03\,\text{mm}^2$ in the biasing are due to the very accurate full-scale current programming in the range $0 \ldots -20\,\text{dB}$ in 32 steps. This feature may not be needed in many applications.

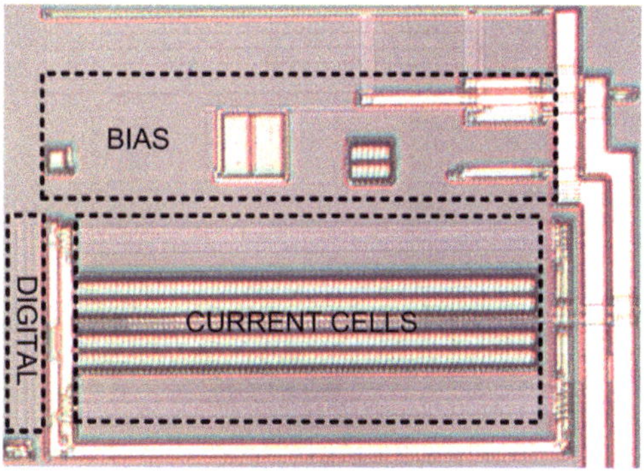

Fig. 6.40 Die photograph of converter core

Table 6.3 Area contribution of the major building blocks

Building block	Area (mm^2)	Percentage (%)
Current-cell array	0.43	44.8
Biasing + calibration	0.24	25
Digital calibration control, decoders	0.035	3.6
Supply, data, and clock routing	0.15	15.6
Output stage	0.105	11
Overall converter	0.96	100

Figure 6.40 shows a die photograph of the converter core. It corresponds to the layout shown in Fig. 6.39, except that it does not include the output stage located farther to the right. Due to the mandatory metal filling in modern CMOS processes, only the uppermost layers of the metallization structures are clearly discernible in such photographs.

Table 6.3 summarizes the area contributions of the major building blocks. The current-cell array occupies almost half of the total silicon area, one quarter is used by the biasing, including the analog calibration circuitry. The digital calibration control together with the data decoders and multiplexers consumes less than 5 %. Interestingly, the top-level voltage supply routing plus the data and clock wiring ($\approx$ 16 %) is larger than the entire output stage (11 %).

6.2.8 *Experimental Results*

The prototype converter is tested with a sampling rate of 130 MS/s in Nyquist-mode, and with 300 MS/s using an on-chip interpolation filter with an upsampling factor of 3. Although in the latter case the signal bandwidth is limited to 50 MHz, the spectral images due to the sampling process only appear above 250 MHz. This greatly relaxes the reconstruction filter order but at the cost of higher power drain and slightly reduced linearity performance of the converter. Thus, the optimum transmitter architecture cannot be determined straightforwardly but also depends on a variety of system considerations.

Table 6.4 Power drawn from 1.5 V supply

Building block	CLK=130 MHz (mW)	CLK=300 MHz (mW)
Converter core—static	8.7	8.7
Converter core—dynamic	13.8	31.8
Converter core—total	22.5	40.5
Output stage	29	29
Digital	1.5	3.5
Total analog + digital	53	73

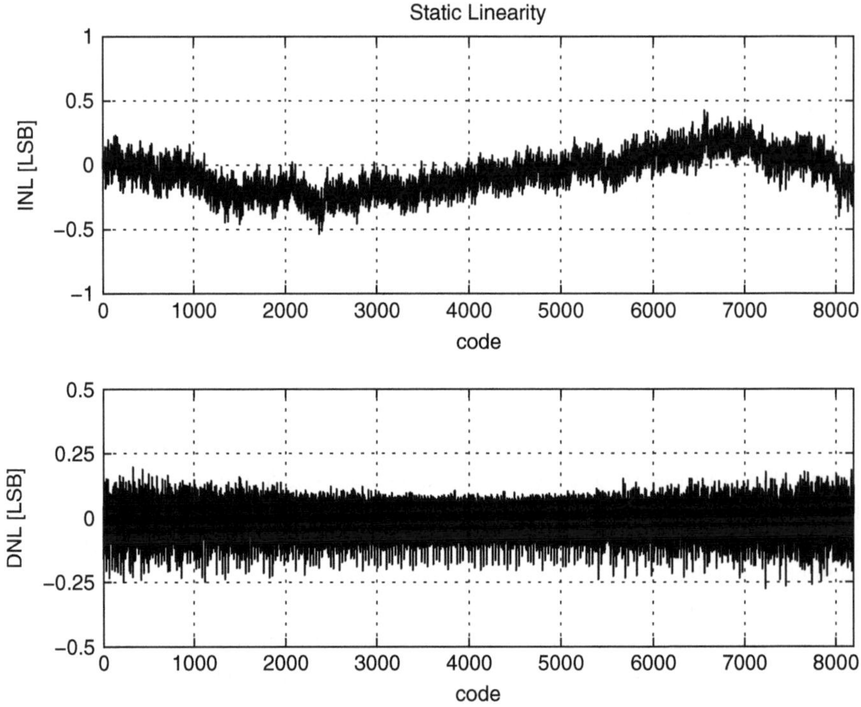

Fig. 6.41 Calibrated static linearity

Table 6.4 reports the measured power dissipation of the converter for the two operating modes when synthesizing a 1 MHz full-scale sine wave with the interleaved RZ-mode activated. At 130 MS/s the total power drawn from the 1.5 V supply is 53 mW. When the clock rate is increased to 300 MHz, the converter dissipates 73 mW. In both modes, the active output stage consumes 30 mW, while the static power dissipation of the converter core including the bias circuit amounts to 8.7 mW. The rest is dominated by the dynamic switching of the current cells with a slope of approximately 0.106 mW/MHz. At chip level a split analog/digital 1.5 V voltage supply pair is used for the converter.

Figure 6.41 shows the static linearity of the converter with the background calibration activated. The maximum INL and DNL are −0.53/0.42 LSB and −0.27/0.19 LSB, respectively.

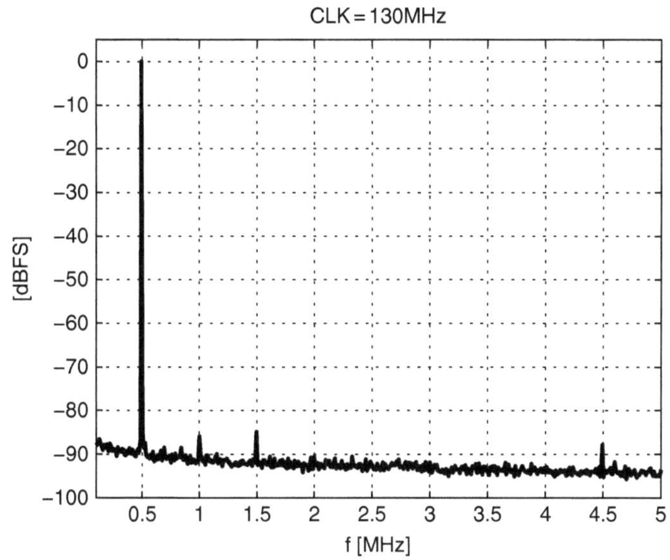

Fig. 6.42 500 kHz single tone at 130 MS/s

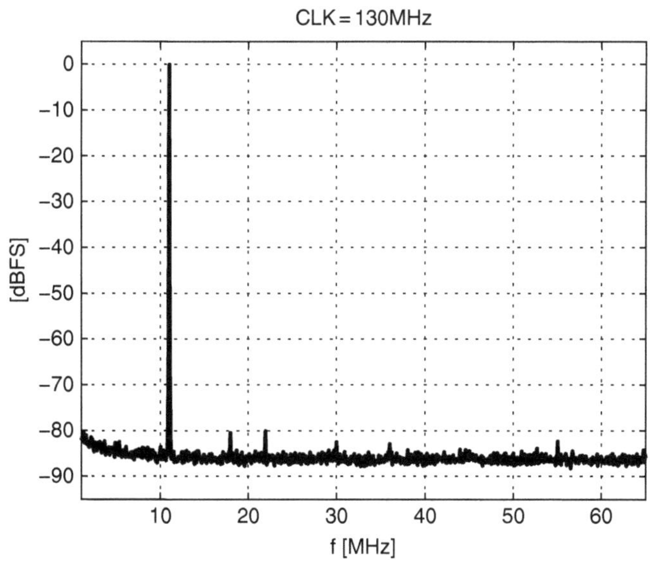

Fig. 6.43 11 MHz single tone at 130 MS/s

Figures 6.42–6.44 show measured output spectra for a single-tone full-scale signal at 500 kHz, 11 MHz, and 64 MHz, respectively, synthesized with a sampling rate of 130 MS/s. Even close to the Nyquist frequency the linearity stays above 73 dB. The optimized feedback network of the transimpedance stage (see Sect. 6.2.6) consists of a 4 pF capacitor in parallel to the 400 Ω resistor. The resulting

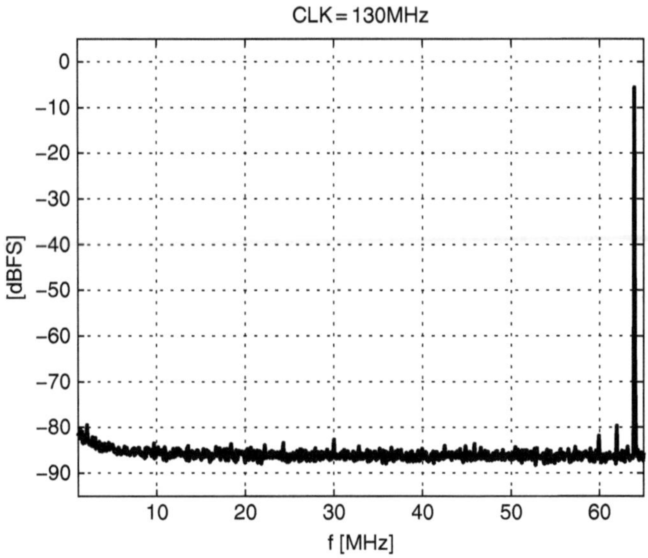

Fig. 6.44 64 MHz single tone at 130 MS/s

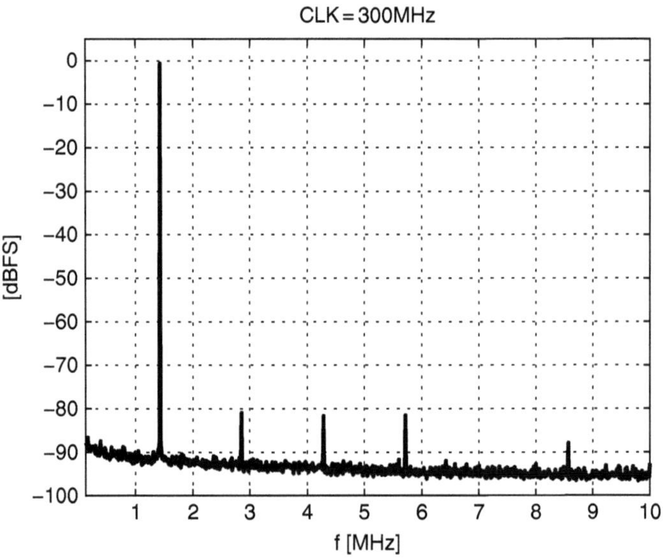

Fig. 6.45 1.4 MHz single tone at 300 MS/s

LHP real pole of the output stage is placed at 100 MHz. It contributes an additional attenuation of 1.5 dB at the Nyquist frequency. Together with the attenuation of the sinc function of 3.9 dB a total maximum in-band attenuation of 5.4 dB results. All dynamic measurements are done with a full-scale output voltage of 1.5 Vpp.

Figure 6.45 shows the measured output spectrum of a 1.4 MHz full-scale signal sampled at 300 MHz. A preemphasis filter is included into the digital interpolation

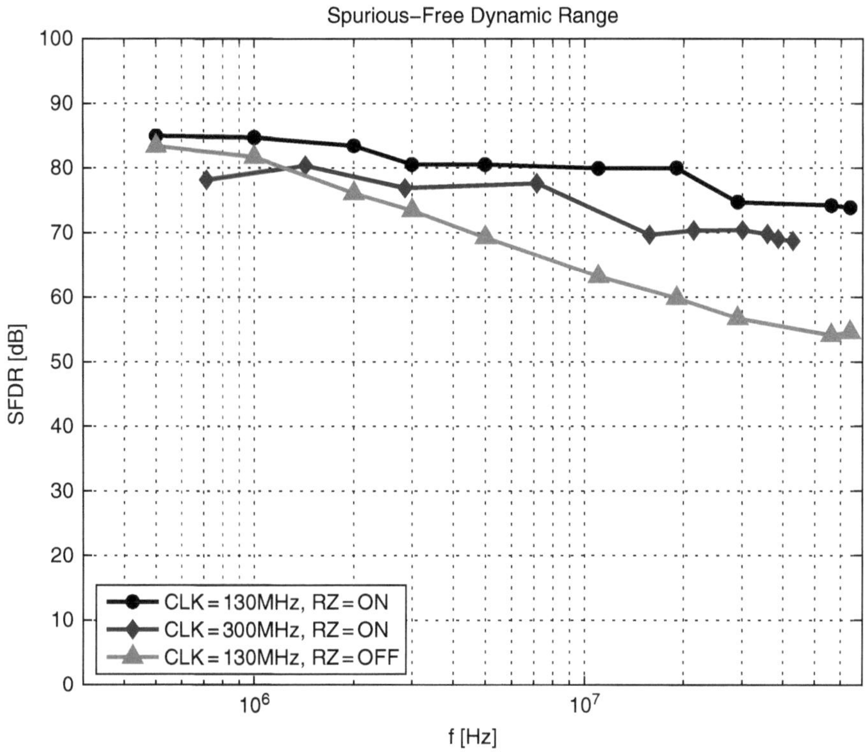

Fig. 6.46 SFDR at 130 and 300 MS/s

such that the overall amplitude response remains flat over the whole signal frequency range.

The SFDR-performance as a function of the signal frequency for a sampling rate of 130 and 300 MS/s is shown in Fig. 6.46. At 130 MS/s, with interleaving enabled (RZ=ON), the SFDR reaches 85 dB at low signal frequencies. It then remains above 80 dB for signal frequencies up to nearly 20MHz and drops to 73.8 dB close to the Nyquist frequency. At 300 MS/s, we notice a degradation of a few dB for low signal frequencies. Above 10 MHz the SFDR remains slightly above 70 dB and finally drops to 68.7 dB close to the upper signal frequency band edge of 50 MHz. With the interleaving turned off (RZ=OFF), i.e., using only one half of the dual current cells in the MSB- and ULSB-segments in NRZ-mode, the SFDR at 130 MS/s still approaches 85 dB for low signal frequencies. Close to Nyquist, instead, the SFDR drops to 54 dB. The effective RZ realized with interleaved current cells thus improves the dynamic linearity of the converter by about 20 dB close to the Nyquist frequency.

Two-tone output spectra at 130 and 300 MS/s are shown in Figs. 6.47 and 6.48. The third-order intermodulation distortion for a closely spaced two-tone input signal as a function of the center frequency is shown in Fig. 6.49. Again, at

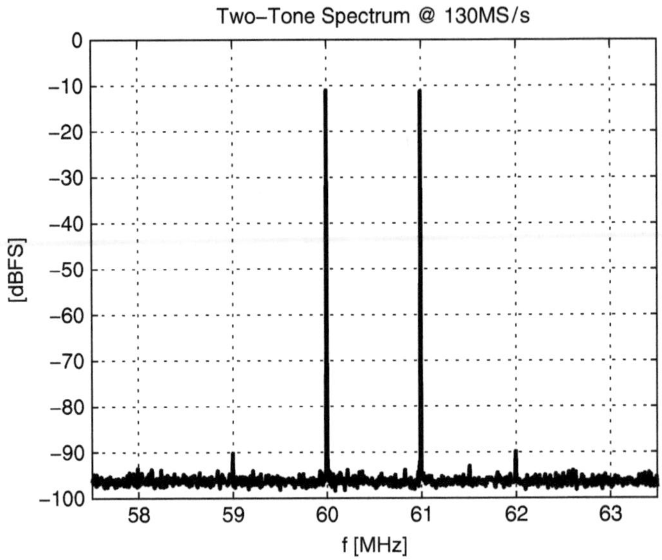

Fig. 6.47 60 + 61 MHz two-tone at 130 MS/s

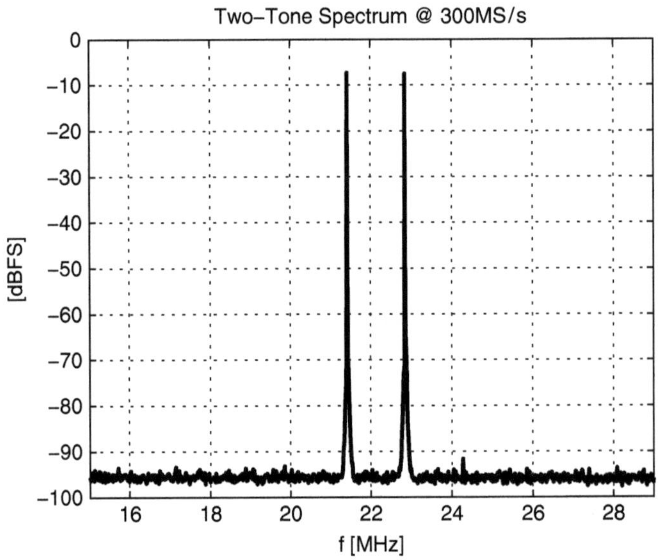

Fig. 6.48 21 + 23 MHz two-tone at 300 MS/s

300 MS/s we notice a slight degradation of the linearity, but in both cases with interleaving enabled (RZ=ON), the IM3-products stay above 75 dBc over the entire signal frequency band. With interleaving disabled (RZ=OFF), the third-order intermodulation at 130 MS/s steadily drops to 56 dBc around 60 MHz.

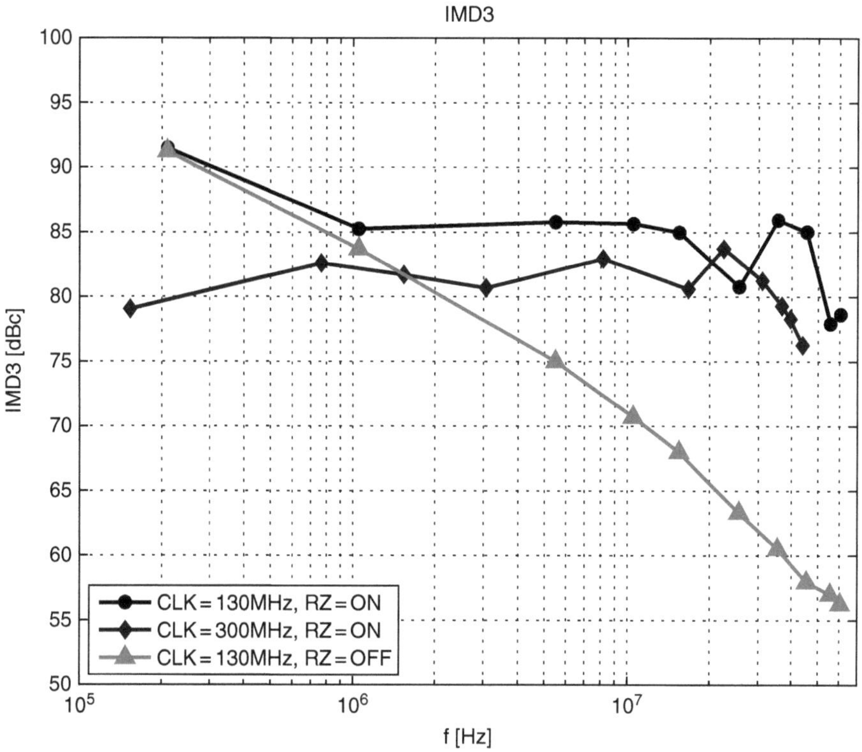

Fig. 6.49 Third-order intermodulation at 130 and 300 MS/s

Figure 6.50 shows a four-band DMT-signal synthesized at 300 MS/s. It has a bandwidth of 43 MHz, a carrier spacing of 50 kHz, and a Crest Factor of 5.4. Five different symbols with the same Crest Factor have been evaluated and the resulting MTPR/MBPR averaged. The worst-case MTPR is better than 62 dB and the MBPR better than 65 dB. Also shown is a seventh-order polynomial fit for the multitone nonlinearity, which can be used for system modeling. In such a model, the MTPR and MBPR would remain above 66 dB over the whole signal band.

Of interest in a 2× time-interleaved converter architecture is the image tone at $\frac{1}{2} f_{\mathrm{CLK}} - f_{\mathrm{SIG}}$, generated by the aggregate gain and phase mismatch between the interleaved current cells. Figure 6.51 shows the measured magnitude of the image tone for a full-scale sine wave synthesized at 130 MS/s. Without calibration the in-band image appears between $-65 \cdots -55$ dBc. Applying background calibration without randomization pushes the image tone down by almost 20 dB. Finally, with randomized calibration time slot the image tone is further suppressed and remains below -100 dBc, very close to the level of the non-interleaved converter (RZ = OFF). In both cases the image tone is still visible above the noise floor, indicating a finite amount of digital crosstalk. With randomization the residual gain mismatch stemming from dynamic effects of the calibration process is no longer constant over

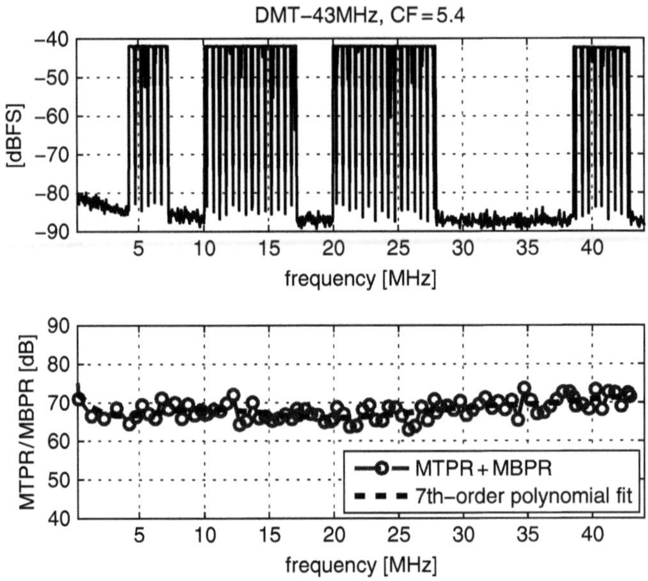

Fig. 6.50 Multiband-DMT with CF=5.4 at 300 MS/s

time, and the energy residing in the image tone is spectrally expanded into noise, like the calibration refresh tones [93].

The above measurement results demonstrate an adequate performance level for a next-generation digital communication system with a signal bandwidth up to 65 MHz, while using a dual-polarity current-steering D/A-converter with active output stage and maximized output swing. This advancement is mostly due to the full-clock interleaved current-cell architecture, realizing an effective return-to-zero (RZ) with NRZ output waveform. Besides, this performance level also requires a high-linearity operational amplifier (using a push-pull output stage for high efficiency) with optimized feedback network. The correct choice of the feedback capacitor relaxes the slew-rate specification of the amplifier considerably but still contributes only a small additional in-band attenuation of 1dB for an analog bandwidth of 50 MHz. However, in an oversampled application, the low-pass frequency response of the output stage can easily be equalized by including a digital preemphasis filter. Alternatively, the LHP real pole of the converter output stage can be included into the overall transfer function of the reconstruction filter connected to the DAC.

Time-domain randomization of the calibration slot not only suppresses the low-frequency calibration refresh tones but is also effective in suppressing the image tone resulting from the residual dynamic mismatch of the time-interleaved current-cell architecture after calibration.

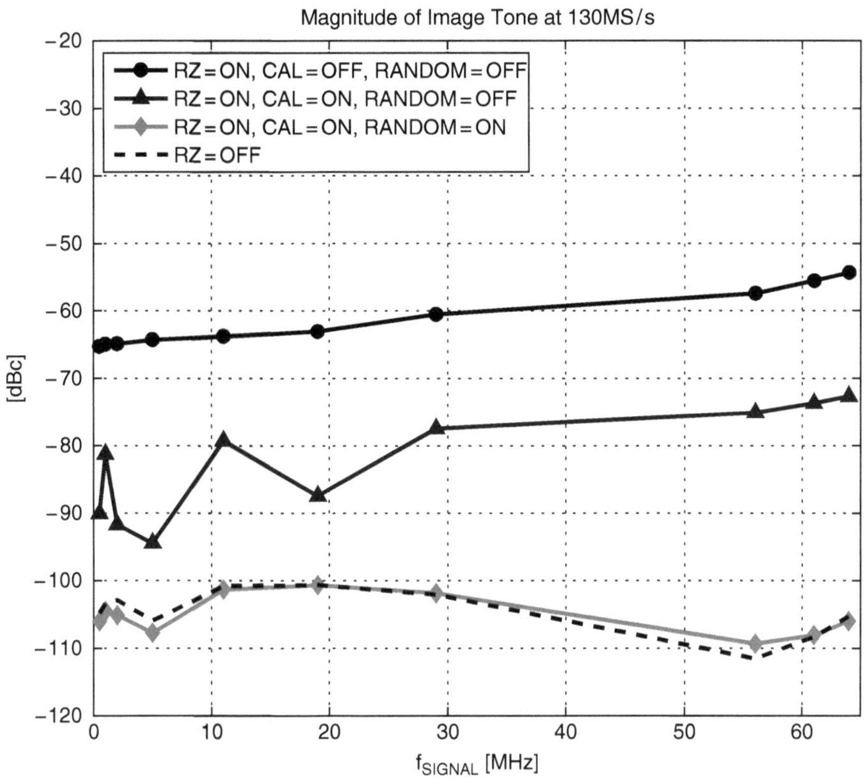

Fig. 6.51 Image tone magnitude at 130 MS/s

6.3 A Figure-of-Merit for Nyquist D/A-Converters

The spectral performance of Nyquist-rate D/A-converters is in most cases characterized in terms of the SFDR as a function of the signal frequency. A figure-of-merit (FOM) definition for Nyquist D/A-converters that puts equal weight on the static and dynamic linearity, bandwidth, and power consumption is found in [178]:

$$\mathrm{FOM}_2 = \frac{2^{\frac{\mathrm{SFDR(DC)}-1.76}{6.02}} \cdot 2^{\frac{\mathrm{SFDR}(0.5\cdot f_{\mathrm{CLK}})-1.76}{6.02}} \cdot f_{\mathrm{CLK}}}{P - \frac{1}{2} I_{\mathrm{FS}}^2 R_L D}. \tag{6.14}$$

As an approximation for the static linearity, the SFDR for a low-frequency single tone is used, while the dynamic performance is represented by the SFDR for a signal located close to the Nyquist frequency. The third factor in the numerator is the sampling frequency. In the denominator we find the power dissipated as heat in the chip. It results by subtracting the power delivered to the load from the total power

Table 6.5 FOM_2 comparison

Label	Ref.	Tech. [μm]	Area [mm²]	P [mW]	CLK [MHz]	SFDR @ DC [dB]	SFDR @ $\frac{f_{CLK}}{2}$ [dB]	FOM_2 [10³/pJ]
[Lin98]	[33]	0.35	0.6	125	500	66	51	2
[Bug00]	[104]	0.35	5.1	100	100	82	72	19
[VdB01]	[117]	0.35	0.35	110	1,000	74	61	36
[AvB01]	[179]	0.35	1	111	300	80	42	2.2
[Sco03]	[76]	0.25	1.95	400	400	98	75	306
[Hyd03]	[132]	0.25	0.44	53	250	75	76	114
[Hua04,1]	[92]	0.18	1	97	200	76	61	9.9
[Hua04,2]	[92]	0.18	1	97	200	85	44	4.0
[Tii01]	[101]	0.18	1	20	100	84	34	2.9
[Cha06]	[180]	0.18	3.2	177	100	75	75	12.1
[Cha07]	[181]	0.18	3	127	150	85	83	203
[Dev06]	[34]	0.18	0.35	22	250	74	62.5	57
[Dor05]	[125]	0.18	1.13	216	350	80	65	19.7
[OSu04]	[127]	0.18	0.44	82	320	95	40	16.7
[Con03]	[103]	0.13	0.1	16.7	100	82	62	74.5
[Cla07,1]	[42]	0.13	0.49	25	200	83.7	54.5	46
[Cla07,2]	[42]	0.13	0.49	19	100	83.7	67.5	136
[Cla08]	[93]	0.13	0.96	53	130	85	73.8	144
[Uen05]	[182]	0.09	0.75	21.6	200	68	56	10.7
[Tan11]	[183]	0.14	2.4	270	200	83	78.5	60.5
[Tse11]	[184]	0.09	0.825	128	1250	75	66	75
[Lin09]	[128]	0.065	0.31	188[a]	1,600	74	52.5	14

[a] Power dissipation is reported only for 2.9 GS/s in [128]. Therefore, FOM_2 @ 1.6 GS/s is actually higher than reported in the table and figure above

drawn from the supply.[10] The duty factor D indicates the duty cycle of the output current, i.e., $D = 1$ for a NRZ-DAC, while $D = 0.5$ for a half-clock RZ-DAC.

FOM_2 also has the form Speed · Accuracy^2/Power with the dimension of 1/energy. It is therefore inversely proportional to the energy required for the conversion of one data sample, weighted with the product of the low-frequency and high-frequency spectral performance of the DAC.

A comparison of recently published Nyquist-rate D/A-converters with FOM_2 is summarized in Table 6.5 and displayed graphically in Fig. 6.52. Each design is represented by a two-dimensional number given by ($x = 2^B$/area, $y = FOM_2$). The x-coordinate is the inverse silicon area, normalized to the physical resolution in binary exponential form. The y-coordinate is the FOM defined in Eq. (6.14), normalized to the inverse energy quantum of 1/pJ. Thus, again, along the x-axis

[10] A closer analysis reveals that in most current-steering DAC designs the power delivered to the load is much smaller than the total power drawn from the supply. Therefore, the numerator in Eq. (6.14) can be approximated by the total power dissipation P with only a small difference in the resulting FOM.

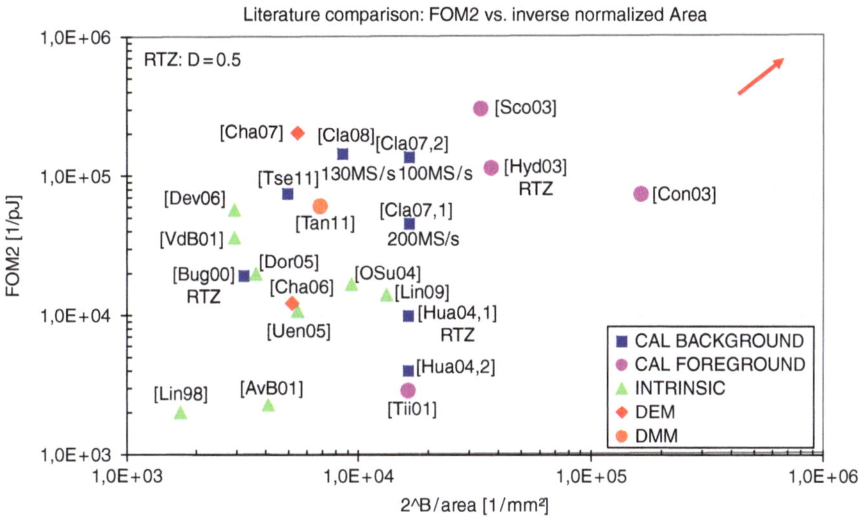

Fig. 6.52 Comparison of Nyquist-rate DACs with FOM_2

the converter area efficiency increases, while along the y-axis the converter performance efficiency, represented by FOM_2 improves.

As shown in Fig. 6.52, calibration naturally improves the area efficiency of D/A-converters. Except for [101], the converters calibrated in foreground do not include all the required trimming infrastructure on-chip; therefore the area efficiency of these designs is overestimated. Uncalibrated designs, on the other hand, seem to achieve the highest FOM_2 with a resolution around 10 bits, showing also a clear advancement with shorter gate lengths. The best power efficiency in this resolution range, due to the lowest dynamic power dissipation, is achieved with a fully binary architecture [34].

The implementation of DEM-techniques in Nyquist-rate D/A-converters is a relatively new development [180, 181]. Traditionally, element-shuffling is only used in oversampled converters, leading to a gain in resolution after baseband filtering. This advantage is clearly not available in Nyquist-rate converters, such that no improvement in terms of dynamic range is possible. This is not reflected in the definition of FOM_2 in Eq. (6.14). However, when used in conjunction with current-source calibration, DEM-techniques could help to make the element transitions more signal-independent, leading to an increase in dynamic linearity, i.e., improved SFDR at high signal frequencies.

Dynamic mismatch mapping (DMM) [183] attempts to reorder the switching sequence in the unary MSB-array, with the goal to minimize static *and* dynamic errors contemporaneously. This is done by comparing, e.g., at start-up, the static and dynamic differences of the unit current pulses, as delivered by each individual MSB-cell and a unique reference cell and, consequently, by remapping the MSB switching sequence. The remapping algorithm used in this example pursues the objective to minimize the error energy for two consecutive transitions, i.e., it assumes a digital input ramp. Nevertheless, it achieves a considerable boost in dynamic linearity for single-tone and two-tone signals [183]. The implemented DMM can be classified as a foreground calibration technique, because each MSB-cell must first be characterized independently, before the optimum MSB-mapping is fully established and data processing can commence. Due to the use of a not power-optimized converter core in the demonstrator testchip, FOM_2 seems to suffer somewhat from increased power dissipation at the reported sampling rate.

The design described in Sect. 6.1 and [42] has a "sweet spot" for the performance efficiency at around 100 MS/s, with a FOM_2 almost a factor 3 higher than at 200 MS/s. Using an active output stage and interleaved-RZ, the converter described in Sect. 6.2 and [93] achieves a 6 % higher FOM_2 at 130 MS/s, although the dynamic linearity has improved by more than 6 dB and the signal bandwidth by 30 %. The reason is the increased power consumption due to the interleaved architecture and the active output stage. On the other hand, the latter allows to increase the full-scale range by at least 50 % and provides a buffered voltage output. These system-level advantages are not reflected at all in the power efficiency definition of FOM_2, solely defined by the static and dynamic linearity, as well as the power dissipation at a given clock rate.

For an SFDR-measurement, as required by (6.14), we need to monitor the full Nyquist bandwidth, which tends to be time consuming for converters operating at high clock rates. Also, it does not distinguish between spectral components related to actual nonlinearities and, for example, disturbances injected via parasitic paths. The latter are more related to on-chip isolation issues and may depend heavily on the actual environment the converter is embedded in.

Alternatively, as a measure for the linearity at low and high frequencies in (6.14) we could also use the "subtractive" third-order intermodulation products IM_3^- (see the definition (D.42) in Sect. D.7), obtained from two-tone measurements. Since these intermodulation products always appear in close proximity[11] to the carriers, this would eliminate the ambiguity with respect to the origin of spectral impurities, as well as the need to monitor the full Nyquist-spectrum.

Regardless of the nonlinearity measure used, a completely fair comparison with a FOM based on very few and general parameters is only possible among sufficiently similar converters, e.g., having the same physical resolution, load resistance, sampling rate, and full-scale output voltage. The inclusion of more parameters into the FOM-definition to cover more concurrent design aspects is in principle possible, but this also tends to reduce the physical insight. Also, the proper choice of the necessary weighting factors is not always obvious and may depend on the scope of the definition.

[11] Of course, assuming a reasonably small carrier spacing.

In this specific example, the FOM defined by Eq. (6.14) seems to favor designs operating at "moderate" sampling frequencies with respect to the speed capability of digital circuits fabricated in the given process technology, whereas converters targeting very high signal bandwidths inevitably suffer from an overproportional power penalty. Of course, this could eventually be equalized by choosing a different weighting of the contributing factors in (6.14) but inevitably also leads to a FOM-definition with a different scope.

Chapter 7
Conclusion and Outlook

7.1 Conclusions

High-performance CMOS D/A-converters suitable for SoC-integration in a digital transceiver face stringent silicon area and power consumption constraints. The resolution of these converters is fundamentally limited by the matching and the finite output impedance of the current sources, circuit and jitter noise, as well as a number of dynamic effects that dominate the spectral purity of the analog output at higher signal frequencies.

The noise generated in the biasing of a current-steering DAC appears as amplitude modulation on the output signal and adds to the noise-floor given by the current sources and the I/V-converter. It is shown in Sect. 2.3.2 that multitone signals are much less affected by the bias noise than a single carrier with the same peak amplitude, because the modulated noise is distributed across a large number of carriers with accordingly lower absolute value.

Any spectral impurity of the sampling clock also generates noise in the analog output signal. While the NRZ-DAC shows a jitter-limited SNR equal to the ideal sampler, for an RZ-DAC it approaches a finite value for low frequencies. This makes the standard RZ-architecture very inconvenient for high-resolution converters. Again, the jitter specification for multitone systems is shown (Sect. 2.3.3) to be considerably relaxed, due to the distribution of the noise across a large number of carriers. The developed analytical formulas are also verified experimentally.

Random mismatch affects the accuracy of the static converter characteristic. Excluding all systematic effects that arise from the fabrication process, packaging stress and thermal gradients, the static resolution is solely given by the sizing and the chosen bias point of the current source transistors. A statistical formulation of the yield as a function of the unit-cell matching for different converter segmentation, calculated in Sect. 3.1, allows to calculate the minimum required current source area for given linearity specification and desired fabrication yield.

M. Clara, *High-Performance D/A-Converters*, Springer Series in Advanced Microelectronics 36, DOI 10.1007/978-3-642-31229-8_7,

The output resistance of the current sources limits the INL of the current-steering DAC. The derived analytical expressions for the single-ended and fully differential converter in Sect. 3.1.5 show that the latter is much less sensitive (with a factor inversely proportional to the number of unit cells) toward this effect. Although in today's technologies the output resistance can still be kept sufficiently high, eventually this will be one of the major limiting factors for the static linearity in much more advanced CMOS processes.

In an idealized converter with perfectly matched current cells and instantaneous charge redistribution at the beginning of the sampling period, three basic effects limit the dynamic linearity (Chap. 4.1). Two of these are related to signal-dependent error charges injected into the current outputs, either by asymmetries in the current pulse shape or by a signal-dependent charging of the parasitic capacitance connected to the source node of the current switches. The third effect is the frequency-dependent output impedance of the unit current sources. While the signal dependency of the switching errors can, at least in principle, be eliminated, the output impedance of the current sources is a fundamental limit for the achievable dynamic linearity. However, at least for still some technology generations to come, it will be possible to make the output impedance of the DAC-elements reasonably large, so as not to limit the dynamic linearity in fully differential designs.

Oversampled converters traditionally use dynamic element matching (DEM) techniques to improve the static linearity. With large oversampling ratios, the simple CLA can give an adequate improvement in resolution, provided that the tonal behavior of this algorithm can be tolerated by the system. The design example in Sect. 5.1 describes the implementation of a $\Sigma\Delta$-DAC with CLA. The converter has a signal bandwidth of 2.2 MHz and achieves a dynamic range of 87 dB with a power consumption of only 9 mW from a 1.5 V supply.

For low oversampling ratio the simple, but very effective DWA algorithm cannot be used in a current-steering D/A-converter, because the unavoidable switching errors are highly signal correlated and cause large, frequency-independent distortion. Return-to-zero (RZ) can in principle eliminate the direct relation between the input signal and the switching errors, allowing the usage of DWA in a current-steering converter. But the inherent 6 dB signal loss and the increased jitter sensitivity make it impractical for high-resolution converters. With a time-interleaved architecture an effective RZ can be implemented, while still preserving the desired NRZ-waveform. The design example described in Sect. 5.2 is a $\Sigma\Delta$-DAC with a signal bandwidth close to 30 MHz and a DR of nearly 12 bits. The converter can also be used in a low-power mode with reduced bandwidth (2.2 MHz), but with a resolution increased to 14 bits.

Another method to increase the static linearity of a current-steering D/A-converter is calibration. Dynamic current calibration is shown to be a suitable method to achieve 14-bit resolution and beyond. Since the single DAC-elements are trimmed in the background during normal operation, this method also adapts to slowly changing environmental conditions. The design described in Sect. 6.1 introduces a segment boundary trimming that iteratively matches the sum of the lower segments to the unit elements of the higher segments. The converter is supplied with 1.5 V and achieves a low-frequency SFDR close to 84 dB with a power consumption of 25 mW at 200 MS/s.

A persistent problem of dynamic background calibration is the periodicity of the calibration process itself. The introduction of a random calibration slot length distributes the energy of the calibration refresh tones across a certain bandwidth, achieving a spur attenuation of more than 20 dB and eventually merging the calibration noise with the circuit-noise floor. The method is basically limited by the leakage currents that discharge the storage capacitor in the DAC-cell over time.

The design described in Sect. 6.2 uses a self-calibrated variable reference cell that allows direct trimming of DAC-elements in differently weighted segments. Only for the lowest binary-coded segment a boundary match is performed. The interleaved current-cell architecture together with an optimized transimpedance output stage enables the converter to achieve a SFDR > 70 dB for signal frequencies up to 65 MHz. In an interleaved RZ-architecture with dynamic current calibration, the residual dynamic mismatch due to the background calibration causes an in-band image tone at $\frac{1}{2} f_{\mathrm{CLK}} - f_{\mathrm{SIG}}$. It is shown that the randomization of the calibration slot length also eliminates the image tone in the same way as the calibration refresh tones.

7.2 Outlook

It is expected that future research on current-steering D/A-converters will be focused into the following directions:

- With the migration to CMOS technologies featuring smaller minimum gate length, the "cost" of digital circuitry in terms of silicon area and power consumption will continue to drop. This, in turn, will make the introduction of more advanced element shuffling techniques, which, at the same time, provide (high-order) mismatch shaping and constant switching activity, increasingly attractive. Such advanced DEM-algorithms can also be used in Nyquist-converters to increase the dynamic linearity. First design examples have been published in [180, 181]. A potential problem with Nyquist-rate DEM is that the "mismatch noise" is completely in-band. At least for full-Nyquist operation, there is thus no inherent gain in dynamic range possible.
  Nevertheless, an attractive combination could be the implementation of calibrated current sources together with special DEM-algorithms, which are optimized to eliminate only the signal dependency of the DAC-element transitions. In such an arrangement, the calibration would be responsible for the static linearity, while the element shuffling provides the necessary decorrelation of the element switching with respect to the input signal to achieve good dynamic linearity.
- In newer CMOS technologies the accuracy that can be achieved with dynamic current calibration will be increasingly limited by the gate leakage current, unless a thick gate oxide device is available and can be used within the available voltage

headroom. Ultimately, dynamic current calibration will be replaced by discrete current source trimming, e.g., using a local CALDAC in every current source. In terms of area efficiency, a binary CALDAC with redundant coding might be the optimum solution, although this necessitates a more sophisticated digital calibration algorithm. Also the storage of the digital trimming value will require increasingly less silicon area, as CMOS technologies advance.

- Direct current-source calibration, performed by comparing each DAC-cell with a single reference element, is basically a "DNL-calibration". This method is found in most of today's D/A-converter designs that employ self-calibration. The main drawback of this method is that it does not take into account the additional error of the code-dependent output impedance of the DAC-array. As the output resistance of the current sources decreases with advancing technologies, this factor might become the major limitation for future current-steering DAC designs.
  An "INL-calibration" technique, on the other hand, tries to trim the static DAC-characteristic code by code, independent of the type of static nonlinearity error. Therefore, not only a single reference element is needed for comparison, but the complete ideal characteristic must be available. This is best done in the digital domain by measuring the actual output code with an A/D-converter. This CALADC must have a resolution and linearity that exceeds the DAC-resolution, albeit only at DC. This could be achieved quite efficiently with a single-bit $\Sigma\Delta$-ADC at the expense of some silicon area.
  Because the complete converter is needed for the output voltage measurement, INL-calibration must be performed in foreground. If the absolute value of the single DAC-elements, e.g., the difference to a stable reference element, is additionally measured and stored, continuous calibration can still be performed element-wise in the background. In this case the element trimming must also keep track of the overall converter characteristic in relation to the value of the single element. The cheaply available digital storage and computing power in future CMOS technologies could make this approach feasible very soon.
- A global CALDAC seems to be the most area-efficient discrete calibration device. Since it is a dynamic circuit, it is far more critical with respect to the dynamic performance of the overall converter. The difficulty for a modeling approach is that, after calibration, the switching activity of the global CALDAC depends on the input code *and* on the initial mismatch within both converters, different for each fabricated device.
- Calibration of dynamic effects will eventually become very important, especially for wideband signal synthesis. A first implementation of a foreground technique, based on a dynamic characterization of unary DAC-cells and reordering of the switching sequence according to a minimum error energy criterion, has recently been described in [183]. A dynamic error calibration that runs continuously in the background, though, will be much more difficult to achieve, but it is the only method that allows to track, independently from the actual system requirements, also slowly changing environmental parameters, such as supply voltage and on-chip temperature distribution. Since dynamic errors are related more or less to

the behavior of the current switching and retiming circuitry in the DAC-cells, the real challenge of such an approach will be the rerouting of the cell current to an appropriate on-chip sensor, without altering the dynamic properties that characterize the DAC-cell during normal operation.

- digital pre-distortion (DPD) to improve the dynamic linearity of a current-steering D/A-converter is a technique not yet broadly established.[1] Nonideal effects that occur only at the moment of the switching transition, e.g., switching asymmetries and charge sharing, could be described by frequency-dependent nonlinearities. Once a suitable nonlinear model of the converter, e.g., based on Volterra kernels, is established, the digital input signal of the DAC can in principle be predistorted by a (nonlinear) digital filter and thereby cancel the nonlinear distortion generated by the converter core.
  One of the problems, when using DPD to linearize a D/A-converter, will be the reliable identification of the nonlinear coefficients. To allow for an easy measurement of the error energy during the identification process, the level of the harmonic distortion due to the switching asymmetry and the charge-sharing effect could be increased considerably by applying a suitable switching algorithm, e.g., a standard DWA. In this case a single-tone measurement at a very low signal frequency would be sufficient to determine the nonlinear coefficients, e.g., using an eventually available CALADC. The question remains, however, whether the dynamic error sources affecting the converter during normal wideband signal synthesis truly follow the same, simple model. Also, the whole DAC is required for the characterization, i.e., again we are at first stuck with a true foreground calibration technique. Since we can, eventually, only monitor the dynamic behavior of *single* DAC-elements in the background, the extension to an adaptive tracking of the nonlinear coefficients related to the DPD will only be possible, if we are able to extrapolate their relative change from the individually measured dynamic behavior of the single DAC-cells.

[1] DPD is currently a very active area of research in the context of RF power amplifiers [185].

Appendix A
DAC Bias Noise Model

A.1 Bias Noise Model Without 1/f-Noise

We approximate the correlated bias noise of Fig. 2.14 using a white current noise source with rms-value σ_{IB} that is filtered by a first-order low-pass with corner frequency $\omega_{p,\mathrm{Bias}}$. Consequently, 1/f-noise is neglected in this model. The bias noise has a (one-sided) PSD $S_B(\omega)$:

$$\overline{I_{n,\mathrm{Bias}}(\omega)^2} \approx S_B(\omega) = \frac{4\sigma_{\mathrm{IB}}^2}{\omega_{p,\mathrm{Bias}}} \cdot \frac{1}{1+\left(\dfrac{\omega}{\omega_{p,\mathrm{Bias}}}\right)^2},$$

$$\frac{1}{2\pi}\int_0^\infty S_B(\omega)\mathrm{d}\omega = \sigma_{\mathrm{IB}}^2. \tag{A.1}$$

With a sine-wave input $\mathrm{d}(t) = A_{\sin}\sin(\omega_{\sin}t)$ the (one-sided) output noise PSD becomes

$$S_{i,\mathrm{SIN}}(\omega,\omega_{\sin}) \approx \overline{I_{n,\mathrm{Array}}^2} + \overline{I_{n,\mathrm{OS}}^2} + A_{\sin}^2 \frac{S_B(\omega-\omega_{\sin}) + S_B(\omega+\omega_{\sin})}{4}. \tag{A.2}$$

The low-frequency bias noise is translated to the carrier frequency due to the multiplication with the signal (see Fig. 2.16). Because the sampling operation of the data converter is not considered, Eq. (A.2) is only valid for frequencies sufficiently below Nyquist, where aliasing effects occur. The maximum noise density occurs in the vicinity of the carrier frequency. For $\omega_{\sin} >> \omega_{p,\mathrm{Bias}}$ we get

$$\max\left(S_{i,\mathrm{SIN}}(\omega)\right) = S_{i,\mathrm{SIN}}(\omega_{\sin}) \approx \overline{I_{n,\mathrm{Array}}^2} + \overline{I_{n,\mathrm{OS}}^2} + \frac{A_{\sin}^2\sigma_{\mathrm{IB}}^2}{\omega_{p,\mathrm{Bias}}}. \tag{A.3}$$

M. Clara, *High-Performance D/A-Converters*, Springer Series in Advanced Microelectronics 36, DOI 10.1007/978-3-642-31229-8,

If we synthesize a homogeneous multitone signal (see Sect. 2.2.5), the total output noise is given by the sum of all carrier contributions:

$$S_{i,\mathrm{DMT}}(\omega) \approx \overline{I^2_{n,\mathrm{Array}}} + \overline{I^2_{n,\mathrm{OS}}} + A^2_{\mathrm{DMT}} \sum_{k=0}^{N_C} \frac{S_B(\omega-\omega_k)+S_B(\omega+\omega_k)}{4}. \tag{A.4}$$

With $\omega_k = \omega_{\min} + k\Delta\omega$ for $k = [0; N_C]$ and $\omega_{N_C} = \omega_{\min} + N_C \cdot \Delta\omega = \omega_{\max}$. Using (2.7) we can write

$$\begin{aligned} S_{i,\mathrm{DMT}}(\omega) \approx{} & \overline{I^2_{n,\mathrm{Array}}} + \overline{I^2_{n,\mathrm{OS}}} \\ & + \frac{2}{\mathrm{CF}^2} \cdot \frac{A^2_{\mathrm{peak}}\sigma^2_{\mathrm{IB}}}{\omega_{p,\mathrm{Bias}}\left(\omega_{\max}-\omega_{\min}\right)} \\ & \times \sum_{k=0}^{N_C} \left(\frac{\Delta\omega}{1+\left(\frac{\omega-\omega_k}{\omega_{p,\mathrm{Bias}}}\right)^2} + \frac{\Delta\omega}{1+\left(\frac{\omega+\omega_k}{\omega_{p,\mathrm{Bias}}}\right)^2} \right). \end{aligned} \tag{A.5}$$

When the carrier spacing $\Delta\omega$ is much smaller than $\omega_{p,\mathrm{Bias}}$, then the noise density does not change significantly when we increase the number of carriers and keep the Crest Factor constant. Although the amplitude of the carriers A_{DMT} decreases according to (2.7), increasingly more carriers contribute to the noise.[1] In the limit $\Delta\omega \to 0$ the discrete sum can be transformed into an integral and solved:

$$\begin{aligned} & \lim_{\Delta\omega\to 0} \sum_{k=0}^{N_C} \frac{\Delta\omega}{1+\left(\frac{\omega\mp\omega_k}{\omega_{p,\mathrm{Bias}}}\right)^2} \\ & = \int_{\omega_{\min}}^{\omega_{\max}} \frac{\mathrm{d}\nu}{1+\left(\frac{\omega\mp\nu}{\omega_{p,\mathrm{Bias}}}\right)^2} \\ & = \omega_{p,\mathrm{Bias}} \left[\arctan\left(\frac{\omega_{\max}\mp\omega}{\omega_{p,\mathrm{Bias}}}\right) - \arctan\left(\frac{\omega_{\min}\mp\omega}{\omega_{p,\mathrm{Bias}}}\right) \right]. \end{aligned} \tag{A.6}$$

In the case of a sufficiently dense multitone signal with small carrier spacing the total (one-sided) output noise PSD can thus be approximated by

[1] It can be shown that the average signal power per unit bandwidth remains constant.

$$S_{i,\mathrm{DMT}}(\omega) \approx \overline{I^2_{n,\mathrm{Array}}} + \overline{I^2_{n,\mathrm{OS}}} + \frac{2}{\mathrm{CF}^2} \cdot \frac{A^2_{\mathrm{peak}}\sigma^2_{\mathrm{IB}}}{\omega_{\max} - \omega_{\min}} \left(B_n(\omega) + B_n(-\omega)\right),$$

$$B_n(\omega) = \arctan\left(\frac{\omega_{\max} - \omega}{\omega_{p,\mathrm{Bias}}}\right) - \arctan\left(\frac{\omega_{\min} - \omega}{\omega_{p,\mathrm{Bias}}}\right). \tag{A.7}$$

The maximum of the PSD occurs at $\frac{1}{2}(\omega_{\max} + \omega_{\min})$, and, in case this frequency is sufficiently larger than $\omega_{p,\mathrm{Bias}}$, we can approximate the maximum noise density by

$$\begin{aligned}\max\left(S_{i,\mathrm{DMT}}(\omega)\right) &= S_{i,\mathrm{DMT}}\left(\frac{\omega_{\max} + \omega_{\min}}{2}\right) \\ &= \overline{I^2_{n,\mathrm{Array}}} + \overline{I^2_{n,\mathrm{OS}}} \\ &\quad + \frac{4}{\mathrm{CF}^2} \cdot \frac{A^2_{\mathrm{peak}}\sigma^2_{\mathrm{IB}}}{(\omega_{\max} - \omega_{\min})} \arctan\left(\frac{\omega_{\max} - \omega_{\min}}{2\omega_{p,\mathrm{Bias}}}\right).\end{aligned} \tag{A.8}$$

For wideband systems with $\omega_{\max} - \omega_{\min} >> 2\omega_{p,\mathrm{Bias}}$, the arctangent converges toward $\pi/2$ and (A.8) can be approximated by

$$\max\left(S_{i,\mathrm{DMT}}(\omega)\right) \approx \overline{I^2_{n,\mathrm{Array}}} + \overline{I^2_{n,\mathrm{OS}}} + \frac{A^2_{\mathrm{peak}}}{\mathrm{CF}^2}\sigma^2_{\mathrm{IB}} \frac{2\pi}{(\omega_{\max} - \omega_{\min})} \quad \text{for } \omega_{\max} - \omega_{\min} >> 2\omega_{p,\mathrm{Bias}}. \tag{A.9}$$

If the noise injected by the DAC-biasing is dominant, then the ratio of the maximum output noise PSD for a single-tone signal and a multitone signal is given by

$$\begin{aligned}\frac{\max\left(S_{i,\mathrm{SIN}}(\omega)\right)}{\max\left(S_{i,\mathrm{DMT}}(\omega)\right)} &\approx \frac{\mathrm{CF}^2}{4\omega_{p,\mathrm{Bias}}} \cdot \frac{\omega_{\max} - \omega_{\min}}{\arctan\left(\frac{\omega_{\max} - \omega_{\min}}{2\omega_{p,\mathrm{Bias}}}\right)} \\ &\approx \frac{\mathrm{CF}^2}{2\pi} \cdot \frac{\omega_{\max} - \omega_{\min}}{\omega_{p,\mathrm{Bias}}} \quad \text{for } \omega_{\max} - \omega_{\min} >> 2\omega_{p,\mathrm{Bias}}.\end{aligned} \tag{A.10}$$

Equation (A.10) shows that especially wideband multitone systems are much less sensitive to noise injected from the bias circuit. Therefore, a noise specification on the basis of a full-scale single-tone assumption can lead to significant overdesign of the DAC-biasing.

A.2 DMT Synthesis with Correlated 1/f-Noise

To also include 1/f-noise in the noise model of the DAC-biasing, we extend Eq. (A.1) by including an additional term that is inversely proportional to frequency:

$$S_B(\omega) = \frac{4\sigma_F^2}{\omega_{p,\mathrm{Bias}}} \cdot \frac{1}{1 + \left(\frac{\omega}{\omega_{p,\mathrm{Bias}}}\right)^2} \cdot \left(1 + \frac{\omega_{1/f}}{\omega}\right) \tag{A.11}$$

$\omega_{1/f}$ is the corner frequency of the 1/f-noise (see Fig. 2.14). First we determine the total noise power by integrating (A.11). Because the flicker noise component in (A.11) would have infinite power at DC, we must introduce a lower limit for the integration bandwidth, $\omega_{F\min}$:

$$\begin{aligned} \frac{1}{2\pi} \int_{\omega_{F\min}}^{\infty} S_B(\omega) \mathrm{d}\omega &= \frac{1}{2\pi} \int_{\omega_{F\min}}^{\infty} \frac{4\sigma_F^2}{\omega_{p,\mathrm{Bias}}} \cdot \frac{1}{1 + \left(\frac{\omega}{\omega_{p,\mathrm{Bias}}}\right)^2} \cdot \left(1 + \frac{\omega_{1/f}}{\omega}\right) \mathrm{d}\omega \\ &= \sigma_F^2 \cdot \left[1 - \frac{2}{\pi} \frac{\omega_{1/f}}{\omega_{p,\mathrm{Bias}}} \cdot \ln\left(\frac{\omega_{F\min}}{\sqrt{\omega_{p,\mathrm{Bias}}^2 + \omega_{F\min}^2}}\right)\right]. \end{aligned} \tag{A.12}$$

The value of $\omega_{F\min}$ depends on the minimum frequency of interest, e.g., the symbol rate in a DSL-system or an appropriate fraction thereof. Since we want the total power of the correlated bias noise to be equal to σ_{IB}^2, the constant σ_F is given by

$$\sigma_F = \sigma_{\mathrm{IB}} \cdot \frac{1}{\sqrt{1 - \frac{2}{\pi} \frac{\omega_{1/f}}{\omega_{p,\mathrm{Bias}}} \cdot \ln\left(\frac{\omega_{F\min}}{\sqrt{\omega_{p,\mathrm{Bias}}^2 + \omega_{F\min}^2}}\right)}}. \tag{A.13}$$

Neglecting aliasing effects, the (one-sided) output noise PSD when synthesizing a homogeneous multitone signal becomes

$$\begin{aligned} S_{i,\mathrm{DMT}}(\omega) \approx\ & \overline{I_{n,\mathrm{Array}}^2} + \overline{I_{n,\mathrm{OS}}^2} \\ & + \frac{2}{\mathrm{CF}^2} \cdot \frac{A_{\mathrm{peak}}^2 \sigma_F^2}{\omega_{p,\mathrm{Bias}} (\omega_{\max} - \omega_{\min})} \sum_{k=0}^{N_C} \frac{1 + \frac{\omega_{1/f}}{|\omega \mp \omega_k|}}{1 + \left(\frac{\omega \mp \omega_k}{\omega_{p,\mathrm{Bias}}}\right)^2} \Delta\omega. \end{aligned} \tag{A.14}$$

Assuming a sufficiently dense multitone signal we can again approximate (A.14) by performing the limit transition $\Delta\omega \to 0$ and replacing the sum with an integral:

$$
\begin{aligned}
S_{i,\mathrm{DMT}}(\omega) \approx\; & \overline{I_{n,\mathrm{Array}}^2} + \overline{I_{n,\mathrm{OS}}^2} \\
& + \frac{2}{\mathrm{CF}^2} \cdot \frac{A_{\mathrm{peak}}^2 \sigma_F^2}{\omega_{p,\mathrm{Bias}} \left(\omega_{\max} - \omega_{\min}\right)} \int_{\omega_{\min}}^{\omega_{\max}} \frac{1 + \dfrac{\omega_{1/f}}{|\omega \mp \nu|}}{1 + \left(\dfrac{\omega \mp \nu}{\omega_{p,\mathrm{Bias}}}\right)^2} \Delta\nu.
\end{aligned}
\tag{A.15}
$$

Unfortunately, the integral in (A.15) does not exist for $\omega_{\min} < \omega < \omega_{\max}$. Instead, we can solve Eq. (A.15) for the regions $\omega < \omega_{\min}$ and $\omega > \omega_{\max}$, i.e., outside of the frequency band occupied by the carriers. For $\omega < \omega_{\min}$, we get

$$
\begin{aligned}
S_{i,\mathrm{DMT}}(\omega) \approx\; & \overline{I_{n,\mathrm{Array}}^2} + \overline{I_{n,\mathrm{OS}}^2} \\
& + \frac{2}{\mathrm{CF}^2} \cdot \frac{A_{\mathrm{peak}}^2 \sigma_F^2}{\omega_{p,\mathrm{Bias}} \left(\omega_{\max} - \omega_{\min}\right)} \left[B_{n1}(\omega) + B_{n1}(-\omega)\right] \\
B_{n1}(\omega) =\; & \arctan\left(\frac{\omega_{\max} - \omega}{\omega_{p,\mathrm{Bias}}}\right) - \arctan\left(\frac{\omega_{\min} - \omega}{\omega_{p,\mathrm{Bias}}}\right) \\
& - \frac{\omega_{1/f}}{2\omega_{p,\mathrm{Bias}}} \cdot \ln\left(\frac{\omega_{p,\mathrm{Bias}}^2 + (\omega - \omega_{\max})^2}{\omega_{p,\mathrm{Bias}}^2 + (\omega - \omega_{\min})^2}\right) + \frac{\omega_{1/f}}{\omega_{p,\mathrm{Bias}}} \cdot \ln\left(\frac{\omega - \omega_{\max}}{\omega - \omega_{\min}}\right).
\end{aligned}
\tag{A.16}
$$

For $\omega > \omega_{\max}$, instead:

$$
\begin{aligned}
S_{i,\mathrm{DMT}}(\omega) \approx\; & \overline{I_{n,\mathrm{Array}}^2} + \overline{I_{n,\mathrm{OS}}^2} \\
& + \frac{2}{\mathrm{CF}^2} \cdot \frac{A_{\mathrm{peak}}^2 \sigma_F^2}{\omega_{p,\mathrm{Bias}} \left(\omega_{\max} - \omega_{\min}\right)} \left[B_{n2}(\omega) + B_{n2}(-\omega)\right] \\
B_{n2}(\omega) =\; & \arctan\left(\frac{\omega_{\max} - \omega}{\omega_{p,\mathrm{Bias}}}\right) - \arctan\left(\frac{\omega_{\min} - \omega}{\omega_{p,\mathrm{Bias}}}\right) \\
& + \frac{\omega_{1/f}}{2\omega_{p,\mathrm{Bias}}} \cdot \ln\left(\frac{\omega_{p,\mathrm{Bias}}^2 + (\omega - \omega_{\max})^2}{\omega_{p,\mathrm{Bias}}^2 + (\omega - \omega_{\min})^2}\right) - \frac{\omega_{1/f}}{\omega_{p,\mathrm{Bias}}} \cdot \ln\left(\frac{\omega - \omega_{\max}}{\omega - \omega_{\min}}\right).
\end{aligned}
\tag{A.17}
$$

Figure A.1 shows a comparison of the calculated bias noise PSD and the result of a time-domain behavioral simulation for a homogeneous multitone signal sampled at 200 MS/s. The following parameters are used in this example:

- Crest Factor: $\mathrm{CF} = 4.1$
- DMT bandwidth: $f_{\min} = 10\,\mathrm{MHz} \ldots f_{\max} = 30\,\mathrm{MHz}$
- Number of carriers: $N_C + 1 = 925$
- Corner frequency of bias noise filter: $\omega_{p,\mathrm{Bias}} = 3\,\mathrm{MHz}$
- Flicker noise corner frequency: $\omega_{1/f} = 0.5\,\mathrm{MHz}$
- Minimum integration frequency: $\omega_{F\min} = 2\,\mathrm{kHz}$

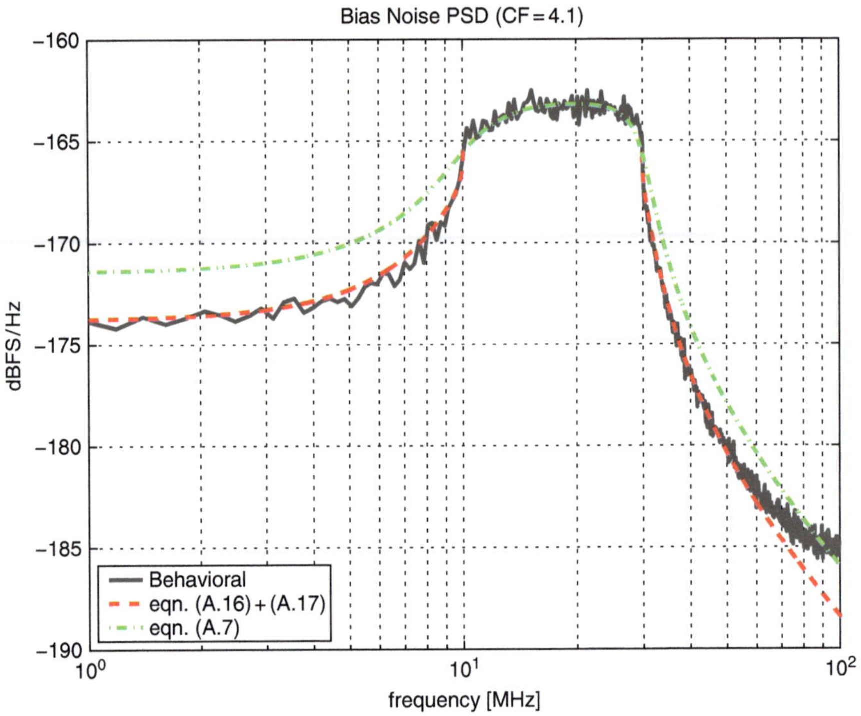

Fig. A.1 DMT-signal with correlated 1/f-noise

Also shown in Fig. A.1 is the theoretical result for the same total noise power, but without flicker noise component, as given by Eq. (A.7). Outside of the carrier population the noise is accurately predicted by (A.16) and (A.17). It seems that within the carrier population equation (A.7) can provide a reasonable estimate for the correlated bias noise. This is only true as long as the DMT bandwidth remains larger than the bias noise bandwidth, i.e., $\omega_{\max} - \omega_{\min} > \omega_{p,\text{Bias}}$. Also, the carrier spacing must be much smaller than the flicker noise corner frequency, $\Delta\omega << \omega_{1/f}$. Very close to the band edges, where 1/f-noise components start to dominate, (A.7) slightly underestimates the total output noise.

A.3 Maximum SNR Limited by Correlated Bias Noise

Let the noise voltage spectral density at the output of the D/A-converter for the midscale and full-scale input code be given by $\overline{V_{n,\text{MS}}(f)}$ and $\overline{V_{n,\text{FS}}(f)}$, respectively. The correlated bias noise PSD is then the difference $\overline{V_{n,\text{FS}}(f)^2} - \overline{V_{n,\text{MS}}(f)^2}$. Using (A.2), the output noise power density when synthesizing a sine wave with maximum amplitude 1 is given by

$$\overline{V_{n,\mathrm{DAC}}(f, f_{\mathrm{sin}})^2} = \overline{V_{n,\mathrm{MS}}(f)^2} + \frac{1}{4}\left(\overline{V_{n,\mathrm{FS}}(f \mp f_{\mathrm{sin}})^2} - \overline{V_{n,\mathrm{MS}}(f \mp f_{\mathrm{sin}})^2}\right). \tag{A.18}$$

The total noise power N_{DAC} is calculated by integrating (A.18). Suppose that the correlated bias noise is completely within the bandwidth of interest f_B. Then we can write

$$N_{\mathrm{DAC}} = \int_{f_B} \overline{V_{n,\mathrm{MS}}(f)^2}\mathrm{d}f + \frac{1}{2}\int_0^{\infty}\left(\overline{V_{n,\mathrm{FS}}(f)^2} - \overline{V_{n,\mathrm{MS}}(f)^2}\right)\mathrm{d}f. \tag{A.19}$$

With the full-scale voltage range V_{FS} the maximum signal power is given by $\frac{1}{2}V_{\mathrm{FS}}^2$. The maximum possible SNR due to circuit noise, including the correlated bias noise, is therefore

$$\mathrm{SNR}_{\mathrm{max}} = \frac{\frac{1}{2}V_{\mathrm{FS}}^2}{\int_{f_B} \overline{V_{n,\mathrm{MS}}(f)^2}\mathrm{d}f + \frac{1}{2}\int_0^{\infty}\left(\overline{V_{n,\mathrm{FS}}(f)^2} - \overline{V_{n,\mathrm{MS}}(f)^2}\right)\mathrm{d}f}. \tag{A.20}$$

Finally, with (A.1) and assuming a load resistance R_L:

$$\mathrm{SNR}_{\mathrm{max}} = \frac{\frac{1}{2}V_{\mathrm{FS}}^2}{\int_{f_B} \overline{V_{n,\mathrm{MS}}(f)^2}\mathrm{d}f + \frac{1}{2}\sigma_{\mathrm{IB}}^2 R_L^2}. \tag{A.21}$$

The dynamic range (DR) of the converter (see Sect. 2.2.4) is defined as the resolution for small signals, extrapolated to the full-scale value. Therefore, the correlated bias noise is not limiting the DR of the converter:

$$\mathrm{DR} = \frac{\frac{1}{2}V_{\mathrm{FS}}^2}{\int_{f_B} \overline{V_{n,\mathrm{MS}}(f)^2}\mathrm{d}f}. \tag{A.22}$$

Appendix B
Jitter Noise

B.1 Sampling Jitter Model

In general, we can describe the sampling clock by a sine wave with a random phase term $\varphi(t)$ [186]:

$$v(t) = V_0 \sin(\omega_0 t + \varphi(t))$$
$$\omega_0 = \frac{2\pi}{T} \tag{B.1}$$

$\varphi(t)$ is a stationary random process with zero mean and represents the momentary deviation from the linear phase behavior of an ideal clock signal. It is therefore commonly called "phase noise". For a first-order PLL with noiseless input the (two-sided) PSD of the phase noise can be modeled by a second-order Butterworth low-pass spectrum [51, 54]:

$$S_\varphi(\omega) = \left(\frac{2\pi}{T}\right)^2 \frac{\sigma_{\mathrm{LT}}^2}{\omega_{\mathrm{PLL}}} \frac{1}{1 + \left(\frac{\omega}{\omega_{\mathrm{PLL}}}\right)^2}$$
$$\frac{1}{2\pi} \int_{-\infty}^{\infty} S_\varphi(j\omega) \mathrm{d}\omega = \left(\frac{2\pi}{T}\right)^2 \frac{\sigma_{\mathrm{LT}}^2}{2} \tag{B.2}$$

$S_\varphi(\omega)$ in Eq. (B.2) is also called a Lorentzian spectrum. $\omega_{\mathrm{PLL}} = 2\pi f_{\mathrm{PLL}}$ is a good approximation of the PLL-bandwidth. In contrast to "white" clock jitter we now speak of "colored" or "pink" jitter. For our purposes, the phase noise model of (B.2) applies, with reasonable accuracy, also to higher-order PLL's [51].

The sampling instants are given by the zero crossings of $v(t)$ in Eq. (B.1). The deviations ΔT_k from the ideal sampling instants kT can be obtained by sampling

M. Clara, *High-Performance D/A-Converters*, Springer Series in Advanced Microelectronics 36, DOI 10.1007/978-3-642-31229-8,

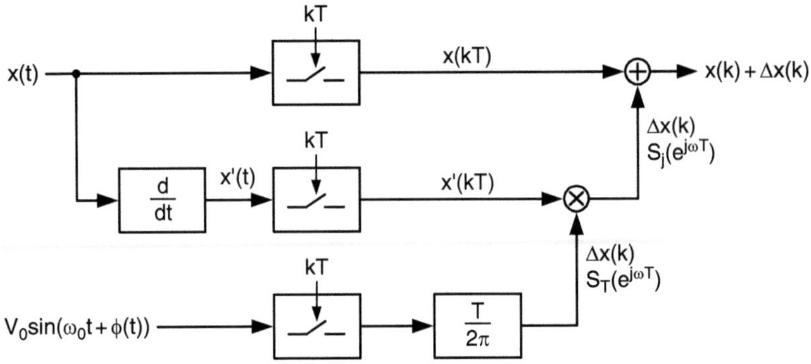

Fig. B.1 Sampling jitter model

$v(t)$ with an ideal clock $v_{\text{ideal}}(t) = V_0 \sin(\omega_0 t)$ having the same average (linear) phase, but no random phase excursions. Assuming that the phase deviations are small, $\varphi(t) << 2\pi$, we can write [53]

$$\Delta T_k = \frac{T}{2\pi}\varphi(kT). \tag{B.3}$$

The stationary, zero-mean random process ΔT_k has a (one-sided) PSD and a total power given by

$$S_T(\mathrm{e}^{j\omega T}) = 2T \cdot \frac{1-\mathrm{e}^{-2\omega_{\text{PLL}}T}}{|1-\mathrm{e}^{-\omega_{\text{PLL}}T}\mathrm{e}^{-j\omega T}|^2} \cdot \frac{1}{2}\sigma_{\text{LT}}^2$$

$$E\left(\Delta T_k^2\right) = \frac{1}{2\pi}\int_0^{\frac{\pi}{T}} S_T(\mathrm{e}^{j\omega T})\mathrm{d}\omega = \frac{\sigma_{\text{LT}}^2}{2}. \tag{B.4}$$

The model for the sampling process is shown in Fig. B.1. The analog input signal $x(t)$ is ideally sampled at instants $t = k \cdot T$, along with the jittered sampling clock and the first time derivative of the signal itself. The latter is then multiplied with the timing error sequence ΔT_k to yield the linearly approximated amplitude error sequence Δx_k.
The sampling process applied to a sine wave $x(t) = A_{\sin}\sin(\omega_{\sin}t)$ translates the PSD of the timing jitter sequence ΔT_k to the frequency of the sampled sine wave, $\omega_{\sin}$. Because the amplitude error is proportional to the slew rate, i.e., the first time derivative of the signal, the power of the resulting jitter noise is proportional to the square of the signal frequency.

The (one-sided) jitter noise PSD $S_j(\mathrm{e}^{j\omega T})$ and the total jitter noise power N_j, obtained by integration over the first Nyquist band, are given by

$$S_j(\mathrm{e}^{j\omega T}) = A_{\sin}^2 \omega_{\sin}^2 \frac{S_T\left(\mathrm{e}^{j(\omega-\omega_{\sin})T}\right) + S_T\left(\mathrm{e}^{j(\omega+\omega_{\sin})T}\right)}{4}$$

$$N_j = \frac{1}{2\pi} \int_0^{\frac{\pi}{T}} S_j(\mathrm{e}^{j\omega T}) \mathrm{d}\omega = \frac{1}{2} A_{\sin}^2 \omega_{\sin}^2 \frac{\sigma_{\mathrm{LT}}^2}{2}. \tag{B.5}$$

With the signal power of the sine wave, $P_{\sin} = \frac{1}{2} A_{\sin}^2$, directly the jitter-limited SNR of Eq. (2.39) results.

In the vicinity of the carrier frequency $\omega_{\sin}$ the (one-sided) jitter noise PSD can be approximated by

$$S_j(\omega, \omega_{\sin}) \approx \frac{A_{\sin}^2 \omega_{\sin}^2 \frac{1}{2} \sigma_{\mathrm{LT}}^2}{\omega_{\mathrm{PLL}}} \left[\frac{1}{1 + \left(\frac{\omega - \omega_{\sin}}{\omega_{\mathrm{PLL}}}\right)^2} + \frac{1}{1 + \left(\frac{\omega + \omega_{\sin}}{\omega_{\mathrm{PLL}}}\right)^2} \right]$$

$$\frac{1}{2\pi} \int_0^{+\infty} S_j(\omega, \omega_{\sin}) \mathrm{d}\omega = \frac{1}{2} A_{\sin}^2 \omega_{\sin}^2 \frac{\sigma_{\mathrm{LT}}^2}{2}. \tag{B.6}$$

B.2 Non Return-to-Zero DAC

When the NRZ-DAC processes a sine wave $d_k = A_{\sin} \sin(k\omega_{\sin} T)$, then the normalized amplitude error sequence becomes

$$e_k = \frac{A_{\sin}}{T} \cdot [\sin(k\omega_{\sin} T) - \sin((k-1)\omega_{\sin} T)] \, \Delta T_k. \tag{B.7}$$

Using the identity

$$\sin(a) - \sin(b) = 2 \sin\left(\frac{a-b}{2}\right) \cos\left(\frac{a+b}{2}\right) \tag{B.8}$$

we can write

$$e_k = 2 \frac{A_{\sin}}{T} \sin\left(\frac{\omega_{\sin} T}{2}\right) \cdot \cos\left(k\omega_{\sin} T - \frac{\omega_{\sin} T}{2}\right) \Delta T_k. \tag{B.9}$$

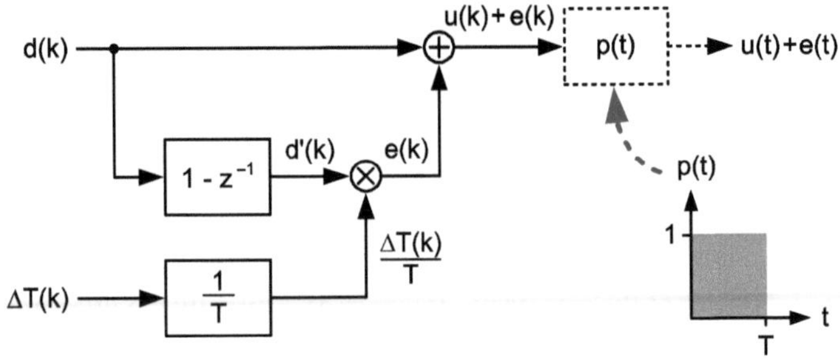

Fig. B.2 Simplified timing error model for DAC

The power of the error sequence e_k can be calculated as

$$E\left(e_k^2\right) = \frac{1}{2}A_{\sin}^2\omega_{\sin}^2\mathrm{sinc}^2\left(\frac{\omega_{\sin}T}{2}\right)E\left(\Delta T_k^2\right)$$

$$\mathrm{sinc}(x) = \frac{\sin(x)}{x}. \tag{B.10}$$

The (one-sided) phase noise PSD and total jitter noise power of the NRZ-DAC are given by

$$S_{j,\mathrm{NRZ}}(\mathrm{e}^{j\omega T}) = A_{\sin}^2\omega_{\sin}^2\mathrm{sinc}^2\left(\frac{\omega_{\sin}T}{2}\right)\frac{S_T\left(\mathrm{e}^{j(\omega-\omega_{\sin})T}\right)+S_T\left(\mathrm{e}^{j(\omega+\omega_{\sin})T}\right)}{4}$$

$$N_{j,\mathrm{NRZ}} = \frac{1}{2\pi}\int_0^{\frac{\pi}{T}} S_{\mathrm{jitter}}(\mathrm{e}^{j\omega T})\mathrm{d}\omega = \frac{1}{2}A_{\sin}^2\omega_{\sin}^2\mathrm{sinc}^2\left(\frac{\omega_{\sin}T}{2}\right)\frac{\sigma_{\mathrm{LT}}^2}{2}. \tag{B.11}$$

The signal power residing in the first Nyquist band of an NRZ-DAC is

$$P_{\sin} = \frac{1}{2}A_{\sin}^2\mathrm{sinc}^2\left(\frac{\omega_{\sin}T}{2}\right). \tag{B.12}$$

Combining (B.11) and (B.12), the generally valid expression for the SNR in Eq. (2.39) results.

In case the maximum frequency of interest is sufficiently smaller than the Nyquist frequency, such that the spectral shaping of the output signal due to the zero-order hold can be neglected, then the simplified timing error model of Fig. B.2 is sufficient [58, 187]. The jitter noise PSD at the DAC output can then also be approximated by Eq. (2.42), at least with respect to the bulk of the jitter noise power residing around the carrier frequency.

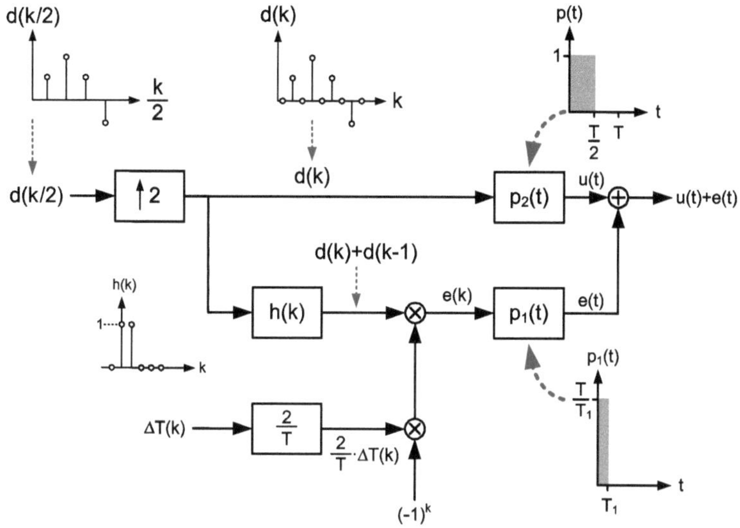

Fig. B.3 Model for RZ-DAC with sampling jitter

B.3 Return-to-Zero DAC

The Return-to-Zero behavior is controlled by two clock edges in every sampling period. We assume that both clock edges have the same statistical properties and are generated from the same clock source by ideal frequency division. The clock source, usually an integrated VCO, runs at a much higher frequency. Because the PLL-bandwidth and the total power of the jitter sequence do not change, we can write

$$S_{T,\mathrm{RZ}}(\mathrm{e}^{j\omega\frac{T}{2}}) = T \cdot \frac{1 - \mathrm{e}^{-2\omega_{\mathrm{PLL}}\frac{T}{2}}}{\left|1 - \mathrm{e}^{-\omega_{\mathrm{PLL}}\frac{T}{2}}\mathrm{e}^{-j\omega\frac{T}{2}}\right|^2} \cdot \frac{\sigma_{\mathrm{LT}}^2}{2}$$

$$E\left(\Delta T_k^2\right) = \frac{1}{4\pi}\int_0^{\frac{2\pi}{T}} S_{T,\mathrm{RZ}}(\mathrm{e}^{j\omega\frac{T}{2}})\mathrm{d}\omega = \frac{\sigma_{\mathrm{LT}}^2}{2}. \tag{B.13}$$

In Eq. (B.13) we assume that the timing jitter ΔT_k is sampled at a rate of $2/T$. Therefore, also the integration for the total jitter power is performed from 0 to $2\pi/T$.

We can model the RZ-signal generation at twice the sampling rate, as shown in Fig. B.3. First we perform an upsampling with zero-insertion by a factor of two.[1]

[1] We only consider half-clock Return-to-Zero.

The resulting code sequence d_k, now at twice the sampling rate $2/T$, is then applied to the half-clock RZ unit impulse response $p_2(t)$, having a duration of $T/2$. The ideal output signal of the RZ-DAC can be written as

$$u_{RZ}(t) = \sum_k d_k \cdot \frac{1-(-1)^k}{2} \cdot \mathrm{rect}\left(\frac{t - k \cdot T - \frac{T}{4}}{\frac{T}{2}}\right). \tag{B.14}$$

The term $\frac{1}{2}\left(1-(-1)^k\right)$ allows us to represent the input code sequence at twice the sampling rate $2/T$, because every second sample (even k) is automatically set to zero. The nonzero samples (odd k) represent the original input data sampled with the clock period T.

Assuming a sine-wave input, we have after the upsampling and zero-insertion process:

$$d_k = A_{\sin} \sin\left(k\omega_{\sin}\frac{T}{2}\right) \cdot \frac{1-(-1)^k}{2}. \tag{B.15}$$

The effect of the clock jitter on the falling edge of the output pulses inherently contains a sign inversion. Instead of performing this sign inversion on the input code sequence after the upsampling process, in the model of Fig. B.3, we perform a modulation of the timing error sequence with $(-1)^k$. This is equivalent to shifting the jitter noise PSD of Eq. (B.13) from 0 to the clock frequency $\omega_0 = \frac{2\pi}{T}$. The modulated timing error is then multiplied with the sequence $d_k + d_{k-1}$, generated by the hold function $h(k) = \delta(k) + \delta(k-1)$. The error pulse sequence e_k can be written as

$$e_k = (d_k + d_{k-1}) \cdot \frac{2}{T} \Delta T_k \cdot (-1)^k. \tag{B.16}$$

Using (B.15), the error pulse sequence e_k can be expanded into

$$\begin{aligned} e_k &= \frac{2}{T} A_{\sin}\left[\sin\left(k\omega_{\sin}\frac{T}{2}\right) \cdot \frac{1-(-1)^k}{2}\right. \\ &\qquad \left. + \sin\left((k-1)\omega_{\sin}\frac{T}{2}\right) \cdot \frac{1-(-1)^{k-1}}{2}\right] \cdot \Delta T_k \cdot (-1)^k \\ &= \frac{2}{T} A_{\sin} \cos\left(\frac{\omega_{\sin} T}{4}\right) \sin\left(\frac{(2k-1)\omega_{\sin} T}{4}\right) \Delta T_k \cdot (-1)^k \\ &\qquad - \frac{2}{T} A_{\sin} (-1)^k \sin\left(\frac{\omega_{\sin} T}{4}\right) \cos\left(\frac{(2k-1)\omega_{\sin} T}{4}\right) \cdot \Delta T_k \cdot (-1)^k. \end{aligned} \tag{B.17}$$

Finally the (one-sided) phase noise PSD of the half-clock RZ-DAC processing a sine-wave signal is given by

$$S_{j,\mathrm{RZ}}(\mathrm{e}^{j\omega\frac{T}{2}}) = \frac{1}{T^2}A_{\sin}^2\sin^2\left(\frac{\omega_{\sin}T}{4}\right)\left[S_{T,\mathrm{RZ}}\left(\mathrm{e}^{j(\omega-\omega_{\sin})\frac{T}{2}}\right)+S_{T,\mathrm{RZ}}\left(\mathrm{e}^{j(\omega+\omega_{\sin})\frac{T}{2}}\right)\right]$$
$$+\frac{1}{T^2}A_{\sin}^2\cos^2\left(\frac{\omega_{\sin}T}{4}\right)\left[S_{T,\mathrm{RZ}}\left(\mathrm{e}^{j(\omega-\omega_0-\omega_{\sin})\frac{T}{2}}\right)\right.$$
$$\left.+S_{T,\mathrm{RZ}}\left(\mathrm{e}^{j(\omega-\omega_0+\omega_{\sin})\frac{T}{2}}\right)\right]. \tag{B.18}$$

To derive the low-frequency limit of the jitter noise power $N_{j,\mathrm{RZ}}$, we use the following approximation for the (one-sided) jitter noise PSD when the signal frequency converges toward zero:

$$\lim_{\omega_{\sin}\to 0} S_{j,\mathrm{RZ}}(\omega) \approx S_{j,\mathrm{RZ},0}(\omega)$$

$$S_{j,\mathrm{RZ},0}(\omega) = \frac{4}{T^2}A_{\sin}^2\frac{\sigma_{\mathrm{LT}}^2}{\omega_{\mathrm{PLL}}}\left(\frac{1}{1+\left(\frac{\omega-\omega_0}{\omega_{\mathrm{PLL}}}\right)^2}+\frac{1}{1+\left(\frac{\omega+\omega_0}{\omega_{\mathrm{PLL}}}\right)^2}\right). \tag{B.19}$$

By integrating $S_{j,\mathrm{RZ},0}(\omega)$ over the first Nyquist band we get an approximation $N_{j,\mathrm{RZ},0}$ for the total jitter noise power of a RZ-DAC processing a very-low-frequency sine wave:

$$N_{j,\mathrm{RZ},0} = \frac{1}{2\pi}\int_0^{\frac{1}{2}\omega_0} S_{j,\mathrm{RZ},0}(\omega)\mathrm{d}\omega$$
$$= \frac{4}{T^2}A_{\sin}^2\frac{\sigma_{\mathrm{LT}}^2}{2\pi\omega_{\mathrm{PLL}}}\left[\int_0^{\frac{1}{2}\omega_0}\frac{\mathrm{d}\omega}{1+\left(\frac{\omega-\omega_0}{\omega_{\mathrm{PLL}}}\right)^2}+\int_0^{\frac{1}{2}\omega_0}\frac{\mathrm{d}\omega}{1+\left(\frac{\omega+\omega_0}{\omega_{\mathrm{PLL}}}\right)^2}\right]$$
$$= \frac{4}{T^2}A_{\sin}^2\frac{\sigma_{\mathrm{LT}}^2}{2\pi}\left[\int_0^{\frac{-\omega_0}{2\omega_{\mathrm{PLL}}}}\frac{\mathrm{d}\nu}{1+\nu^2}+\int_0^{\frac{3\omega_0}{2\omega_{\mathrm{PLL}}}}\frac{\mathrm{d}\nu}{1+\nu^2}\right]$$
$$= \frac{4}{T^2}A_{\sin}^2\frac{\sigma_{\mathrm{LT}}^2}{2\pi}\left[\arctan\left(\frac{3\omega_0}{2\omega_{\mathrm{PLL}}}\right)-\arctan\left(\frac{\omega_0}{2\omega_{\mathrm{PLL}}}\right)\right]. \tag{B.20}$$

Using the identity

$$\arctan(x)+\arctan(y) = \arctan\left(\frac{x+y}{1-xy}\right) \quad \text{for } x\cdot y<1 \tag{B.21}$$

Eq. (B.20) becomes

$$N_{j,\mathrm{RZ},0} = \frac{4}{T^2} A_{\sin}^2 \frac{\sigma_{\mathrm{LT}}^2}{2\pi} \arctan\left(\frac{4\omega_0 \omega_{\mathrm{PLL}}}{3\omega_0^2 + 4\omega_{\mathrm{PLL}}^2}\right). \tag{B.22}$$

Since $\omega_{\mathrm{PLL}} << \omega_0$, we can use the approximation:

$$\arctan(x) \approx x \quad \text{for small } x \tag{B.23}$$

and (B.22) becomes

$$N_{j,\mathrm{RZ},0} = \frac{4}{T^2} A_{\sin}^2 \frac{\sigma_{\mathrm{LT}}^2}{2\pi} \frac{4\omega_{\mathrm{PLL}}}{3\omega_0}. \tag{B.24}$$

According to Eq. (1.9), the baseband amplitude of a half-clock RZ-DAC is reduced by the duty factor $D = T_S/T$. Therefore, the signal power residing in the first Nyquist band with $D = 0.5$ is given by

$$P_{S,\mathrm{RZ}} = \frac{1}{2}\left(\frac{A_{\sin}}{2}\right)^2 = \frac{A_{\sin}^2}{8}. \tag{B.25}$$

In contrast to an NRZ-implementation, the jitter-limited SNR of a RZ-DAC, evaluated over the first Nyquist band, approaches a constant value for low signal frequencies:

$$\lim_{\omega_{\sin}\to 0} \mathrm{SNR}_{j,\mathrm{RZ}} \approx \mathrm{SNR}_{j,\mathrm{RZ}}(0) = -20\log_{10}\left(4\sigma_{\mathrm{LT}}\sqrt{\frac{\omega_{\mathrm{PLL}}\omega_0}{3\pi^3}}\right). \tag{B.26}$$

B.4 Jitter in Multitone Systems

In a multitone signal, the individual carriers are independently affected by sampling jitter. The jitter noise PSD's of the single carriers $S_j(\omega, \omega_k)$ can be summed to yield the total jitter noise PSD of a sampled multitone signal:

$$S_{j,\mathrm{DMT}}(\omega) = \sum_{k=0}^{N_C} S_j(\omega, \omega_k). \tag{B.27}$$

When we use a PLL-clock for the data converters, then the phase noise contribution of each carrier can be described by Eq. (B.5) or, at least in the vicinity of the carrier, by the approximation (B.6). In an NRZ-DAC, the resulting jitter noise PSD generated around each individual carrier k at frequency ω_k is given by

$$S_j(\omega,\omega_k) \approx \frac{A_{\mathrm{DMT}}^2 \frac{1}{2}\sigma_{\mathrm{LT}}^2}{\omega_{\mathrm{PLL}}} \left[\frac{\omega_k^2}{1+\left(\frac{\omega-\omega_k}{\omega_{\mathrm{PLL}}}\right)^2} + \frac{\omega_k^2}{1+\left(\frac{\omega+\omega_k}{\omega_{\mathrm{PLL}}}\right)^2} \right]$$

$$\omega_k = \omega_{\min} + k\Delta\omega \qquad \text{for } k = 0,1,\ldots,N_C = \frac{\omega_{\max}-\omega_{\min}}{\Delta\omega}. \tag{B.28}$$

Using (2.7), the total jitter noise PSD, obtained by summing the contributions of all carriers, can be written as

$$S_{j,\mathrm{DMT}}(\omega) \approx \frac{A_{\mathrm{peak}}^2 \sigma_{\mathrm{LT}}^2}{\mathrm{CF}^2 \omega_{\mathrm{PLL}}(\omega_{\max}-\omega_{\min})} \sum_{k=0}^{N_C} \left[\frac{\omega_k^2 \Delta\omega}{1+\left(\frac{\omega-\omega_k}{\omega_{\mathrm{PLL}}}\right)^2} + \frac{\omega_k^2 \Delta\omega}{1+\left(\frac{\omega+\omega_k}{\omega_{\mathrm{PLL}}}\right)^2} \right]$$

$$\omega_k = \omega_{\min} + k\Delta\omega \qquad \text{for } k = 0,1,\ldots,N_C = \frac{\omega_{\max}-\omega_{\min}}{\Delta\omega}. \tag{B.29}$$

When the carrier spacing $\Delta\omega$ is much smaller than ω_{PLL}, then the noise density does not change significantly when we increase the number of carriers and keep the Crest Factor constant. Although the amplitude of the carriers A_{DMT} decreases according to (2.7), increasingly more carriers contribute to the noise. The average in-band signal power in a given bandwidth that is larger than the carrier spacing remains constant, even if carriers are added or removed. The same is true for the average jitter noise power obtained by integration of (B.29) over the same bandwidth. To first order, the overall jitter noise PSD does not depend on the number of carriers, as long as the condition $\Delta\omega << \omega_{\mathrm{PLL}}$ is fulfilled, i.e., for a sufficiently dense multitone signal [60].

In the limit $\Delta\omega \to 0$ the discrete sum can be transformed into an integral and solved:

$$\lim_{\Delta\omega\to 0} \sum_{k=0}^{N_C} \frac{(\omega_{\min}+k\Delta\omega)^2}{1+\left(\frac{\omega \mp \omega_{\min} \mp k\Delta\omega}{\omega_{\mathrm{PLL}}}\right)^2}$$

$$\Delta\omega = \int_{\omega_{\min}}^{\omega_{\max}} \frac{\nu^2}{1+\left(\frac{\omega\mp\nu}{\omega_{\mathrm{PLL}}}\right)^2} \mathrm{d}\nu. \tag{B.30}$$

The total jitter noise PSD of a homogeneous multitone signal with sufficiently high carrier density and $\Delta\omega << \omega_{\mathrm{PLL}}$ can thus be approximated by

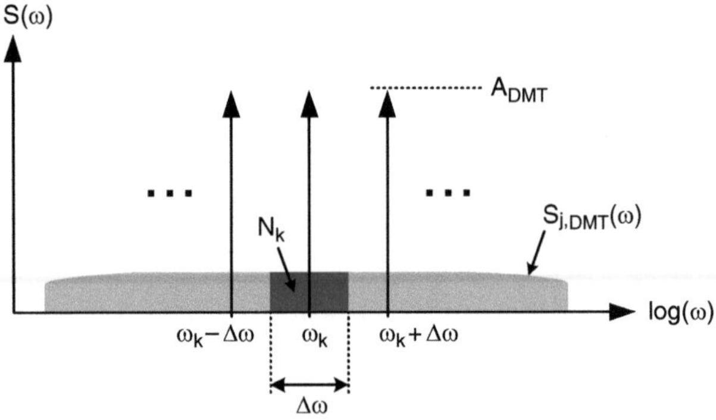

Fig. B.4 SNR per carrier

$$S_{j,\mathrm{DMT}}(\omega) \approx \frac{A_{\mathrm{peak}}^2}{\mathrm{CF}^2}\sigma_{\mathrm{LT}}^2 \cdot (B_j(\omega) + B_j(-\omega))$$

$$\begin{aligned} B_j(\omega) &= \omega_{\mathrm{PLL}} + \frac{\omega \cdot \omega_{\mathrm{PLL}}}{\omega_{\max} - \omega_{\min}} \cdot \ln\left(\frac{\omega_{\mathrm{PLL}}^2 + (\omega - \omega_{\max})^2}{\omega_{\mathrm{PLL}}^2 + (\omega - \omega_{\min})^2}\right) \\ &\quad + \frac{\omega^2 - \omega_{\mathrm{PLL}}^2}{\omega_{\max} - \omega_{\min}}\left[\arctan\left(\frac{\omega_{\max} - \omega}{\omega_{\mathrm{PLL}}}\right) - \arctan\left(\frac{\omega_{\min} - \omega}{\omega_{\mathrm{PLL}}}\right)\right]. \end{aligned} \tag{B.31}$$

To calculate the frequency-dependent SNR per carrier, we observe that in a homogeneous multitone signal, each carrier occupies a frequency band of $\Delta\omega$, equal to the carrier spacing. We are interested in the ratio of the local signal power to the jitter noise power falling into the carrier bandwidth. As shown in Fig. 2.28, repeated here for convenience as Fig. B.4, if the carrier bandwidth $\Delta\omega$ is small, we can approximate the total jitter noise power for carrier k as

$$N_k \approx \frac{A_{\mathrm{peak}}^2}{\mathrm{CF}^2}\sigma_{\mathrm{LT}}^2(B_j(\omega_k) + B_j(-\omega_k))\frac{\Delta\omega}{2\pi}. \tag{B.32}$$

The signal power of each carrier in a homogeneous multitone signal is equal to

$$P_{\mathrm{carrier}} = \frac{1}{2}A_{\mathrm{DMT}}^2 \approx \frac{A_{\mathrm{peak}}^2}{\mathrm{CF}^2}\frac{\Delta\omega}{\omega_{\max} - \omega_{\min}}. \tag{B.33}$$

By combining (B.33) and (B.32) the SNR/carrier at frequency ω_k is obtained [60]:

$$\mathrm{SNR}(\omega_k) = \frac{2\pi}{\sigma_{\mathrm{LT}}^2(\omega_{\max} - \omega_{\min})(B_j(\omega_k) + B_j(-\omega_k))}. \tag{B.34}$$

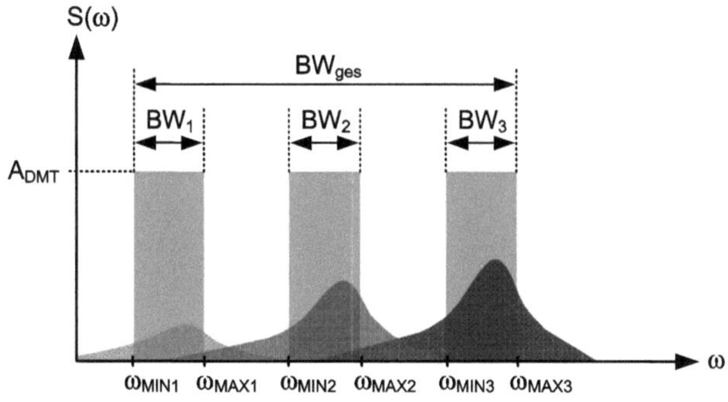

Fig. B.5 Multiband DMT-signal affected by sampling jitter

Figure B.5 shows how a multiband DMT-signal is affected by sampling jitter. The worst noise performance is expected in the highest band, while the lower bands are gradually less affected by jitter.

Using (B.29) through (B.31), the jitter noise PSD of a homogeneous multiband DMT-signal can be written as

$$S_{j,\mathrm{DMT}}(\omega) \approx \frac{A_{\mathrm{peak}}^2}{\mathrm{CF}^2}\sigma_{\mathrm{LT}}^2 \sum_{n=1}^{L} \frac{\omega_{\max,n} - \omega_{\min,n}}{\sum\limits_{k=1}^{L}(\omega_{\max,k} - \omega_{\min,k})} (B_{j,n}(\omega) + B_{j,n}(-\omega)). \quad \text{(B.35)}$$

In the multiband case the PSD shape of each of the subbands must be taken into account with $B_{j,n}(\omega) = B_j(\omega, \omega_{\max,n}, \omega_{\min,n})$. By assuming again $\Delta\omega << \omega_{\mathrm{PLL}}$, we can calculate from the carrier amplitude given by Eq. (2.8) and the noise PSD (B.35) the jitter-limited SNR/carrier for the tone located at ω_k in a homogeneous multiband multitone signal as

$$\mathrm{SNR}(\omega_k) = \frac{2\pi}{\sigma_{\mathrm{LT}}^2 \sum\limits_{n=1}^{L}(\omega_{\max,n} - \omega_{\min,n}) \cdot (B_{j,n}(\omega_k) + B_{j,n}(-\omega_k))}. \quad \text{(B.36)}$$

Figure B.6 shows a comparison of the simulated and calculated jitter noise PSD for a homogeneous multiband DMT-signal. Expect for frequencies close to the Nyquist frequency, the jitter noise prediction according to Eq. (B.35) is reasonably accurate. An experimental verification of the multitone jitter formulae derived above is found in Sect. 2.3.3.

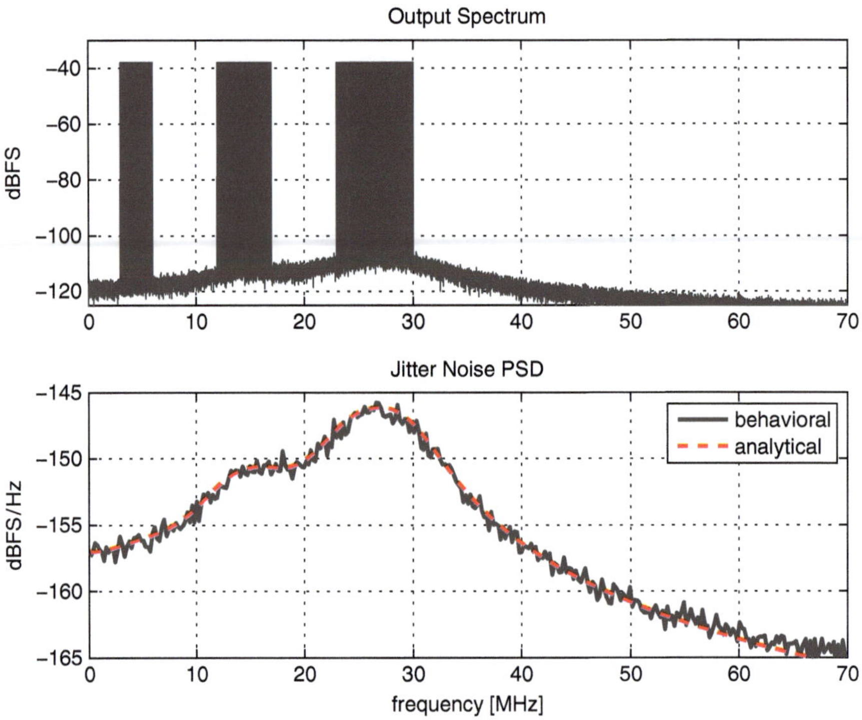

Fig. B.6 Simulation and analytical approximation

Appendix C
Code-Dependent Output Resistance

In a current-steering D/A-converter, the cumulative output resistance of the current cells connected to the same output node forms a current divider with the load resistor. Because the number of current sources switched to the positive and negative output depends on the digital input code, this current division becomes code dependent and distorts the output characteristic of the converter. In the following we will derive the integral nonlinearity (INL) of a single-ended and a fully differential converter affected by the finite output resistance of its unit current cells.

C.1 Single-Ended Converter

Referring to Fig. 3.6, we assume that the DAC consists of N identical unit current sources with current I_{unit} and output resistance R_{unit}. With the input code $k \in [0; N]$ the output currents delivered to the single-ended load resistors R_L are

$$\begin{aligned} I_{\text{OUTP}}(k) &= kI_{\text{unit}} \cdot \frac{R_{\text{unit}}}{R_{\text{unit}} + kR_L} \\ I_{\text{OUTN}}(k) &= (N-k)I_{\text{unit}} \cdot \frac{R_{\text{unit}}}{R_{\text{unit}} + (N-k)R_L}. \end{aligned} \tag{C.1}$$

The single-ended output voltage is therefore

$$V_{\text{OUTP}}(k) = kI_{\text{unit}} \cdot \frac{R_L R_{\text{unit}}}{R_{\text{unit}} + kR_L}. \tag{C.2}$$

The ideal output ramp having the same endpoints, i.e., the same value for $k = 0$ and $k = N$, is given by

M. Clara, *High-Performance D/A-Converters*, Springer Series in Advanced Microelectronics 36, DOI 10.1007/978-3-642-31229-8,

$$V_{\mathrm{OUTP,ideal}}(k) = kI_{\mathrm{unit}} \cdot \frac{R_L R_{\mathrm{unit}}}{R_{\mathrm{unit}} + NR_L}. \tag{C.3}$$

The difference $\Delta V(k)$ between the actual output characteristic and the ideal output ramp is proportional to the INL of the converter. Using (C.2) and (C.3), we get

$$\Delta V(k) = kR_L R_{\mathrm{unit}} I_{\mathrm{unit}} \cdot \left(\frac{1}{R_{\mathrm{unit}} + kR_L} - \frac{1}{R_{\mathrm{unit}} + NR_L} \right). \tag{C.4}$$

First we calculate the derivative of $\Delta V(k)$ with respect to k:

$$\frac{\mathrm{d}}{\mathrm{d}k}\Delta V(k) = R_L R_{\mathrm{unit}} I_{\mathrm{unit}} \cdot \left(\frac{R_{\mathrm{unit}}}{(R_{\mathrm{unit}} + kR_L)^2} - \frac{1}{R_{\mathrm{unit}} + NR_L} \right). \tag{C.5}$$

Equating (C.5) to zero and solving for k, we get the code with the maximum deviation of the actual output characteristic from the ideal output ramp:

$$R_L R_{\mathrm{unit}} I_{\mathrm{unit}} \cdot \left(\frac{R_{\mathrm{unit}}}{(R_{\mathrm{unit}} + kR_L)^2} - \frac{1}{R_{\mathrm{unit}} + NR_L} \right) = 0,$$

$$R_{\mathrm{unit}}\left(R_{\mathrm{unit}} + NR_L\right) - \left(R_{\mathrm{unit}} + kR_L\right)^2 = 0,$$

$$k^2 R_L^2 + 2kR_L R_{\mathrm{unit}} - NR_L R_{\mathrm{unit}} = 0,$$

$$k = -\frac{R_{\mathrm{unit}}}{R_L} \pm \sqrt{\left(\frac{R_{\mathrm{unit}}}{R_L}\right)^2 + N\frac{R_{\mathrm{unit}}}{R_L}}. \tag{C.6}$$

Because, by definition, k must be a positive integer, we have only one valid solution:

$$k_{\mathrm{INL,max}} = -\frac{R_{\mathrm{unit}}}{R_L} + \sqrt{\left(\frac{R_{\mathrm{unit}}}{R_L}\right)^2 + N\frac{R_{\mathrm{unit}}}{R_L}}. \tag{C.7}$$

Since in practice the output impedance of the current sources is always much larger than the load impedance, we can find an approximation for k_{max} by letting the ratio R_{unit}/R_L go toward infinity:

$$k_{\mathrm{INL,max}} \approx \lim_{R_{\mathrm{unit}}/R_L \to \infty} \left[-\frac{R_{\mathrm{unit}}}{R_L} + \sqrt{\left(\frac{R_{\mathrm{unit}}}{R_L}\right)^2 + N\frac{R_{\mathrm{unit}}}{R_L}} \right] = \frac{N}{2}. \tag{C.8}$$

As expected, the maximum INL for the single-ended output will occur around midscale. By inserting (C.8) into (C.4), we can calculate the maximum deviation of the single-ended output from the ideal ramp:

$$\Delta V_{\max} \approx \Delta V\left(\frac{N}{2}\right) = R_L R_{\text{unit}} I_{\text{unit}} \cdot \left(\frac{\frac{N}{2}}{R_{\text{unit}} + \frac{N}{2} R_L} - \frac{\frac{N}{2}}{R_{\text{unit}} + N R_L} \right). \tag{C.9}$$

Assuming $R_{\text{unit}} >> N \cdot R_L$, Eq. (C.9) reduces to [74]

$$\Delta V_{\max} \approx \frac{I_{\text{unit}} R_L^2 N^2}{4 R_{\text{unit}}}. \tag{C.10}$$

If $\Delta V_{\max}$ is small, then we can approximate the actual single-ended LSB-voltage by $V_{\text{LSB}} \approx I_{\text{unit}} R_L$. The achievable worst-case INL for a single-ended current-steering D/A-converter with finite current source output resistance is then given by

$$\text{INL}_{\max,s} \approx \frac{\Delta V_{\max}}{I_{\text{unit}} R_L} = \frac{N^2}{4} \cdot \frac{R_L}{R_{\text{unit}}}. \tag{C.11}$$

C.2 Fully Differential Converter

Referring to Fig. 3.6, the differential output current for input code k can be written as

$$\begin{aligned} I_{\text{OUT}}(k) &= I_{\text{OUTP}}(k) - I_{\text{OUTN}}(k) \\ &= k I_{\text{unit}} \cdot \frac{R_{\text{unit}}}{R_{\text{unit}} + k R_L} - (N-k) I_{\text{unit}} \cdot \frac{R_{\text{unit}}}{R_{\text{unit}} + (N-k) R_L}. \end{aligned} \tag{C.12}$$

The differential output ramp is described by

$$V_{\text{OUT}}(k) = \frac{k R_L R_{\text{unit}} I_{\text{unit}}}{R_{\text{unit}} + k R_L} - \frac{(N-k) R_L R_{\text{unit}} I_{\text{unit}}}{R_{\text{unit}} + (N-k) R_L}. \tag{C.13}$$

The *ideal* output ramp, fitted to the endpoints of Eq. (C.13), is therefore

$$V_{\text{OUT,ideal}}(k) = (2k - N) I_{\text{unit}} \cdot \frac{R_L R_{\text{unit}}}{R_{\text{unit}} + N R_L}. \tag{C.14}$$

Subtracting (C.14) from (C.13) we get the deviation of the differential output characteristic from the ideal output ramp:

$$\Delta V(k) = \frac{k R_L R_{\text{unit}} I_{\text{unit}}}{R_{\text{unit}} + k R_L} - \frac{(N-k) R_L R_{\text{unit}} I_{\text{unit}}}{R_{\text{unit}} + (N-k) R_L} - \frac{(2k-N) R_L R_{\text{unit}} I_{\text{unit}}}{R_{\text{unit}} + N R_L}. \tag{C.15}$$

Differentiation of (C.15) with respect to k, solving for the zeros and performing the limit transition $R_{\text{unit}}/R_L \rightarrow \infty$, gives the approximate code positions of the maximum deviation:

$$k_{\text{INL,max}} \approx N \cdot \frac{3 \pm \sqrt{3}}{6} = \frac{N}{2} \pm \frac{\sqrt{3}}{6} N. \quad \text{(C.16)}$$

MAPLE™ has been used to obtain (C.16) from (C.15). Reinserting (C.16) into (C.15) and assuming once again $R_{\text{unit}} >> N \cdot R_L$, we obtain the maximum deviation of the output characteristic from the ideal ramp in a fully differential implementation with perfectly matched load resistors:

$$\Delta V_{\max} \approx \frac{\sqrt{3} I_{\text{unit}} R_L^3 N^3}{18 R_{\text{unit}}^2}. \quad \text{(C.17)}$$

With the approximate differential LSB-size $V_{\text{LSB}} \approx 2 I_{\text{unit}} R_L$, the achievable worst-case INL for a fully differential current-steering D/A-converter with finite current source output resistance becomes

$$\text{INL}_{\max,d} \approx \frac{\Delta V_{\max}}{2 I_{\text{unit}} R_L} = \frac{\sqrt{3} N^3}{36} \cdot \frac{R_L^2}{R_{\text{unit}}^2}. \quad \text{(C.18)}$$

Because in calculating (C.11) and (C.18) we have performed the limit transition $R_{\text{unit}}/R_L \rightarrow \infty$, we are also interested in the prediction accuracy of the worst-case INL as a function of the unit current-cell resistance. Shown in Figs. C.1 and C.2 are the relative error of the estimated worst-case INL for the single-ended and fully differential DAC as a function of the ratio R_{unit}/R_L with the resolution as a parameter. The relative prediction error is defined as

$$\text{Error} = \frac{\text{INL}_{\text{calc}} - \text{INL}_{\text{sim}}}{\text{INL}_{\text{sim}}} \cdot 100\,\%, \quad \text{(C.19)}$$

where INL_{calc} is the calculated worst-case INL according to (C.11) and (C.18), respectively. INL_{sim} is the result of a behavioral simulation.

Interestingly, the prediction error is always positive, which means that the estimated INL-error is an upper bound for the actual INL-error. Recall that for the derivation of (C.11) and (C.18) we assumed an ideal converter with infinitely large output resistance compared to the load resistor. If $R_{\text{unit}} > 3.2 \times 10^5 \cdot R_L$, the prediction accuracy for a single-ended converter is better than 10 % for resolutions up to 16 bits. In a differential implementation we require for the same prediction accuracy $R_{\text{unit}} > 6.9 \times 10^5 \cdot R_L$. These values are easily achieved in practical MOS-implementations. Therefore, in most cases, Eqs. (C.11) and (C.18) will yield pretty accurate estimations of the achievable worst-case INL-error due to the finite output resistance of the current sources.

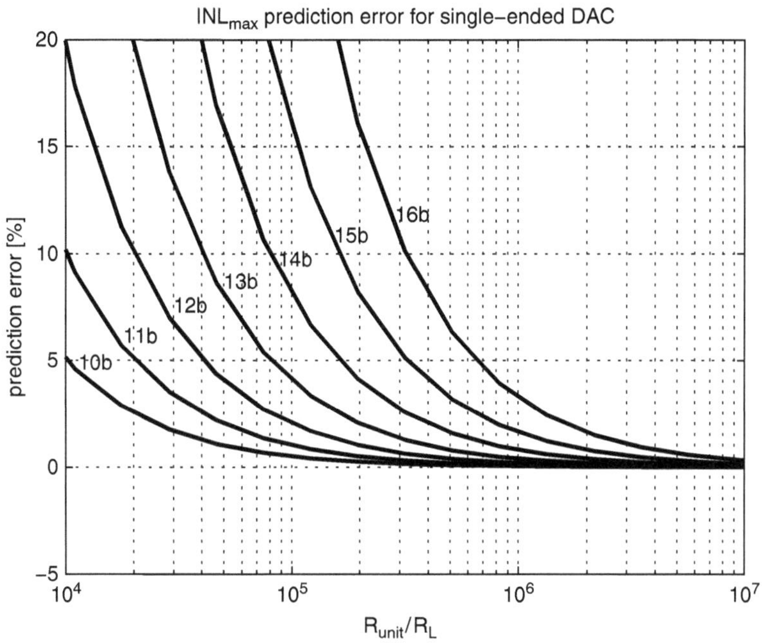

Fig. C.1 INL prediction error for single-ended DAC

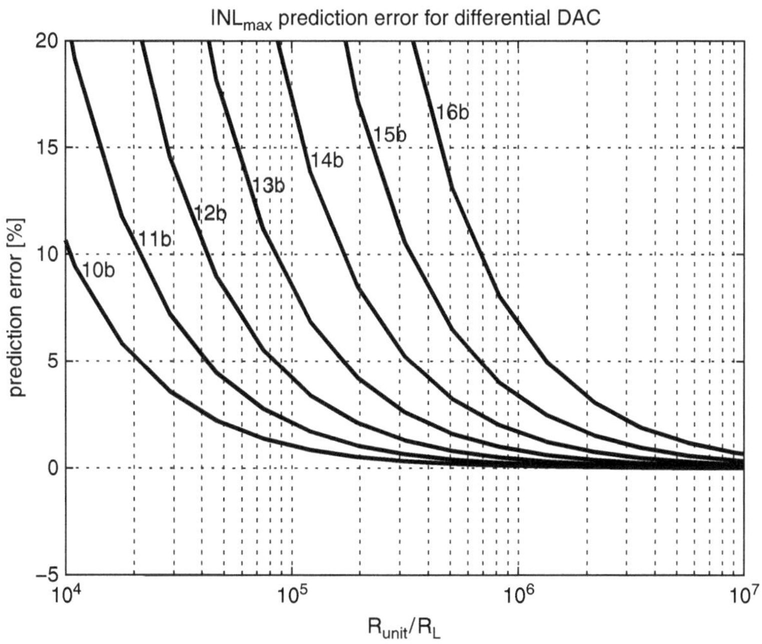

Fig. C.2 INL prediction error for differential DAC

Appendix D
Switching errors

D.1 Generalized Switching Error Model

In the general model of a unary NRZ D/A-converter affected by transitional errors (see Fig. 4.9), the sampled and quantized data $d(k) + q(k)$ is applied to the ideal unit pulse $p_{\mathrm{NRZ}}(t)$. The quantization error is represented by $q(k)$, while $d(k)$ has infinite resolution in this model. The ideal output signal of the NRZ-DAC is

$$u(t) = \sum_k (d(k) + q(k)) \cdot p_{\mathrm{NRZ}}(t - kT). \tag{D.1}$$

The switching functions $s_{\mathrm{ON}}(\cdot)$ and $s_{\mathrm{OFF}}(\cdot)$, fed with the quantized input data $d(k) + q(k)$, deliver the switching sequences $w_{\mathrm{ON}}(k)$ and $w_{\mathrm{OFF}}(k)$. These represent the number of unit cells that are being switched on and switched off at the beginning of each sample k. The switching sequences are applied to the unit error pulse shapes $p_{\mathrm{ON}}(t)$ and $p_{\mathrm{OFF}}(t)$. Their impulse responses correspond to the error made when switching a unit current cell in the respective direction. The combined output $e(t)$ represents the error signal that is finally added to the ideal DAC output $u(t)$:

$$e(t) = \sum_k \left[w_{\mathrm{ON}}(k) p_{\mathrm{ON}}(t - kT) - w_{\mathrm{OFF}}(k) p_{\mathrm{OFF}}(t - kT) \right] \tag{D.2}$$

which can also be written as

$$\begin{aligned} e(t) &= \sum_k (w_{\mathrm{ON}}(kT)\delta(t - kT)) * p_{\mathrm{ON}}(t) \\ &\quad - \sum_k (w_{\mathrm{OFF}}(kT)\delta(t - kT)) * p_{\mathrm{OFF}}(t). \end{aligned} \tag{D.3}$$

M. Clara, *High-Performance D/A-Converters*, Springer Series in Advanced Microelectronics 36, DOI 10.1007/978-3-642-31229-8,

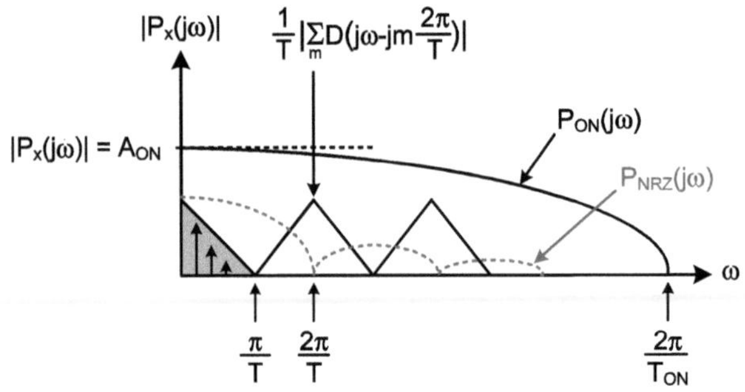

Fig. D.1 Spectrum of short error pulse

The number of switching transitions in ON- and OFF-direction for each data sample k is obtained by sampling the continuous-time functions $w_{\mathrm{ON}}(t)$ and $w_{\mathrm{OFF}}(t)$ at integer multiples of the sampling period T. These functions are thus the continuous-time representation of the switching sequences $w_{\mathrm{ON}}(k)$ and $w_{\mathrm{OFF}}(k)$. Specifically, we can obtain $w_{\mathrm{ON}}(t)$ and $w_{\mathrm{OFF}}(t)$ from $w_{\mathrm{ON}}(k)$ and $w_{\mathrm{OFF}}(k)$ by ideal reconstruction according to the sampling theorem.

The Fourier transform of (D.3) delivers

$$E\left(\mathrm{e}^{j\omega T}\right) = \frac{1}{T}\sum_m W_{\mathrm{ON}}\left(j\omega - jm\frac{2\pi}{T}\right) P_{\mathrm{ON}}(j\omega) - \frac{1}{T}\sum_m W_{\mathrm{OFF}}\left(j\omega - jm\frac{2\pi}{T}\right) P_{\mathrm{OFF}}(j\omega). \tag{D.4}$$

If we approximate the unit error pulses $p_{\mathrm{ON}}(t)$ and $p_{\mathrm{OFF}}(t)$ with rectangular functions of area A_{ON} and A_{OFF} then we can write Eq. (D.4) as

$$E\left(\mathrm{e}^{j\omega T}\right) = \frac{1}{T}\sum_m W_{\mathrm{ON}}\left(j\omega - jm\frac{2\pi}{T}\right)\cdot A_{\mathrm{ON}}\cdot \mathrm{sinc}\left(\frac{\omega T_{\mathrm{ON}}}{2}\right) - \frac{1}{T}\sum_m W_{\mathrm{OFF}}\left(j\omega - jm\frac{2\pi}{T}\right)\cdot A_{\mathrm{OFF}}\cdot \mathrm{sinc}\left(\frac{\omega T_{\mathrm{OFF}}}{2}\right). \tag{D.5}$$

The first zeros of the error pulse spectra are found at $1/T_{\mathrm{ON}}$ and $1/T_{\mathrm{OFF}}$, respectively. If we assume that the switching transitions are very fast and the associated error signals die out in a very short time span compared to the clock period, then the spectra of the unit error pulses are approximately constant over the first Nyquist band $0\cdots\pi/T$, the frequency range we are primarily interested in (see also Fig. D.1). In this case $p_{\mathrm{ON}}(t)$ and $p_{\mathrm{OFF}}(t)$ are completely characterized by their area, and, by reconstruction with an ideal brickwall-filter of bandwidth π/T, Eq. (D.5) becomes

$$E(j\omega) = E\left(e^{j\omega T}\right) \cdot \mathrm{rect}\left(\frac{\omega T}{\pi}\right) \approx \frac{A_{\mathrm{ON}}}{T} \cdot W_{\mathrm{ON}}(j\omega) - \frac{A_{\mathrm{OFF}}}{T} \cdot W_{\mathrm{OFF}}(j\omega). \quad \text{(D.6)}$$

According to (D.6), the baseband error spectrum $E(j\omega)$ is obtained by evaluating the spectra of the continuous-time representations of the switching sequences, $w_{\mathrm{ON}}(t)$ and $w_{\mathrm{OFF}}(t)$. The scaling factors are given by the respective error pulse area divided by the clock period. This indicates the averaging of the transition effect over the whole sampling period.

The corresponding time-domain representation of the "reconstructed" error signal is obtained by the inverse Fourier transform of (D.6):

$$e(t) \approx \frac{A_{\mathrm{ON}}}{T} \cdot w_{\mathrm{ON}}(t) - \frac{A_{\mathrm{OFF}}}{T} \cdot w_{\mathrm{OFF}}(t). \quad \text{(D.7)}$$

If the continuous-time representations of the switching sequences, $w_{\mathrm{ON}}(t)$ and $w_{\mathrm{OFF}}(t)$, are given analytically and the condition $T_{\mathrm{ON}}, T_{\mathrm{OFF}} << T$ holds, we can approximate the spectral content of the error signal in the first Nyquist band by the Fourier transform of (D.7).

The described "Dirac" approximation[1] for the switching error pulses allows a unified treatment for different shapes of the switching error, since its effect is fully described by the area of the error in relation to the unit pulse area.

D.2 Switching Transition Mismatch with Thermometer Coding

In a unary array with N elements, let the decoded input data sequence for a single tone with normalized amplitude $A_{\sin} \in [0, 1]$ and normalized signal frequency $\omega_{\sin}$, according to Eq. (4.14), be given by

$$d(k) = \frac{N}{2}\left(1 + A_{\sin}\cos(kT)\right). \quad \text{(D.8)}$$

With thermometer coding, the DAC-cells are switched on sequentially, always starting from the first element in the array. If the signal gradient $d(k) - d(k-1)$ is positive, then DAC-cells are only switched ON, whereas for negative signal gradients, DAC-cells are only switched OFF. The switching sequences $w_{\mathrm{ON}}(k)$ and $w_{\mathrm{OFF}}(k)$ are therefore

[1] In the time-domain representation, a constant spectrum means that the unit error pulses must be of the form $p_{\mathrm{ON}}(t) = A_{\mathrm{ON}}\delta(t)$ and $p_{\mathrm{OFF}}(t) = A_{\mathrm{OFF}}\delta(t)$, with $\delta(t)$ being the Dirac function.

$$
\begin{aligned}
& d(k) - d(k-1) > 0: \\
& w_{\mathrm{ON}}(k) = \frac{N \cdot A_{\sin}}{2}\left[\cos(kT) - \cos((k-1)T)\right] + q(k) - q(k-1) \\
& w_{\mathrm{OFF}}(k) = 0 \\
& d(k) - d(k-1) < 0: \\
& w_{\mathrm{ON}}(k) = 0 \\
& w_{\mathrm{OFF}}(k) = -\frac{N \cdot A_{\sin}}{2}\left[\cos(kT) - \cos((k-1)T)\right] - (q(k) - q(k-1)). \qquad \text{(D.9)}
\end{aligned}
$$

The minus sign for $w_{\mathrm{OFF}}(k)$ is necessary because the number of switching cells must be a positive number, but the differentiated signal is negative.

The differentiated quantization noise $q(k)-q(k-1)$ also appears in the switching error path and will introduce excess quantization noise. After filtering $w_{\mathrm{ON}}(k)$ and $w_{\mathrm{OFF}}(k)$ with the unit error pulses $p_{\mathrm{ON}}(t)$ and $p_{\mathrm{OFF}}(t)$, the excess quantization noise has the (one-sided) PSD:

$$
S_{\mathrm{eq}}(\omega) = \frac{\Delta^2}{12} \cdot \frac{2}{f_{\mathrm{CLK}}} \cdot \frac{A_{\mathrm{ON}}^2 + A_{\mathrm{OFF}}^2}{2\Delta^2 T^2} \cdot \left|1 - \mathrm{e}^{-j\omega T}\right|^2 \qquad \text{(D.10)}
$$

Δ is the LSB-size, i.e., the value of the unit DAC-elements and thus the height of the unit pulse $p_{\mathrm{NRZ}}(t)$. By integrating (D.10) over the frequency band of interest, $0 \ldots f_B$, the excess quantization noise power due to the switching error can be calculated:

$$
N_{\mathrm{eq}}(f_B) = \frac{A_{\mathrm{ON}}^2 + A_{\mathrm{OFF}}^2}{2\Delta^2 T^2} \cdot \frac{\Delta^2}{12} \cdot \frac{2}{\pi} \cdot \left[\frac{2\pi f_B}{f_{\mathrm{CLK}}} - \sin\left(\frac{2\pi f_B}{f_{\mathrm{CLK}}}\right)\right]. \qquad \text{(D.11)}
$$

As long as the average energy of the error pulses for the ON- and OFF-transition is small compared to the unit pulse energy ($\Delta^2 T^2$), the excess quantization noise added to the output signal is negligibly small. Additionally, its power spectrum has a zero at DC due to the inherent differentiation.

The deterministic part of the reconstructed error signal $e(t)$ is given by:

$$
\begin{aligned}
e(t) &= \frac{A_{\mathrm{OFF}}}{T} \cdot \frac{N}{2} A_{\sin}\left[\cos(t) - \cos(t-T)\right] \quad \text{for } \frac{T}{2} \le t < \pi + \frac{T}{2}, \\
e(t) &= \frac{A_{\mathrm{ON}}}{T} \cdot \frac{N}{2} A_{\sin}\left[\cos(t) - \cos(t-T)\right] \quad \text{for } \pi + \frac{T}{2} \le t < 2\pi + \frac{T}{2}.
\end{aligned} \qquad \text{(D.12)}
$$

A Fourier series expansion of (D.12) with the aid of MAPLE™ yields the following expression for the harmonic distortion products relative to the fundamental tone with amplitude $\frac{1}{2} N \Delta A_{\sin} \mathrm{sinc}\left(\frac{\pi}{2\mathrm{OSR}}\right)$:

$$\text{for } n = 1, 2 \ldots :$$
$$\mathrm{HD}_{2n} = 2 \cdot \frac{|A_{\mathrm{ON}} - A_{\mathrm{OFF}}|}{\Delta T} \cdot \frac{\sqrt{2 - 2\cos\left(\frac{\pi}{\mathrm{OSR}}\right)}}{\pi(2n+1)(2n-1)} \cdot \frac{\frac{\pi}{2\mathrm{OSR}}}{\sin\left(\frac{\pi}{2\mathrm{OSR}}\right)}$$
$$\mathrm{HD}_{2n+1} = 0. \tag{D.13}$$

In the case of a rise and fall time mismatch the area of the unit error pulses is given by

$$A_{\mathrm{ON}} = \frac{T_{\mathrm{rise}}\Delta}{2}$$
$$A_{\mathrm{OFF}} = \frac{T_{\mathrm{fall}}\Delta}{2}. \tag{D.14}$$

And Eq. (D.13) becomes

$$\text{for } n = 1, 2 \ldots :$$
$$\mathrm{HD}_{2n} = \frac{|T_{\mathrm{rise}} - T_{\mathrm{fall}}|}{T} \cdot \frac{\sqrt{2 - 2\cos\left(\frac{\pi}{\mathrm{OSR}}\right)}}{\pi(2n+1)(2n-1)} \cdot \frac{\frac{\pi}{2\mathrm{OSR}}}{\sin\left(\frac{\pi}{2\mathrm{OSR}}\right)}$$
$$\mathrm{HD}_{2n+1} = 0. \tag{D.15}$$

D.3 Switching Transition Mismatch with Data Weighted Averaging

With data weighted averaging (DWA) the DAC-elements for a given sample k are switched on beginning from the first switched-off element of sample $k-1$. When the last element of the array is reached, the process wraps around to the first element in the array. Neglecting a possible $\Sigma\Delta$-modulation, the switching sequences can be written as

$$\begin{aligned}
&\text{for } d(k) > \frac{N}{2} \quad \begin{cases} w_{\mathrm{ON}}(k) = N - d(k-1) - q(k-1) \\ w_{\mathrm{OFF}}(k) = N - d(k) - q(k) \end{cases} \\
&\text{for } d(k) < \frac{N}{2} \quad \begin{cases} w_{\mathrm{ON}}(k) = d(k) + q(k) \\ w_{\mathrm{OFF}}(k) = d(k-1) + q(k-1). \end{cases}
\end{aligned} \tag{D.16}$$

Again, the switching sequences also contain the differentiated quantization error $q(k) - q(k-1)$. Therefore, the Eqs. (D.10) and (D.11) still hold. Except for the highest performance applications (e.g., [188]) the excess quantization noise due to the unit current pulse asymmetry will be negligible.

With a sinusoidal input signal of the form

$$d(k) = \frac{N}{2}\left(1 + A_{\sin}\sin(kT)\right) \tag{D.17}$$

the reconstructed error signal $e(t)$ can be written as

$$\begin{aligned}
&\text{for } \frac{T}{2} \le t < \pi + \frac{T}{2}: \\
&e(t) = \frac{N}{2}\left[\frac{A_{\mathrm{ON}}}{T}\left(1 - A_{\sin}\sin(t-T)\right) - \frac{A_{\mathrm{OFF}}}{T}\left(1 - A_{\sin}\sin(t)\right)\right]. \\
&\text{for } \frac{T}{2} \le t < \pi + \frac{T}{2}: \\
&e(t) = \frac{N}{2}\left[\frac{A_{\mathrm{ON}}}{T}\left(1 + A_{\sin}\sin(t)\right) - \frac{A_{\mathrm{OFF}}}{T}\left(1 + A_{\sin}\sin(t-T)\right)\right].
\end{aligned} \tag{D.18}$$

Again, $e(t)$ is periodic with the same period 2π as the input signal. A Fourier series expansion of (D.18) with the aid of MAPLE^{TM} yields the following expression for the harmonic distortion products relative to the fundamental tone with amplitude $\frac{1}{2}N\Delta A_{\sin}\mathrm{sinc}\left(\frac{\pi}{2\mathrm{OSR}}\right)$:

$$\begin{aligned}
&\text{for } n = 1, 2\ldots: \\
&\mathrm{HD}_{2n} = 2\cdot\frac{|A_{\mathrm{ON}} - A_{\mathrm{OFF}}|}{\Delta\cdot T}\cdot\frac{\sqrt{2 + 2\cos\left(\frac{\pi}{\mathrm{OSR}}\right)}}{\pi(2n+1)(2n-1)}\cdot\frac{\frac{\pi}{2\mathrm{OSR}}}{\sin\left(\frac{\pi}{2\mathrm{OSR}}\right)} \\
&\mathrm{HD}_{2n+1} = 0.
\end{aligned} \tag{D.19}$$

With a rise and fall time mismatch, we obtain by inserting (D.14) into (D.19):

$$\begin{aligned}
&\text{for } n = 1, 2\ldots: \\
&\mathrm{HD}_{2n} = \frac{|T_{\mathrm{rise}} - T_{\mathrm{fall}}|}{T}\cdot\frac{\sqrt{2 + 2\cos\left(\frac{\pi}{\mathrm{OSR}}\right)}}{\pi(2n+1)(2n-1)}\cdot\frac{\frac{\pi}{2\mathrm{OSR}}}{\sin\left(\frac{\pi}{2\mathrm{OSR}}\right)} \\
&\mathrm{HD}_{2n+1} = 0.
\end{aligned} \tag{D.20}$$

Equations (D.19) and (D.20) are also valid in case a digital noiseshaper is used, as long as the number of cells that switch does not depend significantly on the noiseshaper activity. For a sufficiently large number of unit cells and sufficiently large input signal amplitude, this is approximately fulfilled in practical situations.

D.4 Charge Sharing with Thermometer Coding

With an input code sequence given by Eq. (D.8) and assuming perfect settling, the output voltages at the end of sample $k-1$ are given by

$$V_{\text{out},p}(k-1) = \frac{N}{2}\left[1 + A_{\text{sin}}\cos((k-1)T)\right] I_{\text{unit}} R_L,$$
$$V_{\text{out},n}(k-1) = \frac{N}{2}\left[1 - A_{\text{sin}}\cos((k-1)T)\right] I_{\text{unit}} R_L. \tag{D.21}$$

The tail nodes of the current cells follow the respective output voltages, attenuated by the feedthrough factor ρ, for the moment assumed constant. The charges on the tail-node capacitors C_0 switched to the positive and negative outputs at the end of sample $k-1$ are therefore

$$Q_{C0,p}(k-1) = \frac{N}{2}\left[1 + A_{\text{sin}}\cos((k-1)T)\right] I_{\text{unit}} R_L \rho C_0,$$
$$Q_{C0,n}(k-1) = \frac{N}{2}\left[1 - A_{\text{sin}}\cos((k-1)T)\right] I_{\text{unit}} R_L \rho C_0. \tag{D.22}$$

At the beginning of sample k, $w_{\text{ON}}(k)$ and $w_{\text{OFF}}(k)$ represent the number of cells that are switched in the respective direction, given by Eq. (D.9). We assume that the tail nodes of the switched cells immediately undergo charge redistribution, even before the DAC-network starts to settle to the new value for sample k. The total error charge injected into the differential output at the beginning of sample k is thus

$$\begin{aligned} Q_{\text{err}}(k) &= w_{\text{ON}}(k)\left[Q_{C0,n}(k-1) - Q_{C0,p}(k-1)\right] \\ &\quad - w_{\text{OFF}}(k)\left[Q_{C0,p}(k-1) - Q_{C0,n}(k-1)\right]. \end{aligned} \tag{D.23}$$

The corresponding error signal superimposed on the output voltage after baseband filtering can be written as

$$\text{for } \frac{T}{2} \le t < \pi + \frac{T}{2}:$$
$$e(t) = \frac{\rho \cdot N^2 A_{\text{sin}}^2 I_{\text{unit}} R_L^2 C_0}{2T} \cdot \cos(t-T)\left[\cos(t) - \cos(t-T)\right].$$
$$\text{for } \pi + \frac{T}{2} \le t < 2\pi + \frac{T}{2}:$$
$$e(t) = -\frac{\rho \cdot N^2 A_{\text{sin}}^2 I_{\text{unit}} R_L^2 C_0}{2T} \cdot \cos(t-T)\left[\cos(t) - \cos(t-T)\right]. \tag{D.24}$$

A Fourier series expansion of (D.24) delivers, after division by the amplitude of the fundamental tone $\frac{1}{2} N A_{\text{sin}} I_{\text{unit}} R_L \operatorname{sinc}\left(\frac{\pi}{2\text{OSR}}\right)$:

$$
\begin{aligned}
&\text{for } n = 3, 5, 7 \ldots:\\
&\mathrm{HD}_n = \frac{4\rho \cdot C_0 R_L N A_{\sin}}{\pi T n(n+2)(n-2)}\\
&\qquad \cdot \sqrt{4 - (n^2-4)\cos^4\left(\frac{\pi}{2\mathrm{OSR}}\right) + (n^2-8)\cos^2\left(\frac{\pi}{2\mathrm{OSR}}\right)} \cdot \frac{\dfrac{\pi}{2\mathrm{OSR}}}{\sin\left(\dfrac{\pi}{2\mathrm{OSR}}\right)}\\
&\text{for } n = 2, 4, 6 \ldots:\\
&\mathrm{HD}_n = 0.
\end{aligned}
\tag{D.25}
$$

D.5 Charge Sharing with Data Weighted Averaging

With an input signal of the form (D.17), we can write the baseband error signal $e(t)$ as

$$
\begin{aligned}
&\text{for } \frac{T}{2} \le t < \pi + \frac{T}{2}:\\
&e(t) = \frac{\rho \cdot N^2 A_{\sin}^2 I_{\mathrm{unit}} R_L^2 C_0}{2T} \cdot \sin(t-T)\left[2 - \sin(t) - \sin(t-T)\right].\\
&\text{for } \pi + \frac{T}{2} \le t < 2\pi + \frac{T}{2}:\\
&e(t) = -\frac{\rho \cdot N^2 A_{\sin}^2 I_{\mathrm{unit}} R_L^2 C_0}{2T} \cdot \sin(t-T)\left[2 - \sin(t) - \sin(t-T)\right].
\end{aligned}
\tag{D.26}
$$

which, after a Fourier series expansion and division by the amplitude of the fundamental tone, $\frac{1}{2} N A_{\sin} I_{\mathrm{unit}} R_L \operatorname{sinc}\left(\frac{\pi}{2\mathrm{OSR}}\right)$, becomes

$$
\begin{aligned}
&\text{for } n = 3, 5, 7 \ldots:\\
&\mathrm{HD}_n = \frac{4\rho \cdot C_0 R_L N A_{\sin}}{\pi T n(n+2)(n-2)} \sqrt{n^2\cos^2\left(\frac{\pi}{2\mathrm{OSR}}\right) - (n^2-4)\cos^4\left(\frac{\pi}{2\mathrm{OSR}}\right)}\\
&\qquad \cdot \frac{\dfrac{\pi}{2\mathrm{OSR}}}{\sin\left(\dfrac{\pi}{2\mathrm{OSR}}\right)}\\
&\text{for } n = 2, 4, 6 \ldots:\\
&\mathrm{HD}_n = 0.
\end{aligned}
\tag{D.27}
$$

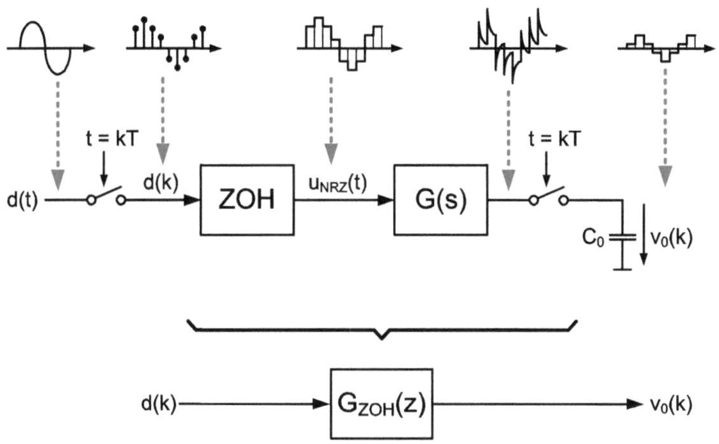

Fig. D.2 Sampled-data model of tail-node voltage

D.6 Output Voltage Feedthrough Factor

The feedthrough factor ρ in Eqs. (D.25) and (D.27) is a measure for the fraction of the output voltage appearing across the tail-node capacitor C_0. In an NRZ-DAC the charge stored on C_0 at the end of the sampling period provokes a signal-dependent error charge injected into the output signal, if the current cell switches at the beginning of the next sample. The aggregate, signal-dependent charge error added by the current-cell array is the cause of odd-order harmonic distortion at the output of the D/A-converter.

The approximations for ρ given in Eqs. (4.23) and (4.24) are only valid if we assume perfect settling of the output voltage in each sample, i.e., all internal time constants of the circuit are much smaller than the clock period T.

In general, we must consider the DAC as a sampled-data system with zero-order hold function feeding an analog output stage. The signal current corresponding to the input data samples is held constant within the clock period, and, starting from the settled end value of the previous sampling period, it generates changing voltages at various nodes in the analog circuit consisting of the current cells and the output stage. To find the correct value for ρ we must determine the settled end values of the dynamic voltage across C_0 under sinusoidal excitation of the DAC.

For an NRZ-DAC we can model the output voltage feedthrough to the tail-node capacitor C_0 as shown in Fig. D.2. The input signal is first sampled by the ideal sampler at times kT and then converted into a pulse train with duration T by the zero-order-hold (ZOH). This is equivalent to the description of the ideal NRZ-DAC in Eq. (1.5).

Let $G(s)$ be the continuous-time transfer function from the signal current of the DAC to the voltage across C_0 in a single unit current cell. Since we assume—in accordance with the simple model described in Sect. 4.1.2—that the behavior of a specific current cell is not influenced by the other cells, i.e., $G(s)$ is not code (transition) dependent and thus time invariant.

The transfer function $G(s)$ thus describes the voltage at C_0 as a function of the normalized signal current, neglecting the sampling operation. The transition from the continuous-time domain to the sampled-data domain can be performed with the ZOH-transform, extensively used in digital control theory [189].

The pulse transfer function $G_{ZOH}(z)$ describes the sampled unit pulse response of an LTI-system $G(s)$, which is preceded by a sampler and a zero-order hold. The pulse transfer function is defined by [189]:

$$\begin{aligned} G_{ZOH}(z) &= \mathscr{Z}\left(\mathscr{L}^{-1}\left(\frac{1-\mathrm{e}^{-Ts}}{s}\cdot G(s)\right)\right) \\ &= \frac{z-1}{z}\cdot\mathscr{Z}\left(\mathscr{L}^{-1}\left(\frac{G(s)}{s}\right)\right). \end{aligned} \tag{D.28}$$

With a steady sinusoidal excitation $\sin(\omega_{\sin}t)$ at the input of the sampler, the amplitude of the normalized signal appearing across C_0 and sampled right at the end of the sampling period, i.e., at $t = kT$, is directly given by $\left|G_{ZOH}\left(\mathrm{e}^{j\omega_{\sin}T}\right)\right|$. The feedthrough factor is therefore

$$\rho(\omega) = \frac{1}{R_L}\left|G_{ZOH}\left(\mathrm{e}^{j\omega T}\right)\right| \tag{D.29}$$

The division by R_L is necessary, because ρ is referred to the full-scale output voltage at DC. The signal current and the charge error packets from the DAC-element transitions are both converted into voltage signals by the output stage of the DAC, represented by a continuous-time filter $T(s)$. For the calculation of the (eventually aliased) odd-order harmonic distortion at frequency ω_{HD}, we can include the filtering effect of the output stage directly in the feedthrough factor:

$$\rho(\omega,\omega_{HD}) = \frac{1}{R_L}\left|G_{ZOH}\left(\mathrm{e}^{j\omega T}\right)\right|\cdot\left|\frac{T(j\omega_{HD})}{T(j\omega)}\right|. \tag{D.30}$$

D.6.1 Single-Polarity DAC with Passive Termination

Neglecting the quantization and the sampling operation, the output voltage of the single-polarity NRZ-DAC shown in Fig. D.3 can be approximated by the continuous-time transimpedance transfer function:

$$\frac{V_{OUT}}{I_{DAC}}(s) \approx T(s) = \frac{R_L}{1+sC_LR_L}. \tag{D.31}$$

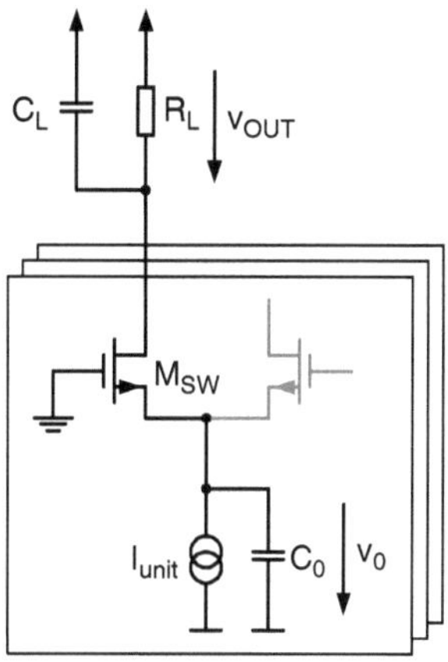

Fig. D.3 Tail-node sampling with passive output stage

The tail node follows the output voltage as

$$\frac{V_0}{V_{OUT}}(s) \approx \frac{gds}{gm} \cdot \frac{1}{1 + s\frac{C_0}{gm}}. \tag{D.32}$$

Multiplying (D.32) with (D.31) the tail-node voltage can be put into relation to the DAC-current:

$$G(s) = \frac{V_0}{I_{DAC}}(s) \approx \frac{gds}{gm} \cdot \frac{R_L}{(1 + sC_L R_L)\left(1 + \frac{C_0}{gm}\right)}. \tag{D.33}$$

The time constant associated with the tail-node capacitance is usually much smaller than the clock period and the time constant of the output node, i.e., $C_0/gm << R_L C_L, T$. We therefore approximate (D.33) by

$$G(s) \approx \frac{gds}{gm} \cdot \frac{R_L}{(1 + sC_L R_L)}. \tag{D.34}$$

Performing the ZOH-transform (D.28), the feedthrough factor becomes

$$\rho(\omega) = \frac{gds}{gm} \cdot \left| \frac{1 - \mathrm{e}^{-\frac{T}{R_L C_L}}}{\mathrm{e}^{j\omega T} - \mathrm{e}^{-\frac{T}{R_L C_L}}} \right|. \tag{D.35}$$

Including the transimpedance transfer function (D.31), we get

$$\rho(\omega, \omega_{HD}) = \frac{gds}{gm} \cdot \left| \frac{1 - \mathrm{e}^{-\frac{T}{R_L C_L}}}{\mathrm{e}^{j\omega T} - \mathrm{e}^{-\frac{T}{R_L C_L}}} \right| \cdot \left| \frac{1 + j\omega C_L R_L}{1 + j\omega_{HD} C_L R_L} \right|. \tag{D.36}$$

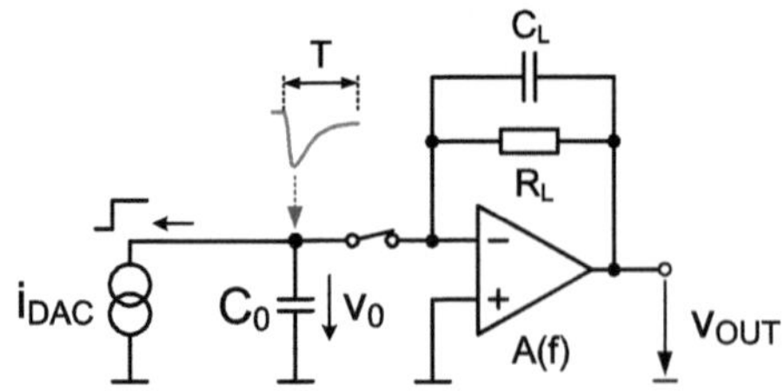

Fig. D.4 Tail-node sampling with active output stage

For sufficiently low frequencies the feedthrough factor can be approximated by (4.23).

D.6.2 Dual-Polarity DAC with Active Termination

We assume that in the dual-polarity DAC of Fig. D.4 the voltage at the tail-node capacitor is given directly by the voltage seen at the virtual ground node. With $A_0 >> 1$ we have

$$G(s) = \frac{V_0}{I_{\mathrm{DAC}}}(s) = \frac{R_L}{A_0} \cdot \frac{1 + s\dfrac{A_0}{2\pi \mathrm{GBW}}}{(1 + sR_L C_L)\left(1 + s\dfrac{1}{2\pi \mathrm{GBW}}\right)}. \tag{D.37}$$

With (4.27) the feedthrough factor becomes

$$\rho(\omega) = \left| 1 + \frac{1 - \dfrac{1}{A_0}}{1 - K_\tau} \cdot \frac{\mathrm{e}^{j\omega T} - 1}{\mathrm{e}^{j\omega T} - \mathrm{e}^{-2\pi \mathrm{GBW} \cdot T}} + \frac{\dfrac{K_\tau}{A_0} - 1}{1 - K_\tau} \cdot \frac{\mathrm{e}^{j\omega T} - 1}{\mathrm{e}^{j\omega T} - \mathrm{e}^{-\frac{T}{R_L C_L}}} \right|. \tag{D.38}$$

Finally, including the transimpedance transfer function $T(j\omega)$ given in (5.4):

$$\rho(\omega, \omega_n) = \left| 1 + \frac{1 - \dfrac{1}{A_0}}{1 - K_\tau} \cdot \frac{\mathrm{e}^{j\omega T} - 1}{\mathrm{e}^{j\omega T} - \mathrm{e}^{-2\pi \mathrm{GBW} \cdot T}} + \frac{\dfrac{K_\tau}{A_0} - 1}{1 - K_\tau} \cdot \frac{\mathrm{e}^{j\omega T} - 1}{\mathrm{e}^{j\omega T} - \mathrm{e}^{-\frac{T}{R_L C_L}}} \right| \cdot \left| \frac{\left(1 + \dfrac{j\omega}{2\pi \mathrm{GBW}}\right)(1 + j\omega R_L C_L)}{\left(1 + \dfrac{j\omega_{\mathrm{HD}}}{2\pi \mathrm{GBW}}\right)(1 + j\omega_{\mathrm{HD}} R_L C_L)} \right|. \tag{D.39}$$

For sufficiently low frequencies the feedthrough factor can again be approximated by (4.24).

D.7 Third-Order Two-Tone Nonlinearity

Following Eq. (4.15), a sampled two-tone signal, scaled to an N-element unary array, can be written as

$$d(k) = \frac{N}{2}\left[1 + \frac{A_{\sin}}{2}\left[\cos\left(2\pi f_1 kT\right) + \cos\left(2\pi f_2 kT\right)\right]\right]. \tag{D.40}$$

With the center frequency of the two-tone signal $f_c = (f_1 + f_2)/2$ and the carrier spacing $\Delta f = |f_2 - f_1|$ we can also write

$$d(k) = \frac{N}{2}\left[1 + \frac{A_{\sin}}{2}\left[\cos\left(2\pi\left(f_c + \Delta f\right)kT\right) + \cos\left(2\pi\left(f_c - \Delta f\right)kT\right)\right]\right]. \tag{D.41}$$

Each carrier has a normalized amplitude $A_{\sin}/2$. The normalized peak value of the two-tone signal is therefore $A_{\sin}$, and we assume of course unclipped operation with $A_{\sin} \in [0, 1]$. In the following, we are primarily interested in the third-order nonlinear products generated by the charge-sharing effect described in Sect. 4.1.4 and D.4. Neglecting aliasing, in the first Nyquist band, the third-order products with an input signal of the form (D.40) will be found at the following frequencies:

$$\begin{aligned}
\mathrm{HD}_{3,L} &\rightarrow 3f_1, \\
\mathrm{HD}_{3,R} &\rightarrow 3f_2, \\
\mathrm{IM}_{3,L}^{+} &\rightarrow (2f_1 + f_2), \\
\mathrm{IM}_{3,R}^{+} &\rightarrow (2f_2 + f_1), \\
\mathrm{IM}_{3,L}^{-} &\rightarrow (2f_1 - f_2), \\
\mathrm{IM}_{3,R}^{-} &\rightarrow (2f_2 - f_1).
\end{aligned} \tag{D.42}$$

If we assume $f_1 < f_2$, the subscripts L and R denote the left- and right-hand position of the corresponding nonlinear spectral component on the frequency axis, at least as long as aliasing is avoided. Ideally, the L and R components have equal magnitude.[2] Normally, the most importance is given to the intermodulation

---

[2]This is always true for memoryless nonlinear systems. Strictly speaking, a D/A-converter is a nonlinear system with memory, even if we assume it rather short (e.g., one sampling period).

contributions $\mathrm{IM}_{3,L}^{-}$ and $\mathrm{IM}_{3,R}^{-}$, since they appear close to the carriers and are thus perceived as in-band distortion. However, also the other third-order products can be of importance, depending on the application.

It is not clear at this point whether a closed-form expression for the third-order products similar to (D.25) for the single-tone case can be obtained, e.g., by performing a Fourier series expansion of an appropriately formulated error signal $e(t)$. Instead of attempting a rigorous calculation, here we try a heuristic approach with the help of a behavioral model according to Fig. 4.9, which can be readily implemented in MATLAB™. By inserting $n = 3$ into Eq. (D.25) and comparing the obtained HD_3 with the simulated third-order nonlinear products using such a model, the following expressions can be established:

$$\mathrm{HD}_{3,LR} \approx \frac{2}{45} \cdot \frac{\rho \cdot C_0 R_L N A_{\sin}}{\pi T} \cdot \sqrt{4-(n^2-4)\cos^4\left(\frac{\pi}{2\mathrm{OSR}}\right)+(n^2-8)\cos^2\left(\frac{\pi}{2\mathrm{OSR}}\right)} \cdot \frac{\frac{\pi}{2\mathrm{OSR}}}{\sin\left(\frac{\pi}{2\mathrm{OSR}}\right)} \tag{D.43}$$

$$\mathrm{IM}_{3,LR}^{+} \approx \frac{12}{165} \cdot \frac{\rho \cdot C_0 R_L N A_{\sin}}{T} \cdot \sqrt{4-(n^2-4)\cos^4\left(\frac{\pi}{2\mathrm{OSR}}\right)+(n^2-8)\cos^2\left(\frac{\pi}{2\mathrm{OSR}}\right)} \cdot \frac{\frac{\pi}{2\mathrm{OSR}}}{\sin\left(\frac{\pi}{2\mathrm{OSR}}\right)} \tag{D.44}$$

$$\mathrm{IM}_{3,LR}^{-} \approx \frac{4}{165} \cdot \frac{\rho \cdot C_0 R_L N A_{\sin}}{T} \cdot \sqrt{1+\left(\frac{3}{\mathrm{OSR}}\right)^2} \cdot \sqrt{4-(n^2-4)\cos^4\left(\frac{\pi}{2\mathrm{OSR}}\right)+(n^2-8)\cos^2\left(\frac{\pi}{2\mathrm{OSR}}\right)} \cdot \frac{\frac{\pi}{2\mathrm{OSR}}}{\sin\left(\frac{\pi}{2\mathrm{OSR}}\right)}. \tag{D.45}$$

In the two-tone case the "oversampling ratio" refers to the center frequency f_c and is defined by $\mathrm{OSR} = f_{\mathrm{CLK}}/(2 \cdot f_c)$. In a first moment, Eqs. (D.43)–(D.45) must be understood as approximations to the exact solutions. For the simulation example

Nonlinear systems with memory can display unequal odd-order distortion and intermodulation products [190]; see also Figs. D.5 and D.6.

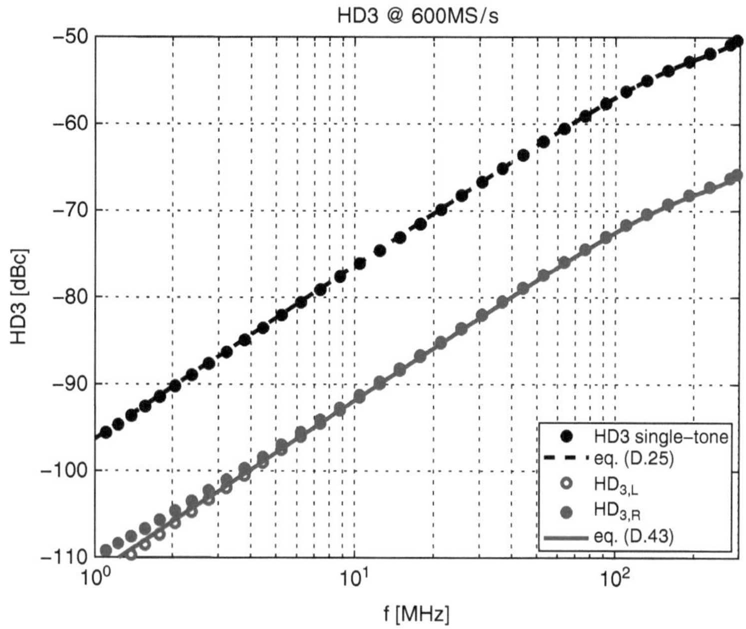

Fig. D.5 HD_3: single tone vs. two tone

described below we use the following model parameters, which are additionally assumed to be frequency independent:

- $N = 127$ (7 bit)
- Ideal LSB-section
- $f_{\text{CLK}} = 600\,\text{MHz}$
- $C_0 = 100\,\text{fF}$
- $\rho = 0.03$
- $R_L = 50\,\Omega$

Figure D.5 shows the comparison of the third-order harmonic distortion for a single-tone and a two-tone signal as a function of the sine-wave frequency f_{sin} and center frequency f_c, respectively. Both signals have the same peak value $A_{\text{sin}} = 1$, corresponding to the converter full-scale. The carrier spacing in this example is $\Delta f = 0.1\,\text{MHz}$. Also shown are Eqs. (D.25), with $n = 3$, and (D.43). In the two-tone case, the third-order harmonic distortion products $\text{HD}_{3,L}$ and $\text{HD}_{3,R}$, as obtained by the behavioral simulation, slightly differ in magnitude at very low frequencies; otherwise, the prediction by Eq. (D.43) is of the same quality as for the single-tone case.

Figure D.6 shows the third-order intermodulation components for the same two-tone signal. Again, left and right components differ slightly in magnitude only at very low frequencies. While the additive intermodulation products $\text{IM}_{3,L}^{+}$ and

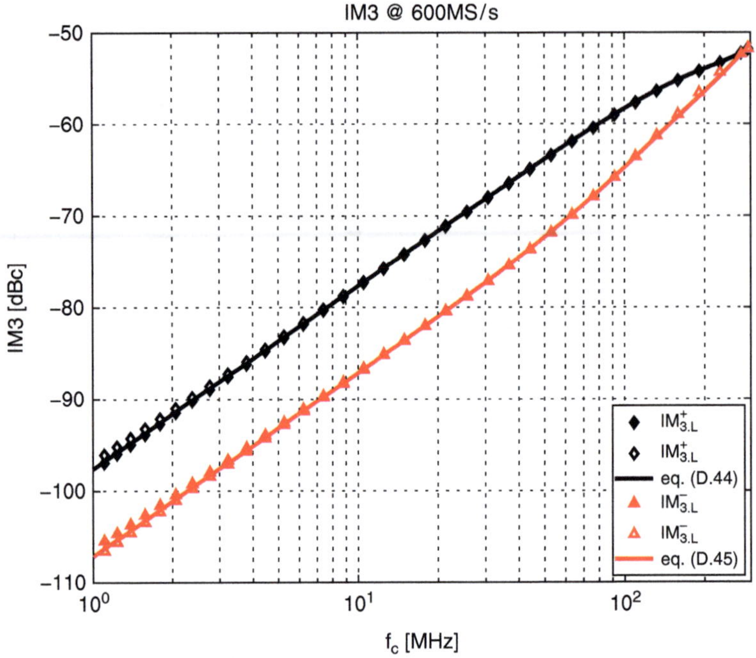

Fig. D.6 IM_3 products for two-tone signal

$IM^+_{3,R}$ display the same frequency behavior as the third-order harmonic distortion, the subtractive components $IM^-_{3,L}$ and $IM^-_{3,R}$ apparently change slope when the center frequency of the two-tone signal surpasses 100 MHz in this example, equal to OSR $= 3$, i.e., $f_c \geq f_{CLK}/6$. Interestingly, around this frequency, the third-order harmonic distortion $HD_{3,LR}$ and the additive intermodulation products $IM^+_{3,LR}$ start to experience aliasing. Error energy seems to be shifted away from the direct cubic and additive intermodulation components to the subtractive intermodulation products at center frequencies larger than $f_{CLK}/6$. This observation is the reason for the deliberately introduced correction factor $\sqrt{1 + (3/\text{OSR})^2}$ in (D.45), which provokes an additional slope increase[3] in the calculated $IM^-_{3,LR}$ for $f_c > f_{CLK}/6$.

A closer inspection also reveals that the prediction accuracy of (D.45) is getting somewhat worse in the frequency range $f_{CLK}/6 < f_c < f_{CLK}/2$, which is clearly due to the heuristic nature of this equation. Of course, a more complex correction function could be introduced into (D.45), in an attempt to provide a tighter fit with the simulated $IM^-_{3,LR}$. However, for all practical purposes, the prediction accuracy of (D.43)–(D.45) seems to be quite sufficient for an estimation of the third-order

[3]It does of course not reach the full 20 dB/decade boost within the limited observation range $f_{CLK}/6 \leq f_c < f_{CLK}/2$.

two-tone nonlinearity, in case the charge-sharing effect, as defined in Sect. 4.1.4, is the dominant distortion mechanism. Of course, this will only remain valid, as long as the prerequisites underlying the model introduced in Sect. 4.1.2, which is the basis for the derivation of the above equations, are applicable.

About the Author

Analog/Mixed-Signal Design Engineer

Martin Clara received the M.Sc. degree in electrical engineering from Vienna University of Technology, Austria, in 1996 and the Ph.D. degree in electronics from Graz University of Technology, Austria, in 2009.

In 1997 he joined Siemens Microelectronics' Design Center in Villach, Austria, as an analog design engineer, mainly working on BiCMOS and CMOS linear circuits.

From 2000 to 2009 he was with Infineon Technologies' Design Center in Villach, Austria, where he designed data converters, linear circuits, and RF building blocks in deep-submicron and nanometer CMOS technologies.

From 2009 to 2012 he was a senior engineer for analog/mixed-signal and RF-design at LANTIQ's design center, based in Villach, Austria. Since 2012 he is with Analog Devices in Wilmington, MA.

M. Clara, *High-Performance D/A-Converters*, Springer Series in Advanced Microelectronics 36, DOI 10.1007/978-3-642-31229-8,

His main interests include the implementation of low-voltage and high dynamic range analog front-ends in advanced CMOS technologies, the concept and design of high-performance data converters, as well as RF-CMOS design.

References

1. F. Maloberti, *Data Converters* (Springer, New York, 2007). ISBN:978-0-387-32485-2
2. F. Maloberti, *Analog Design for CMOS VLSI Systems* (Kluwer Academic Publishers, Boston, MA, 2003). ISBN:0-7923-7550-5
3. P.J.A. Naus, E.C. Dijkmans, E.F. Stikvoort, D.J. Holland, W. Bradinal, A CMOS stereo 16-bit D/A converter for digital audio. IEEE J. Solid-State Circ. **22**(3), 390–395 (1987)
4. R. Wang, S.-H. Kim, S.-H. Lee, S.-B. You, J.-W. Kim, U.-K. Moon, G.C. Temes, A a 100-dB gain-corrected delta-sigma audio DAC with headphone driver. Analog Integr. Circ. Signal Process. **21**(1), 27–31 (2007)
5. D. Johns, K. Martin, Analog Integrated Circuit Design (Wiley, New York, 1997). ISBN:0-471-14448-7
6. F.-J. Wang, G.C. Temes, S. Law, A quasi-passive CMOS pipeline D/A converter. IEEE J. Solid-State Circ. **24**(6), 1752–1755 (1989)
7. K. Khanoyan, F. Behbahanli, A.A. Abidi, A 10b, 400 MS/s Glitch-Free CMOS D/A converter. Symposium on VLSI Circuits Digest of Technical Papers, pp. 73–76, 1999
8. Texas Instruments, DAC8560: 16-bit, ultra-low glitch, voltage output digital-to-analog converter with 2.5V, 2ppm/C internal reference. Datasheet (2006), May 2012. http://www.ti.com
9. Texas Instruments, DAC8820: 16-Bit, Parallel Input Multiplying Digital-to-Analog Converter. Datasheet (2005), May 2012. http://www.ti.com
10. W.A. Kester, *The Data Conversion Handbook* (Elsevier/Newnes, Burlington, MA, 2005). ISBN:0-7506-7841-0
11. B.C. Baker, A DAC for all precision applications. Texas Instrum.: Analog Appl. J. Q3. 5–8 (2008), http://www.ti.com/lit/an/slyt300/slyt300.pdf
12. Analog Devices Inc, AD1862: Ultralow noise 20-bit audio DAC. Datasheet Rev. A (1990), http://www.analog.com/
13. Burr-Brown, PCM1702: BiCMOS advanced sign magnitude 20-bit DIGITAL-TO-ANALOG CONVERTER. Datasheet (1993), www.ti.com
14. Burr-Brown, PCM1704: SoundPlus^{TM} 24-bit, 96 kHz BiCMOS sign-magnitude digital-to-analog converter. Datasheet (1998), www.ti.com
15. M. Egan, The 20-bit DAC is the easiest part of a 1-ppm-accurate precision voltage source. ADI: Analog Dialogue. **44**(2), 3–6 (2010)
16. Analog Devices Inc., AD5791: 1 ppm 20-bit, ± 1 LSB INL, voltage output DAC. Datasheet Rev. C (2010), http://www.analog.com/
17. M. Pelgrom, A 10-bit 50-MHz CMOS D/A-converter with $75-\Omega$ buffer. IEEE J. Solid-State Circ. **25**(6), 1347–1352 (1990)
18. A.B. Grebene, *Bipolar and MOS Analog Integrated Circuit Design* (Wiley, New York, 1984). ISBN:0-471-08529-4

M. Clara, *High-Performance D/A-Converters*, Springer Series in Advanced Microelectronics 36, DOI 10.1007/978-3-642-31229-8,

19. P.R. Gray, R.G. Meyer, *Analysis and Design of Analog Integrated Circuits*, 3rd edn. (Wiley, New York, 1993). ISBN:0-471-57495-3
20. W. Sansen, *Analog Design Essentials* (Springer, New York, 2006). ISBN:10-387-25746-2
21. J.W. Wang, K.W. Martin, High-resolution and low-power CMOS D/A converter. IEEE J. Solid-State Circ. **24**(5), 1458–1461 (1989)
22. T. Guy, L. Trythall, A. Brodersen, A sixteen-bit monolithic bipolar DAC. IEEE J. Solid-State Circ. **17**(6), 1127–1132 (1982)
23. J.R. Naylor, A complete high-speed voltage output 16-bit monolithic DAC. IEEE J. Solid-State Circ. **18**(6), 729–735 (1982)
24. B. Jewett, J. Liu, K. Poulton, A 1.2GS/s 15b DAC for precision signal generation. IEEE International Solid-State Circuits Conference 2005, Digest of Technical Papers, pp. 110–111, 2005
25. B.M. Gordon, R.P. Talambiras, Signal conversion apparatus, United States Patent 3,108,266, 22 Oct 1963
26. E.A. Lee, D.G. Messerschmitt, *Digital Communication* (Kluwer Academic Publishers, Boston, MA, 1988). ISBN:0-89838-274-2
27. B. Razavi, *RF Microelectronics* (Prentice Hall, Upper Saddle River, New Jersey, 1998). ISBN:0-13-887571-5
28. R. Schreier, N. Abaskharoun, H. Shibata, D. Paterson, S. Rose, I. Mehr, Q. Luu, A 375-mW quadrature bandpass $\Delta\Sigma$ ADC with 8.5-MHz BW and 90-dB DR at 44 MHz. IEEE J. Solid-State Circ. **41**(12), 2632–2640 (2006)
29. S. Wimoesterer, VDSL2. *Funkschau, Heft 17/2005* (2005), pp. 43–44
30. N.M. Blachman, The intermodulation and distortion due to quantization of sinusoids. IEEE Trans. Acoustics, Speech Signal Process. **ASSP-33**, 1417–1426 (1985)
31. R.v.d. Plassche, *Integrated Analog-to-Digital and Digital-to-Analog Converters* (Kluwer Academic Publishers, 1994). ISBN:0-7923-9436-4
32. M. Choe, K. Baek, M. Teshome, A 1.6GS/s 12b return-to-zero GaAs RF DAC for multiple nyquist operation. IEEE J. Solid-State Circ. **40**(12), 2456–2468 (2005)
33. C.-H. Lin, K. Bult, A 10-b, 500-MSample/s CMOS DAC in 0.6 mm^2. IEEE J. Solid-State Circ. **33**(12), 1948–1958 (1998)
34. J. Deveugele, M. Steyaert, A 10-bit 250-MS/s binary-weighted current-steering DAC. IEEE J. Solid-State Circ. **41**(2), 320–329 (2006)
35. A. Van den Bosch, M. Steyaert, W. Sansen; *Static and Dynamic Performance Limitations for High Speed D/A Converters* (Kluwer Academic Publishers, 2004). ISBN:1-4020-7761-0
36. M.J.M. Pelgrom, H.P. Tuinhout, M. Vertregt, Transistor matching in analog CMOS applications. *Technical Digest of the International Electron Devices Meeting IEDM'98*, pp. 915–918, 1998
37. P. Kinget, Device mismatch and tradeoffs in the design of analog circuits. IEEE J. Solid-State Circ. **40**(6), 1212–1224 (2005)
38. J. Croon, W. Sansen, H. Maes, *Matching Properties of Deep Submicron MOS Transistors* (Springer, New York, 2005). ISBN:0-387-24314-3
39. M. Pelgrom, A. Duinmaijer, A. Welbers, Matching properties of mos transistors. IEEE J. Solid-State Circ. **24**(5), 1433–1440 (1989)
40. J. Bastos, A. Marques, M. Steyaert, W. Sansen, A 12-bit intrinsic accuracy high-speed CMOS DAC. IEEE J. Solid-State Circ. **33**(12), 1959–1969 (1998)
41. A. Bugeja, Bang-Sup Song, P. Rakers, S. Gillig, A 14-b, 100-MS/s CMOS DAC designed for spectral performance. IEEE J. Solid-State Circ. **34**(12), 1719–1732 (1999)
42. M. Clara, W. Klatzer, B. Seger, A. Di Giandomenico, L. Gori, A 1.5V 200MS/s 13b 25mW DAC with Randomized Nested Background Calibration in 0.13μm CMOS. IEEE International Solid-State Circuits Conference 2007, Digest of Technical Papers, pp. 250–251, 2007
43. K. Parthasarathy, R. Geiger, Unambiguous characterization and specification of D/A converter performance, in *Proceedings of the 43rd IEEE Midwest Symposium on Circuits and Systems*, vol. 2, pp. 890–894, 2000

44. R. Schreier, G. Temes, *Understanding Delta-Sigma Data Converters* (Wiley, Hoboken, New Jersey, 2004). ISBN:0-471-46585-2
45. B. Boser, B. Wooley, The design of sigma-delta modulation analog-to-digital converters. IEEE J. Solid-State Circ. **23**(6), 1298–1308 (1988)
46. J. Wikner, Studies on CMOS Digital-to-Analog Converters, Ph.D. dissertation, University of Linkping, 2001
47. R. Adams, K.Q. Nguyen, K. Sweetland, A 113-dB SNR oversampling DAC with segmented noise-shaped scrambling. IEEE J. Solid-State Circ. **33**(12), 1871–1878 (1998)
48. K. Laker, W. Sansen, *Design of Analog Integrated Circuits and Systems* (McGraw-Hill, New York, 1994). ISBN:00-703-6060-X
49. B. Razavi, *Design of Analog CMOS Integrated Circuits* (McGraw-Hill, Boston, MA, 2001). ISBN:0-07-118839-8
50. Y. Tsivids, C. McAndrew, *Operation and Modeling of the MOS Transistor*, 3rd edn. (Oxford University Press, New York, 2011). ISBN:978-0-19-517015-3
51. N. Da Dalt, M. Harteneck, C. Sandner, A. Wiesbauer, Numerical modeling of PLL jitter and the impact of its non-white spectrum on the SNR of sampled signals. *Southwest Symposium on Mixed-Signal Design (2001)*, pp. 38–44
52. M. Shinagawa, Y. Akazawa, T. Wakimoto, Jitter analysis of high-speed sampling systems. IEEE J. Solid-State Circ. **25**(1), 220–224 (1990)
53. N. Da Dalt, M. Harteneck, C. Sandner, A. Wiesbauer, On the Jitter requirements of the sampling clock for analog-to-digital converters. IEEE Trans. Circ. Syst. I. **49**(9), 1354–1360 (2002)
54. A. Hajimiri, T.H. Lee, *The Design of Low Noise Oscillators* (Kluwer Academic Publisher, 1999). ISBN:0-7923-8455-5
55. J. Parker, D. Ray, A 1.6-GHz CMOS PLL with on-chip loop filter. IEEE J. Solid-State Circ. **33**(3), 337–343 (1998)
56. E. Drucker, Model PLL dynamics and phase-noise performance. Microwave & RF. **39**, 73–117 (2000)
57. A. Mehrotra, Noise analysis of phase-locked loops. IEEE Trans. Circ. Syst. I: Fundamental Theor. Appl. **49**(9), 1309–1316 (2002)
58. L. Hernndez, A. Wiesbauer, S. Paton, A. Di Giandomenico, Modelling and optimization of low pass continuous-time sigma delta modulators for clock jitter noise reduction, in *Proceedings of the 2004 International Symposium on Circuits and Systems*, vol. 1, pp. I–1072 – I–1075, 2004
59. A. Wiesbauer, M. Clara, M. Harteneck, T. Ptscher, B. Seger, C. Sandner, C. Fleischhacker, A fully integrated analog front-end macro for cable modem applications in 0.18-μm CMOS. IEEE J. Solid-State Circ. **37**(7), 866–873 (2002)
60. M. Clara, N. Da Dalt Jitter noise of sampled multitone signals. IEEE Trans. Circ. Syst. II. **58**(10), 652–656 (2011)
61. H. Pan, M. Segami, M. Choi, J. Cao, F. Hattori, A. Abidi, A 3.3V 12b 50-MS/s A/D-converter in 0.6 micron CMOS with over 80-dB SFDR, in *Analog Circuit Design - High-Speed Analog-to-Digital Converters, Mixed-Signal Design; PLLs and Synthesizers*, ed. by R.v.d. Plassche, J. Huijsing, W. Sansen. (Kluwer Academic Publishers, 2000), pp. 47–73. ISBN:0-7923-7956-X
62. Y. Nakamura, T. Miki, A. Maeda, H. Kondoh, N. Yazawa, A 10-b 70-MS/s CMOS D/A converter. IEEE J. Solid-State Circ. **26**(4), 637–642 (1991)
63. T. Miki, Y. Nakamura, M. Nakaya, S. Asai, Y. Akasaka, Y. Horiba, An 80-MHZ 8-bit CMOS D/A converter. IEEE J. Solid-State Circ. **21**(6), 983–988 (1986)
64. G. Van Der Plas, J. Vandenbussche, W. Sansen, M. Steyaert, G. Gielen, A 14-bit intrinsic accuracy Q^2 random walk CMOS DAC. IEEE J. Solid-State Circ. **34**(12), 1708–1718 (1999)
65. K. Lakshmikumar, R. Hadaway, M. Copeland, Characterization and modeling of mismatch in MOS transistors for precision analog design. IEEE J. Solid-State Circ. **21**(6), 1057–1066 (1986)

66. A. Papoulis, *Probability, Random Variables, and Stochastic Processes*, 3rd edn. (McGraw-Hill, 1991). ISBN 0-07-048477-5
67. G.I. Radulov, M. Heydenreich, R.W. van der Hofstad, J.A. Hegt, A.H.M. van Roermund, Brownian-bridge-based statistical analysis of the DAC INL caused by current mismatch. IEEE Trans. Circ. Syst. II: Analog Digital Signal Process. **54**(2), 146–150 (2007)
68. M. Heydenreich, R. van der Hofstad, G.I. Radulov, Functionals of Brownian bridges arising in the current mismatch in D/A converters. Probab. Eng. Inform. Sci. **23**, 149–172 (2009)
69. W. Feller, On the Kolmogorov-Smirnov limit theorems for empirical distributions. Ann. Math. Stat. **19**(2), 177–189 (1948)
70. J. Pitman, M. Yor, On the distribution of ranked heights of excursions of a Brownian bridge. Ann. Probab. **29**(1), 361–384 (2001)
71. A. Van den Bosch, M. Steyaert, W. Sansen, An accurate statistical Yield Model for CMOS Current-Steering D/A Converters, in *Proceedings of the 2000 International Symposium on Circuits and Systems*, vol. 4, pp. IV–105 – IV–108, 2000
72. K. Lakshmikumar, M. Copeland, R. Hadaway, Reply to "comment on 'characterization and modeling of mismatch in MOS transistors for precision analog design'". IEEE J. Solid-State Circ. **23**(1), 295–296 (1988)
73. N. Smirnov, Table for estimating the goodness of fit of empirical distributions. Ann. Math. Stat. **19**(2), 279–281 (1948)
74. B. Razavi, *Principles of Data Conversion System Design* (Wiley-IEEE Press, New York, 1994). ISBN:978-0-7803-1093-3
75. S. Luschas, H.-S. Lee, Output Impedance Requirements for DACs. *Proceedings of the 2003 International Symposium on Circuits and Systems*, vol. 1, pp. I–861 – I–864, 2003
76. W. Schofield, D. Mercer, L.St. Onge, A 16b 400MS/s DAC with < -80dBc IMD to 300MHz and < -160dBm/Hz noise power spectral density. IEEE International Solid-State Circuits Conference 2003, Digest of Technical Papers, pp. 126–129, 2003
77. J.W. Bruce, Dynamic element matching techniques for data converters, Ph.D. dissertation, University of Nevada, Las Vegas, 2000
78. F. Chen, B. Leung, A high resolution multibit sigma-delta modulator with individual level averaging. IEEE J. Solid-State Circ. **30**(4), 453–460 (1995)
79. S. Norsworthy, R. Schreier, G. Temes, *Delta-Sigma Data Converters: Theory, Design and Simulation* (Wiley-IEEE Press, New York, 1996). ISBN:0-7803-1045-4
80. R. Baird, T. Fiez, Improved $\Delta\Sigma$ DAC linearity using data weighted averaging, in *Proceedings of the 1995 International Symposium on Circuits and Systems*, vol. 1, pp. I–13 – I–16, 1995
81. I. Fujimori, L. Longo, A. Hairapetian, K. Seiyama, S. Kosic, C. Jun, S.L. Chan, A 90-dB SNR 2.5-MHz output-rate ADC using cascaded multibit delta-sigma modulation at 8x oversampling ratio. IEEE J. Solid-State Circ. **35**(12), 1820–1828 (2000)
82. M. Clara, A. Wiesbauer, W. Klatzer, Nonlinear distortion in current-steering D/A-converters due to asymmetrical switching errors, in *Proceedings of the 2004 International Symposium on Circuits and Systems*, vol. 1, pp. I–285 – I–288, 2004
83. B. Leung, S. Sutarja, Multibit Σ-Δ A/D converter incorporating a novel class of dynamic element matching techniques. IEEE Trans. Circ. Syst. II: Analog Digital Signal Process. **39**(1), 35–51 (1992)
84. L. Carley, A noise-shaping coder topology for 15+ bit converters. IEEE J. Solid-State Circ. **24**(2), 267–273 (1989)
85. H. Jensen, I. Galton, A low-complexity dynamic element matching DAC for direct digital synthesis. IEEE Trans. Circ. Syst. II: Analog Digital Signal Process. **45**(1), 13–27 (1998)
86. I. Galton, Spectral shaping of circuit errors in digital-to-analog converters. IEEE Trans. Circ. Syst. II: Analog Digital Signal Process. **44**(10), 808–817 (1997)
87. A. Yasuda, H. Tanimoto, T. Iida, A third-order $\Delta\Sigma$ modulator using second-order noise-shaping dynamic element matching. IEEE J. Solid-State Circ. **33**(12), 1879–1886 (1998)
88. X.-M. Gong, E. Gaalaas, M. Alexander, D. Hester, E. Walburger, J. Bian, A 120 dB multi-bit SC audio DAC with second-order noise shaping. IEEE International Solid-State Circuits Conference 2000, Digest of Technical Papers, pp. 344–345, 2000

89. E. Fogelman, J. Welz, I. Galton, An audio ADC delta-sigma modulator with 100-dB peak SINAD and 102-dB DR using a second-order mismatch-shaping DAC. IEEE J. Solid-State Circ. **36**(3), 339–348 (2001)
90. T. Shui, R. Schreier, F. Hudson, Mismatch shaping for a current-mode multibit delta-sigma DAC. IEEE J. Solid-State Circ. **34**(3), 331–338 (1999)
91. W. Schofield, D. Mercer, L.St. Onge, A 16b 400MS/s DAC with < -80dBc IMD to 300MHz and < -160dBm/Hz noise power spectral density. IEEE International Solid-State Circuits Conference 2007, Visuals Supplement to the Digest of Technical Papers, pp. 90–91, 2003
92. Q. Huang, P.A. Francese, C. Martelli, J. Nielsen, A 200MS/s 14b 97mW DAC in $0.18\mu m$ CMOS. IEEE International Solid-State Circuits Conference 2004, Digest of Technical Papers, pp. 364–532, 2004
93. M. Clara, W. Klatzer, D. Gruber, A. Marak, B. Seger, W. Pribyl, A 1.5V 13bit 130–300MS/s self-calibrated DAC with active output stage and 50MHz signal bandwidth in $0.13\mu m$ CMOS, in *Proceedings of the 34th European Solid-State Circuits Conference*, pp. 262–265, 2008
94. R. Craven, An integrated circuit 12-bit D/A converter. IEEE International Solid-State Circuits Conference 1975, Digest of Technical Papers, pp. 40–41, 1975
95. J. Teichmann, K. Burger, W. Hasche, J. Herrfurth, G. Taschner, One time programming (OTP) with Zener diodes in CMOS processes, in *Proceedings of the 33th European Solid-State Device Research Conference*, pp. 433–436, 2003
96. G. Erdi, A precision trim technique for monolithic analog circuits. IEEE J. Solid-State Circ. **10**(6), 412–416 (1975)
97. D. Comer, A monolithic 12-bit DAC. IEEE Trans. Circ. Syst. II: Analog and Digital Signal Process. **25**(7), 504–509 (1978)
98. C.-H. Yang, C.-M. Su, Method of forming a metal fuse on semiconductor devices, U.S. Patent 6,835,642, 28 Dec 2004
99. J. Fellner, P. Boesmueller, H. Reiter, Lifetime study for a poly fuse in a $0.35\mu m$ polycide CMOS process. *IEEE International Reliability Physics Symposium*, pp. 17–21, 2005
100. P. Pessl, R. Gaggl, J. Hohl, D. Giotta, J. Hauptmann, A four-channel ADSL2+ analog front-end for CO applications with 75 mW per channel, built in $0.13\mu m$ CMOS. IEEE J. Solid-State Circ. **39**(12), 2371–2378 (2004)
101. M. Tiilikainen, A 14-bit 1.8-V 20-mW 1-mm^2 CMOS DAC. IEEE J. Solid-State Circ. **36**(7), 1144–1147 (2001)
102. G.I. Radulov, P. Quinn, J.A. Hegt, A.H.M. van Roermund, An on-chip self-calibration method for current mismatch in D/A converters, in *Proceedings of the 31th European Solid-State Circuits Conference*, pp. 169–172, 2005
103. Y. Cong, R.L. Geiger, A 1.5-V 14-bit 100-MS/s self-calibrated DAC. IEEE J. Solid-State Circ. **38**(12), 2051–2060 (2003)
104. A. Bugeja, B.-S. Song, A Self-Trimming 14-b 100-MS/s CMOS DAC. IEEE J. Solid-State Circ. **34**(12), 1841–1852 (2000)
105. G. Wegmann, E. Vittoz, Very accurate dynamic current mirrors. Electron. Lett. **25**, 644–646 (1989)
106. G. Wegmann, E.A. Vittoz, Analysis and improvements of accurate dynamic current mirrors. IEEE J. Solid-State Circ. **25**(3), 699–706 (1990)
107. W. Groeneveld, H. Schouwenaars, H. Termeer, A self calibration technique for monolithic high-resolution D/A converters. IEEE J. Solid-State Circ. **24**(6), 1517–1522 (1989)
108. E. Vittoz, G. Wegmann, Dynamic current mirrors. in *Analog IC Design: the Current-mode Approach*, ed. by C. Tomazou et al. (Peter Peregrinus, 1990), pp. 297–326. ISBN:0 86341 297 1
109. G. Wegmann, E. Vittoz, F. Rahali, Charge injection in analog MOS switches. IEEE J. Solid-State Circ. **22**(6), 1091–1097 (1987)
110. C.J.J. Dachs, R. Surdeanu, D. Guyot, A. Parlangeli, Y.V. Ponomarev, P.A. Stolk, Junction leakage in advanced CMOS technologies, in *Proceedings of the 31th European Solid-State Device Research Conference*, pp. 175–178, 2001

111. K. Roy, S. Mukhopadhyay, H. Mahmoodi-Meimand, Leakage current mechanisms and leakage reduction techniques in deep-submicrometer CMOS circuits. Proc. IEEE. **91**(2), 305–327 (2003)
112. K.M. Cao, W.-C. Lee, W. Liu, X. Jin, P. Su, S.K.H. Fung, J.X. An, B. Yu, C. Hu, BSIM4 gate leakage model including source-drain partition. *Technical Digest of the International Electron Devices Meeting IEDM'00*, pp. 815–818, 2000
113. R. Suarez, P. Gray, D. Hodges, All-MOS charge-redistribution analog-to-digital conversion techniques. II. IEEE J. Solid-State Circ. **10**(6), 379–385 (1975)
114. K. Falakshahi, C.-K. Yang, B. Wooley, A 14-bit, 10-Msamples/s D/A converter using multi-bit $\Sigma\Delta$ modulation. IEEE J. Solid-State Circ. **34**(5), 607–615 (1999)
115. A. Van den Bosch, M. Steyaert, W. Sansen, SFDR-bandwidth limitations for high speed high resolution current steering CMOS D/A converters, in *Proceedings of the 6th IEEE International Conference on Electronics, Circuits and Systems ICECS'99*, vol. 3, pp. 1193–1196, 1999
116. A. Hastings, *The Art of Analog Layout* (Prentice Hall, Upper Saddle River, NJ, 2001). ISBN:0-13-087061-7
117. A. Van den Bosch, M. Steyaert, W. Sansen, A 10-bit 1-GSample/s nyquist current-steering CMOS D/A converter. IEEE J. Solid-State Circ. **36**(3), 315–324 (2001)
118. J. De Maeyer, P. Rombouts, L. Weyten, Addressing static and dynamic errors in bandpass unit element multibit DAC's, in *Proceedings of the 2004 International Symposium on Circuits and Systems*, vol. 1, pp. I–345 – I–348, 2004
119. K.O. Andersson, J.J. Wikner, Characterization of a CMOS current-steering DAC using state-space models, in *Proceedings of the 43rd IEEE Midwest Symposium on Circuits and Systems*, vol. 2, pp. 668–671, 2000
120. K. Doris, D. Leenarts, A.H.M. van Roermund, Time non linearities in D/A converters, in *Proceedings of European Conference on Circuit Theory and Design, ECCTD'01*, vol. 3, pp. 353–357, 2001
121. M. Kosunen, Digital signal processing and digital-to-analog converters for wide-band transmitters, Ph.D. dissertation, Helsinki University of Technology, 2006
122. J.L. Gonzalez, E. Alarcon, Clock-jitter induced distortion in high speed CMOS switched-current segmented digital-to-analog converters, in *Proceedings of the 2001 International Symposium on Circuits and Systems*, vol. 1, pp. I–512 – I–515, 2001
123. L. Luh, J. Choma, J. Draper, A high-speed fully differential current switch. IEEE Trans. Circ. Syst. II: Analog Digital Signal Process. **47**(4), 358–363 2000
124. N. Ghittori, A. Vigna, P. Malcovati, S. D' Amico, A. Baschirotto, 1.2V low-power multi-mode DAC+filter blocks for reconfigurable (WLAN/UMTS, WLAN/Bluetooth) transmitters. IEEE J. Solid-State Circ. **41**(9), 1970–1982 (2006)
125. K. Doris, J. Briaire, D. Leenaerts, M. Vertregt, A.H.M. van Roermund, A 12b 500MS/s DAC with 70dB SFDR up to 120MHz in $0.18\mu m$ CMOS. IEEE International Solid-State Circuits Conference 2005, Digest of Technical Papers, pp. 116–117, 2005
126. B. Schafferer, R. Adams, A 3V CMOS 400mW 14b 1.4GS/s DAC for multi-carrier applications. IEEE International Solid-State Circuits Conference 2004, Digest of Technical Papers, 360–361, 2004
127. K. O'Sullivan, C. Gorman, M. Hennessy, V. Callaghan, A 12-bit 320-MSample/s current-steering CMOS D/A converter in 0.44 mm^2. IEEE J. Solid-State Circ. **39**(7), 1064–1072 (2004)
128. C.-H. Lin, F.M.L. Van Der Goes, J.R. Westra, J. Mulder, Y. Lin, E. Arslan, E. Ayranci, X. Liu, K. Bult, A 12 bit 2.9 GS/s DAC With IM3 < -60 dbc Beyond 1 GHz in 65 nm CMOS. IEEE J. Solid-State Circ. **44**(12), 3285–3293 (2009)
129. S. Park, G. Kim, S. Park, W. Kim, A digital-to-analog converter based on differential-quad switching. IEEE J. Solid-State Circ. **37**(10), 1335–1338 (2002)
130. D. Mercer, A study of error sources in current steering digital-to-analog converters, in *Proceedings of the IEEE 2004 Custom Integrated Circuits Conference*, pp. 185–190, 2004

131. Xue-Mei Gong, Digital signal processor with reduced pattern dependent noise, U.S. Patent 5,719,572, 17 Feb 1999
132. J. Hyde, T. Humes, C. Diorio, M. Thomas, M. Figueroa, A 300-MS/s 14-bit digital-to-analog converter in logic CMOS. IEEE J. Solid-State Circ. **38**(5), 734–740 (2003)
133. M. Clara, W. Klatzer, A. Wiesbauer, D. Straeussnigg, A 350MHz low-OSR $\Sigma\Delta$ current-steering DAC with active termination in $0.13\mu m$ CMOS. IEEE International Solid-State Circuits Conference 2005, Digest of Technical Papers, pp. 118–119, 2005
134. K. Poulton, R. Neff, A. Muto, W. Liu, A. Burstein, M. Heshami, A 4 Gsample/s 8b ADC in 0.35 μm CMOS. IEEE International Solid-State Circuits Conference 2002, Digest of Technical Papers, pp. 166–167, 2002
135. K. Poulton, R. Neff, B. Setterberg, B. Wuppermann, T. Kopley, R. Jewett, J. Pernillo, C. Tan, A. Montijo, A 20 GS/s 8 b ADC with a 1 MB memory in 0.18 μm CMOS. IEEE International Solid-State Circuits Conference 2003, Digest of Technical Papers, pp. 318–319, 2003
136. D. Draxelmayr, A 6b 600MHz 10mW ADC array in digital 90nm CMOS. IEEE International Solid-State Circuits Conference 2004, Digest of Technical Papers, pp. 264–265, 2004
137. K. Doris, E. Janssen, C. Nani, A. Zanikopoulos, G. van der Weide, A 480 mW 2.6 GS/s 10b time-interleaved ADC With 48.5 dB SNDR up to Nyquist. IEEE J. Solid-State Circ. **46**(12), 2821–2833 (2011)
138. W.F. Ellersick, Data converters for high speed CMOS links, Ph.D. dissertation, Stanford University, 2001
139. K.C.-K. Yang, V. Stojanovic, S. Modjtahedi, M.A. Hororwitz, W.F. Ellersick, A serial-link transceiver based on 8-GSamples/s A/D and D/A converters in $0.25\mu m$ CMOS. IEEE J. Solid-State Circ. **36**(11), 1684–1692, 2001
140. T. Rahkonen, Error correction techniques in high-speed A/D and D/A converters. Lecture Notes in Department of Electrical Engineering and Infotech Oulu, University of Oulu, 2001
141. M.O. Hawksford, Digital-to-analog converter with low intersample transition distortion and low sensitivity to sample jitter and transresistance amplifier slew rate. J. Audio Eng. Soc. **42**(11), 901–917 (1994)
142. J. Steensgard-Madsen, High-performance data converters, Ph.D. dissertation;, Technical University of Denmark, Lyngby, 1999
143. I. Galton, Method and apparatus for cyclic return to zero techniques for digital to analog convertors, U.S. Patent 6,476,748, 5 Nov 2002
144. D. Domanin, U. Gatti, P. Malcovati, F. Maloberti, A multipath polyphase digital-to-analog converter for software radio transmission systems, in *Proceedings of the 2000 International Symposium on Circuits and Systems*, vol. 2, pp. II–361 – II–364, 2000
145. W. Black, D.A. Hodges, Time-interleaved converter arrays. IEEE J. Solid-State Circ. **15**(6), 1022–1029 (1980)
146. N. Kurosawa, H. Kobayashi, K. Maruyama, H. Sugawara, K. Kobayashi, Explicit analysis of channel mismatch effects in time-interleaved ADC systems. IEEE Trans. Circ. Syst. I: Fundamental Theor. Appl. **48**(3), 261–271 (2001)
147. R. Khoini-Poorfard, L.B. Lim, D.A. Johns, Time-interleaved oversampling A/D converters: theory and practice. IEEE Trans. Circ. Syst. II: Analog Digital Signal Process. **44**(8), 634–645 (1997)
148. D. Giotta, P. Pessl, M. Clara, W. Klatzer, R. Gaggl, Low-power 14-bit current steering DAC, for ADSL2+/CO applications in $0.13\mu m$ CMOS, in *Proceedings of the 30th European Solid-State Circuits Conference*, pp. 163–166, 2004
149. J. Kenney, F. Sabouri, V. Leung, J. Guido, E. Zimany, A. Agrillo, J. Trackim, J. Khoury, R. Shariatdoust, A 4 channel analog front end for central office ADSL modems, in *Proceedings of the IEEE 2000 Custom Integrated Circuits Conference*, pp. 307–310, 2000
150. S. Lin, D. Li, W. Chen, 1 V 1.25 GS/s 8 mW D/A converters for MB-OFDM UWB transceivers. *IEEE International Conference on Ultra-Wideband, ICUWB, 2007*, pp. 453–456, 2007
151. C. Sandner, M. Clara, A. Santner, T. Hartig, F. Kuttner, A 6-bit 1.2-GS/s low-power flash-ADC in $0.13\mu m$ digital CMOS. IEEE J. Solid-State Circ. **40**(7), 1499–1505 (2005)

152. L. Wadhwa, Simulation of third-order systems with double-lead using one operational amplifier. Proc. IRE. **50**(6), 1538–1539 (1962)
153. Analog Devices Inc., AD8057: Low cost, single, high performance voltage feedback, 325 MHz amplifier. Datasheet (2003), http://www.analog.com/
154. R. Schreier, Noise-shaped coding, Ph.D. dissertation, University of Toronto, 1991
155. R. Schreier, Delta-Sigma Toolbox, MATLAB^{TM} toolbox (2000), http://www.mathworks.com/matlabcentral/fileexchange
156. M. Clara, W. Klatzer, A. Wiesbauer, Method for digital/analog conversion and corresponding digital/analog converter device, U.S. Patent 7,199,741, 3 Apr 2007
157. S. Rabii, B. Wooley, A 1.8-V digital-audio sigma-delta modulator in $0.8\mu m$ CMOS. IEEE J. Solid-State Circ. **32**(6), 1026–1034 (1997)
158. P. Kinget, M. Steyaert, Impact of transistor mismatch on the speed-accuracy-power trade-off of analog CMOS circuits, in *Proceedings of the IEEE 1996 Custom Integrated Circuits Conference*, pp. 333–336, 1996
159. K. Uyttenhove, M. Steyaert, Speed-power-accuracy trade-off in high-speed ADC's: What about nano-electronics? in *Proceedings of the IEEE 2001 Custom Integrated Circuits Conference*, pp. 341–344, 2001
160. K. Uyttenhove, M. Steyaert, Speed-power-accuracy trade-off in high-speed CMOS ADC's. IEEE Trans. Circuits Syst. II. **49**(4), 280–287 (2002)
161. H. Schouwenaars, W. Groeneveld, C. Bastiaansen, H. Termeer, An oversampled multibit CMOS D/A converter for digital audio with 115-dB dynamic range. IEEE J. Solid-State Circ. **26**(12), 1775–1780 (1991)
162. D.K. Su, B. Wooley, A CMOS oversampling D/A converter with a current-mode semidigital reconstruction filter. IEEE J. Solid-State Circ. 28(12), 1224–1233 (1993)
163. T. Hamasaki, Y. Shinohara, H. Terasawa, K.-I. Ochiai, M. Hiraoka, H. Kanayama, A 3-V, 22-mW multibit current-mode $\Sigma\Delta$ DAC with 100 dB dynamic range. IEEE Journal of Solid-State Circ. **31**(12), 1888–1894 (1996)
164. I. Fujimori, A. Nogi, T. Sugimoto, A multibit delta-sigma audio DAC with 120-dB dynamic range. IEEE J. Solid-State Circ. **35**(8), 1066–1073 (2000)
165. M. Annovazzi, V. Colonna, G. Gandolfi, F. Stefani, A. Baschirotto, A Low-Power 98-dB multibit audio DAC in a standard 3.3-V $0.35\mu m$ CMOS technology. IEEE J. Solid-State Circ. **37**(7), 825–834 (2002)
166. T. Rueger, B. Duewer, S. Hodapp, T. Lei, J. Melanson, B. Trotter, A 110dB ternary PWM current-mode audio DAC with monolithic 2Vrms driver. IEEE International Solid-State Circuits Conference 2004, Digest of Technical Papers, 372–373, 2004
167. P. Francese, P. Ferrat, Q. Huang, A 13-b 1.1-MHz Oversampled DAC with semidigital reconstruction filtering. IEEE J. Solid-State Circ. **39**(12), 2098–2106 (2004)
168. K. Nguyen, A. Bandyopadhyay, B. Adams, K. Sweetland, P. Baginski, A 108 dB SNR, 1.1mW oversampling audio DAC with a three-level DEM technique. IEEE J. Solid-State Circ. **43**(12), 2592–2600 (2008)
169. V. Colonna, M. Annovazzi, G. Boarin, G. Gandolfi, F. Stefani, A. Baschirotto, A 0.22-mm^2 7.25-mW per-channel audio stereo-DAC with 97-dB DR and 39-dB SNR_{OUT}. IEEE J. Solid-State Circ. **40**(7), 1491–1498 (2005)
170. Y.-H. Lee, M.-Y. Choi, S.-B. You, W.-S. Yeum, H.-J. Park, J.-W. Kim, A 4mW per-Channel 101dB-DR stereo audio DAC with transformed quantization structure, in *Proceedings of the IEEE 2006 Custom Integrated Circuits Conference*, pp. 145–148, 2006
171. L. Risbo, R. Hezar, B. Kelleci, H. Kiper, M. Fares, A 108dB-DR 120dB-THD and 0.5Vrms output audio DAC with Inter-symbol-interference-shaping algorithm in 45nm CMOS. IEEE International Solid-State Circuits Conference 2005, Digest of Technical Papers, pp. 484–485, 2011
172. H. Weinrichter, F. Hlawatsch, *Stochastische Grundlagen Nachrichtentechnischer Signale* (Springer, New York, 1991). ISBN:3-211-82303-4
173. M. Clara, W. Klatzer, A. Di Giandomenico, L. Gori, Digital/Analog Wandler mit Selbstkalibrierung, Patentschrift DE102005017304, 2 Nov 2006

174. M. Clara, J. Hauptmann, Digital-to-analog converter arrangement with an array of unary digital-to-analog converting elements useable for different signal types, U.S. Patent 6,831,581, 14 Dec 2004
175. M. Clara, H. Weinberger, C. Fleischhacker, A. Wiesbauer, Digital/Analog converter with programmable gain, U.S. Patent 6,646,580, 11 Nov 2003
176. D. Monticelli, A Quad CMOS single-supply Op Amp with rail-to-rail output swing. IEEE J. Solid-State Circ. **21**(6), 1026–1034 (1986)
177. R. Hogervorst, J. Tero, R. Eschauzier, J.H. Huijsing, A compact power-efficient 3 V CMOS rail-to-rail Input/Output operational amplifier for VLSI cell libraries. IEEE J. Solid-State Circ. **29**(12), 1505–1513 (1994)
178. M. Clara, W. Klatzer, B. Seger, A. Di Giandomenico, L. Gori, A 1.5V 200MS/s 13b 25mW DAC with randomized nested background calibration in $0.13\mu m$ CMOS. IEEE International Solid-State Circuits Conference 2007, Visuals Supplement to the Digest of Technical Papers, pp. 196–197, 2007
179. A. Van den Bosch, M. Borremans, M. Steyaert, W. Sansen, A 12b 500Msample/s current-steering CMOS D/A converter. IEEE International Solid-State Circuits Conference 2001, Digest of Technical Papers, pp. 366–367, 2001
180. K. Chan, I. Galton, A 14b 100MS/s DAC with fully segmented dynamic element matching. IEEE International Solid-State Circuits Conference 2006, Digest of Technical Papers, pp. 390–391, 2006
181. K. Chan, J. Zhu, I. Galton, A 150MS/s 14-bit segmented DEM DAC with greater than 83dB of SFDR across the Nyquist band. Symposium on VLSI Circuits Digest of Technical Papers, pp. 200–201, 2007
182. T. Ueno, T. Yamaji, T. Itakura, A 1.2V, 12-bit, 200MSample/s current-steering D/A converter in 90nm CMOS, in *Proceedings of the IEEE 2005 Custom Integrated Circuits Conference*, pp. 747–750, 2005
183. Y. Tang, J. Briaire, K. Doris, R. van Veldhoven, P.C.W. van Beek, H.J.A. Hegt, A.H.M. Roermund, A 14 bit 200 MS/s DAC With SFDR $>$ 78 dBc, IM3 < -73 dBc and NSD $<$ 163 dBm/Hz across the whole nyquist band enabled by dynamic-mismatch mapping. IEEE J. Solid-State Circ. **46**(6), 1371–1381 (2011)
184. W.-H. Tseng , C.-W. Fan, J.-T. Wu, A 12-Bit 1.25 GS/s DAC in 90nm CMOS With $>$ 70 dBc SFDR up to 500 MHz. IEEE J. Solid-State Circ. **46**(12), 2845–2856 (2011)
185. S.C. Cripps, *RF Power Amplifiers for Wireless Communications*, 2nd edn. (Artech House, Boston, MA, 2006). ISBN:1-59693-018-7
186. B. Razavi, Design of monolithic phase-locked loops and clock recovery circuits-a tutorial, in *Monolithic Phase-Locked Loops and Clock Recovery Circuits : Theory and Design*, ed. by B. Razavi (Wiley-IEEE Press, New York, 1996), pp. 1–39. ISBN:0-7803-1149-3
187. A. Wiesbauer, D. Straeussnigg, R. Gaggl, M. Clara, L. Hernndez, D. Gruber, Clock jitter compensation for current steering DACs, in *Proceedings of the 2006 International Symposium on Circuits and Systems*, pp. 5375–5378, 2006
188. R. Adams, Design and implementation of an audio 18-Bit analog-to-digital converter using oversampling techniques. J. Audio Eng. Soc. **34**(3), 153–166 (1986)
189. G.F. Franklin, J.D. Powell, M.L. Workman, *Digital Control of Dynamic Systems*, 3rd edn. (Addison-Wesley, Menlo Park, CA, 1998). ISBN:0-201-33153-5
190. S.C. Cripps, *Advanced Techniques in RF Power Amplifier Design*, 2nd edn. (Artech House, Boston, MA, 2002). ISBN:1-58053-282-9

Index

M. Clara, *High-Performance D/A-Converters*, Springer Series in Advanced Microelectronics 36, DOI 10.1007/978-3-642-31229-8,

MIX
Papier aus verantwortungsvollen Quellen
Paper from responsible sources
FSC® C105338

If you have any concerns about our products,
you can contact us on
ProductSafety@springernature.com

In case Publisher is established outside the EU,
the EU authorized representative is:
Springer Nature Customer Service Center GmbH
Europaplatz 3, 69115 Heidelberg, Germany

Printed by Libri Plureos GmbH
in Hamburg, Germany